Hazardous Materials

Warren E. Isman

Director, Department of Fire and Rescue Services
Montgomery County, Maryland

Gene P. Carlson

Associate Editor, International Fire Service Training Association
Oklahoma State University

(A Glencoe Book)

MACMILLAN PUBLISHING CO., INC.
New York

COLLIER MACMILLAN PUBLISHERS
London

Glencoe Publishing Co., Inc.
17337 Ventura Boulevard
Encino, California 91316
Collier Macmillan Canada, Ltd.

Library of Congress Catalog Card Number: 78-71739

ISBN   0-02-475020-4

2 3 4 5 6 7 8 9 10   83

Table of Contents

Preface

This text has been written with a dual purpose in mind. Its first use is as the basic text for the college-level course in hazardous materials. In addition, it is a practical source book for field personnel who may not have the opportunity to take a hazardous materials course at a college.

The aim of *Hazardous Materials* is to give the student basic information about the properties of hazardous materials and the methods that are most appropriate for handling the various kinds of accidents that can occur with these substances. At the same time, the text offers the senior officer instruction in managerial techniques that are essential in making intelligent, knowledgeable decisions in directing operations in which hazardous materials are involved.

In general, the text has been prepared for that group of people whom the authors call emergency response personnel. They are all the members of the various agencies who respond to hazardous materials incidents— local and state law enforcement, emergency medical services, civil defense, environmental agencies, and, of course, fire services personnel. It is to their advantage to be familiar with the contents of this book. Such knowledge will ensure their safety as well as that of the general public. Furthermore, because the authority to take charge in an incident is vested in different agencies, personnel in all of these agencies must be ready to meet their responsibilities.

FEATURES

Handling a hazardous materials incident is a very complex job. From the individual on the front line to the first-level supervisor to the senior supervisor, each person requires a different degree of information. Because of its extensive coverage of the subject, this text imparts to each group the kind of knowledge it must have readily at hand.

Each chapter is essentially self-contained and concentrates on one aspect of hazardous materials, such as their chemical properties, modes of transportation, and so forth. Thus, instructors or readers can freely choose the topics which are most important to them. Likewise, a course for emergency response teams can be devised for their special needs and interests.

The tables have been designed so that they provide a quick reference at the incident scene. They are concise and comprehensive, with specific data on label and placard identi-fication, symptoms of poisoning, and treatment to be administered to the injured. Emergency response personnel are advised to carry the text in a command vehicle for reference when needed.

As a visual aid to the reader, both photographs and line art illustrations are used liberally throughout the text. These illustrations are meant to help the student visualize objects or subjects that are difficult to describe but should be known, such as the coupling devices on railroad cars and the common forms of petroleum storage tanks.

Finally, each chapter has a set of study questions which review the main points of the chapter, and many chapters have a case problem concerned with practical application of techniques discussed in the chapter. Also, each chapter has a brief note or reference list citing sources for further study.

ORGANIZATION

This text is composed of sixteen chapters, each of which discusses a particular subject related to hazardous materials. Chapter 1, "The Hazardous Materials Incident," defines the term *hazardous materials,* relates a cross section of serious hazardous materials accidents in recent years, and specifies the agencies by state that either assist in or supply information about handling incidents. Chapter 2, "Review of Basic Chemistry," surveys such topics as temperature–pressure–volume relationships, fire behavior, chemical reactions, and the metric equivalents to and conversions from English measurements.

The next three chapters provide information on specific groups of chemicals. Chapter 3, "Health Hazards," outlines the dangers of agricultural and garden chemicals, corrosives, and poisons. Chapter 4 examines explosive and radioactive materials, their inherent dangers, the safety precautions to observe, and the recommended ways to manage accidents involving these materials. In addition, information about specific types of explosives is included.

Coverage of the types of hazardous materials concludes with Chapter 5, which deals with compressed gases, cryogenics, and flammable products. Like the preceding chapters, this chapter discusses the physical properties, specific chemical hazards, safety devices and practices, and methods of coping with incidents involving these products. Chapters 3, 4, and 5 are primarily directed toward personnel who will be on the scene handling the incident.

Chapters 6 through 10 cover the identification systems and the command and decision-making procedures that should be followed at an incident. Chapter 6 details the techniques to use in identifying the chemical class of a substance and the specific chemical itself. It furthermore explains how to interpret shipping documents and where to go if information is needed to identify an unknown substance.

Emergency response personnel in every community should have a prearranged plan of action to follow in the event of an incident. Chapter 7 looks at what preplanning is all about. For example, the access routes, water supply, layouts of plants, storage facilities, railroad yards, and so forth, should be known to response personnel.

Command operations and communications at a hazardous materials incident are critically important. Chapter 8 describes how to establish and staff a command post. It also examines the responsibilities of the personnel assigned to that staff. In general, logistics come under the direction of the command staff. This aspect of response is also discussed in this chapter.

A hazardous materials incident has the potential to last several days or more. During this extended period of time, agencies must work together in close cooperation. Chapter 9, "Disaster Management," describes the strategy, contingency planning, emergency medical planning, evacuation, and cleanup plans that are part of disaster management.

The last chapter in this group, Chapter 10, points out that decision making means, first and foremost, choosing a course of action and then implementing it. This chapter offers a critical path or decision tree upon which to base the actions which are selected.

There are five major modes of transporting hazardous materials—railroad, truck, pipeline, aircraft, and water. Chapters 11 through 15 describe these modes in detail. Within each chapter, the student is taught about the types of equipment used for transportation, safety devices on this equipment, types of accidents that can occur, examples of commodities carried, and handling incidents with these forms of transport.

The last chapter, Chapter 16, "Fixed

Exterior Storage Facilities," informs the reader on the common types of storage tanks, their internal and external structural supports, techniques for extinguishing fires at storage facilities, and the types of foam that work best in putting out various kinds of fires.

Following Chapter 16, there is a glossary of terms. The glossary allows the student to find, in one place, a source of information on hazardous materials and transportation equipment. Terms for transportation equipment are included because the student, in doing additional research in trade journals, may encounter an unfamiliar technical jargon which the glossary will help one to understand.

ACKNOWLEDGMENTS

In any undertaking as large as this one, there are numerous people to thank. From the entire national fire service we have received input and material for inclusion in the text. For their help we are grateful. We are indebted to so many people that listing them all would be very difficult. However, a few deserve special recognition for the many hours they have spent discussing the tactics, procedures, and methods used in handling hazardous materials incidents. So, special thanks go to Assistant Professor Charles J. Wright, Western Kentucky University; Bruce R. Piringer, Kentucky State Fire Training Director; Chief William J. Patterson, Santa Barbara County (California) Fire Department; Eugen Schlaf, formerly of the Illinois Central Gulf Railroad; R.G. Kuhlman, Burlington Railroad; Chief Robert Shimer, Westernport (Maryland) Fire Department; and Edward Pritchard, Federal Railroad Administration.

Many of the communities where we have conducted our training courses also helped us with this project. We owe a debt of appreciation to William Keller, Bucks County, Pennsylvania Fire School Director; William Porter, Connecticut State Fire Administrator; Chief Richard Lamb, Wayne Township (Indiana) Fire Department; Robert L. Milke, Valencia Community College (Orlando, Florida); and William Hagevig, Alaska Director of Fire Service Training.

Within the Montgomery County (Maryland) Department of Fire/Rescue Service special thanks goes to Lieutenant Leonard King for use of his photographs and Sergeant Richard Long for developing many of the photographs used in the text.

Converting handwritten manuscript to typed pages, while correcting spelling and grammar, is no easy task. This job fell to Mrs. Peggy Holt and Mrs. Anne Carlson, who worked long and tirelessly against difficult deadlines. We will never be able to adequately thank them for their work.

The greatest debt of all, however, is owed to our children who had to put up with many months without either of us being available. So this book has been written for our kids in the hope that as they grow up the problems from hazardous materials will decrease. So to Emily, Susan, Amy, Kenneth, Janet, David, Carla, Theodore, Heather, Kirsten, Aaron, and Kendra, we dedicate this book.

Warren E. Isman
Gene P. Carlson

1
The hazardous materials incident

Among the new materials being produced to satisfy the needs of modern life are many that have dangerous properties. The hazards are all around us—moving along our highways and railways, in large concentrations in processing and manufacturing plants, spread over our fields and farmlands. Although hazardous materials come under federal regulations in packaging and transport, accidents will happen. In recognition of the potential for disaster in the wake of an accident, spill, or leak, various government agencies have programs to study the problems encountered and to develop proper safeguards, and educational materials and reams of helpful information are published by trade associations and fire service periodicals. It becomes an enormous task to absorb the avalanche of information from all sources. To this end, it is the aim of this book to gather together the essential information to enable emergency response personnel to do their job of safeguarding lives and property.

At the scene of a hazardous materials accident or spill, the threat of fire or toxic effects is always present. The emergency forces of the fire service must be ready with current information, knowledge, and skills to deal with the situation. Every town, city, and hamlet has the potential for an incident, even if there are no waterways, airports, or interstate highways nearby. Accidents during transport are just one aspect of the problem. Incidents may occur at storage facilities, manufacturing plants, retail establishments, processing operations—even in homes. Hazardous materials include everything from chemicals used in backyard swimming pools to industrial chemicals carried by special tank trucks or railroad cars.

This text presents step-by-step procedures to analyze the problem, take corrective action, and bring the incident under control. It is not intended to cover all possible solutions to all possible incidents. The reader must learn to take the general information given and adapt it to the specific situation. Neither is this text designed to provide a course in chemistry, but, rather, to make available to emergency services management and response team personnel the information necessary to make intelligent decisions when called to a hazardous materials accident.

WHAT ARE HAZARDOUS MATERIALS?

Hazardous materials can be defined in a number of ways. Here are three common definitions:

1. Any substance or material in a quantity or form that poses an unreasonable risk to safety or health and property when transported in commerce.

2. Any element, compound, or combination thereof, which is flammable, corrosive, etc., and which, because of handling, storage, processing, or packaging, may have detrimental effects on operating and emergency personnel, the public, equipment, and/or the environment.

3. Any substance that must be placarded when moving in interstate commerce.

The first and third definitions deal only with materials during transport. Since emergency response personnel are concerned with hazardous materials at all stages, the second definition will be used for purposes of this text.

This first chapter covers the following:

1. The scope of the problem.
2. Some previous incidents.
3. Modes of transportation.
4. Places where hazardous materials can be found.
5. Sources of information to stay abreast of the hazardous materials problem.

SCOPE OF THE PROBLEM

Almost everyday the newspapers carry a story of a hazardous materials incident somewhere in the United States. A gas leak, a highway tanker accident, an explosion in an industrial plant, in a manufacturing plant, or even in some illegal operation in a private dwelling may be reported. Town size matters little because the locations of these incidents play no favorites. Big or small, all have the potential for disaster. The incident reported may be discussed among emergency response personnel as a piece of news and then dismissed as being the concern of some other jurisdiction. Detailed studies of the incident do not often receive wide circulation. Lessons to be learned from the strategy and tactics used are seldom brought out. What was a disaster one day for one town could become a disaster in another town at some later time, with the same mistakes being made.

To understand the ramifications of the job confronting the emergency response team, let's begin by examining the scope of the problem.

Transportation Problems

During the period 1971 through 1975 there were approximately 32,000 hazardous materials incidents reported to the U.S. Department of Transportation, as tallied in Table 1-1. The trend in every state, as shown by the figures, increases yearly. While a portion of the increase can be ascribed to better reporting procedures, the number of actual incidents occurring is on the rise. In addition, remember that these are only the incidents that were reported. Many more may be assumed to occur, which, for a variety of reasons, go unreported.

In 1971 a nationwide intermodal system for the reporting of hazardous materials incidents was established to provide the Department of Transportation with the

Table 1-1 Tally Of "Incident Locations" (By State) Of Approximately 32,000 Hazardous Materials Incident Reports Received During 1971–1975

Incident Location	1971	1972	1973	1974	1975	Total	Incident Location	1971	1972	1973	1974	1975	Total
Alabama	49	89	73	186	271	668	Nebraska	19	16	38	58	74	205
Alaska	0	1	3	6	11	21	Nevada	4	16	27	19	21	87
Arizona	44	69	92	132	157	494	New Hampshire	1	4	6	14	16	41
Arkansas	13	37	54	114	127	345	New Jersey	59	108	215	272	326	980
California	207	277	357	491	518	1,850	New Mexico	24	42	69	106	114	355
Colorado	47	63	64	95	112	381	New York	57	146	294	422	430	1,349
Connecticut	24	53	84	57	77	295	North Carolina	134	232	104	402	428	1,300
Delaware	1	6	18	29	27	81	North Dakota	1	7	20	26	20	74
Dist. of Col.	2	4	8	9	12	35	Ohio	105	358	415	547	831	2,256
Florida	40	98	139	191	315	783	Oklahoma	29	52	65	87	85	318
Georgia	107	165	237	401	392	1,302	Oregon	49	39	53	47	34	222
Hawaii	0	2	0	2	5	9	Pennsylvania	146	352	573	854	1,200	3,125
Idaho	23	27	26	27	43	146	Rhode Island	3	9	20	21	36	89
Illinois	96	238	316	390	565	1,605	South Carolina	18	61	74	107	90	350
Indiana	60	135	208	266	344	1,013	South Dakota	6	3	52	30	32	123
Iowa	19	34	76	91	152	372	Tennessee	68	146	223	330	407	1,174
Kansas	23	30	46	69	148	316	Texas	154	230	346	457	569	1,756
Kentucky	28	104	113	91	138	474	Utah	27	63	76	49	87	302
Louisiana	64	115	121	179	234	713	Vermont	1	3	6	7	4	21
Maine	0	12	8	12	11	43	Virginia	43	84	149	235	264	775
Maryland	20	65	76	101	126	388	Washington	61	39	43	47	75	265
Massachusetts	29	68	105	159	184	545	West Virginia	13	47	63	62	99	284
Michigan	50	105	157	244	395	951	Wisconsin	54	118	156	230	306	864
Minnesota	41	106	91	165	150	553	Wyoming	7	8	40	60	111	226
Mississippi	27	40	66	57	89	279	American Samoa	0	0	0	1	0	1
Missouri	71	92	158	276	377	974	Puerto Rico	2	0	0	1	0	3
Montana	31	45	33	52	95	256	Canada	0	17	10	25	20	72

SOURCE: U.S. Department of Transportation

factual data necessary to comply with the Hazardous Materials Transportation Control Act of 1970. These reports are analyzed to reveal problem areas, and changes are made in the regulations, where necessary, to alleviate any threat to public safety.

The regulations requiring reporting of hazardous materials incidents are contained in the following sections of the Code of Federal Regulations:

Title 49, Parts 100 to 199 (governing the transport of hazardous materials by *highway and rail*): Section 171.15—Immediate notice of certain hazardous materials incidents;

Section 171.16—Detailed hazardous materials incident reports.

Title 14, Part 103 (governing the transport of hazardous materials *by air*): Section 103.28 —Reporting certain dangerous article incidents.

Title 46, Parts 1 to 65 (covering the transport of hazardous materials *by water*): Section 2.20-65—Immediate notice of certain hazardous materials incidents; Section 2.20-70—Detailed hazardous materials incident reports.

This system is twofold in that an immediate telephone notice is required under certain conditions and a detailed written report is required whenever there is *any* unintentional release of a hazardous material during transportation or temporary storage related to transportation. Figure 1-1 shows a blank copy of the written form that must be filed.

DEPARTMENT OF TRANSPORTATION

Form Approved **OMB No. 04-5613**

HAZARDOUS MATERIALS INCIDENT REPORT

INSTRUCTIONS: Submit this report in duplicate to the Secretary, Hazardous Materials Regulations Board, Department of Transportation, Washington, D.C. 20590, (ATTN: Op. Div.). If space provided for any item is inadequate, complete that item under Section H, "Remarks", keying to the entry number being completed. Copies of this form, in limited quantities, may be obtained from the Secretary, Hazardous Materials Regulations Board. Additional copies in this prescribed format may be reproduced and used, if on the same size and kind of paper.

A	INCIDENT

1. TYPE OF OPERATION

1 ☐ AIR   2 ☐ HIGHWAY   3 ☐ RAIL   4 ☐ WATER   5 ☐ FREIGHT FORWARDER   6 ☐ OTHER *(Identify)* _____

2. DATE AND TIME OF INCIDENT *(Month - Day - Year)*

_____ a.m.
_____ p.m.

3. LOCATION OF INCIDENT

B	REPORTING CARRIER, COMPANY OR INDIVIDUAL

4. FULL NAME

5. ADDRESS *(Number, Street, City, State and Zip Code)*

6. TYPE OF VEHICLE OR FACILITY

C	SHIPMENT INFORMATION

7. NAME AND ADDRESS OF SHIPPER *(Origin address)*

8. NAME AND ADDRESS OF CONSIGNEE *(Destination address)*

9. SHIPPING PAPER IDENTIFICATION NO.

10. SHIPPING PAPERS ISSUED BY

☐ CARRIER   ☐ SHIPPER

☐ OTHER *(Identify)*

D	DEATHS, INJURIES, LOSS AND DAMAGE

DUE TO HAZARDOUS MATERIALS INVOLVED

11. NUMBER PERSONS INJURED

12. NUMBER PERSONS KILLED

13. ESTIMATED AMOUNT OF LOSS AND/OR PROPERTY DAMAGE INCLUDING COST OF DECONTAMINATION *(Round off in dollars)*

14. ESTIMATED TOTAL QUANTITY OF HAZARDOUS MATERIALS RELEASED

$

E	HAZARDOUS MATERIALS INVOLVED

15. CLASSIFICATION *(Sec. 172.4)*	**16. SHIPPING NAME** *(Sec. 172.5)*	**17. TRADE NAME**

F	NATURE OF PACKAGING FAILURE

18. *(Check all applicable boxes)*

(1) DROPPED IN HANDLING	(2) EXTERNAL PUNCTURE	(3) DAMAGE BY OTHER FREIGHT
(4) WATER DAMAGE	(5) DAMAGE FROM OTHER LIQUID	(6) FREEZING
(7) EXTERNAL HEAT	(8) INTERNAL PRESSURE	(9) CORROSION OR RUST
(10) DEFECTIVE FITTINGS, VALVES, OR CLOSURES	(11) LOOSE FITTINGS, VALVES OR CLOSURES	(12) FAILURE OF INNER RECEPTACLES
(13) BOTTOM FAILURE	(14) BODY OR SIDE FAILURE	(15) WELD FAILURE
(16) CHIME FAILURE	(17) OTHER CONDITIONS *(Identify)*	**19.** *SPACE FOR DOT USE ONLY*

Form DOT F 5800.1 (10-70)

Figure 1-1a. DOT hazardous materials incident report

The U.S. Department of Transportation reports that there are more than 100,000 shippers of hazardous materials, which are transported by over 50,000 truck companies, railroads, and airlines. Added to this, are the millions who either store the material or use it

G	PACKAGING INFORMATION - *If more than one size or type packaging is involved in loss of material show packaging information separately for each. If more space is needed, use Section H "Remarks" below keying to the item number.*				
	ITEM		#1	#2	#3
20	TYPE OF PACKAGING INCLUDING INNER RECEPTACLES (*Steel drums, wooden box, cylinder, etc.*)				
21	CAPACITY OR WEIGHT PER UNIT (*55 gallons, 65 lbs., etc.*)				
22	NUMBER OF PACKAGES FROM WHICH MATERIAL ESCAPED				
23	NUMBER OF PACKAGES OF SAME TYPE IN SHIPMENT				
24	DOT SPECIFICATION NUMBER(S) ON PACKAGES (*21P, 17E, 3AA, etc., or none*)				
25	SHOW ALL OTHER DOT PACKAGING MARKINGS (*Part 178*)				
26	NAME, SYMBOL, OR REGISTRATION NUMBER OF PACKAGING MANUFACTURER				
27	SHOW SERIAL NUMBER OF CYLINDERS, CARGO TANKS, TANK CARS, PORTABLE TANKS				
28	TYPE DOT LABEL(S) APPLIED				
29	IF RECONDITIONED OR REQUALIFIED, SHOW	A REGISTRATION NO. OR SYMBOL			
		B DATE OF LAST TEST OF INSPECTION			
30	IF SHIPMENT IS UNDER DOT OR USCG SPECIAL PERMIT, ENTER PERMIT NO.				

H REMARKS - Describe essential facts of incident including but not limited to defects, damage, probable cause, stowage, action taken at the time discovered, and action taken to prevent future incidents. Include any recommendations to improve packaging, handling, or transportation of hazardous materials. Photographs and diagrams should be submitted when necessary for clarification.

31. NAME OF PERSON PREPARING REPORT (*Type or print*)	32. SIGNATURE
33. TELEPHONE NO. (*Include Area Code*)	34. DATE REPORT PREPARED

Reverse of Form DOT F 5800.1 (10-70)

Figure 1-1b. Reverse side—DOT incident report

☆ U. S. GOVERNMENT PRINTING OFFICE: 1974—626-182/8 3-1

in further processing, which gives some idea of the enormity of the problem.

By highway transportation alone, it is estimated that over one billion tons of dangerous cargo are shipped yearly. This includes 600,000 tons (540,000 metric tons) of blasting agents, 300,000 tons (270,000 metric tons) of poisonous substances, and 100,000 shipments of radioactive isotopes.

In the area of explosives, the Associated Press in the May 20, 1977, issue of the *Washington Post* reports that the Bureau of Alcohol, Tobacco, and Firearms (ATF) is urging that there be increased security on the sale and storage of explosives.[1] The Director of ATF, Rex D. Davis, is quoted as saying, "Any missing explosives should be cause for alarm." He went on to say, "Please do not store explosives in places like an attic, barn, or unsecured shed." Out of the three billion pounds of explosives produced in 1976, 200,000 pounds (90,000 kilograms) of blasting agents and 29,000 blasting caps were reported stolen.

In 1976, the ATF investigated 705 incidents involving stolen explosives. There were 69 deaths, 239 injuries, and $12 million in property damage due to explosives reported in that year.

Not only do emergency response personnel have to worry about legal storage, but they also must be prepared for hidden storage of illegal explosives and the possibility of coming upon them when least expected.

Industry Problems

The U.S. Department of Commerce, Bureau of Census, has established a definition for the chemical and allied industries. This text will refer to the classifications in this definition when speaking of this broad industry:

> This major group includes establishments producing basic chemicals and establishments manufacturing products by predominately chemical processes. Establishments classified in this major group manufacture three general classes of products: (1) basic chemicals such as acids, alkalies, salts and organic chemicals; (2) chemical products to be used in further manufacture, such as synthetic fibers, plastics materials, dye colors and pigments; (3) finished chemical products to be used for ultimate consumption, such as drugs, cosmetics and soaps; or to be used as materials or supplies in other industries, such as paints, fertilizers and explosives.[2]

In the text *Hazard Survey of the Chemical and Allied Industries*, the American Insurance Association (AIA) reports that:

> From 1960 through 1967, a total of 205 "large loss" fires and explosions was reported by the National Fire Protection Association in the chemical and allied industries with an aggregate loss of nearly $165,000,000, resulting in 116 fatalities and 1133 nonfatal injuries.[3]

This gives an idea of the extent of the problem, particularly because chemical production since 1967 has increased at a rapid rate.

The AIA text goes on to discuss 12 trends for the future which have a great impact on emergency response personnel. Among the major changes predicted for the future are:

1. Increased production of plastics, synthetic resins, organics, and petroleum chemicals that use flammable and unstable chemicals in processing.
2. Operation at extremely high pressures and temperatures.
3. Decrease in number of plant personnel as automation of plant continues.
4. Increase in bulk handling of chemicals.
5. Decrease in quality control during construction of the chemical plants.
6. Need for highly skilled operators for the chemical plant.

7. Need for improved preventive maintenance programs.

The top 25 chemicals manufactured in 1973 are listed in Table 1-2. Since the table shows just the top 25 chemicals, it does not even begin to scratch the surface.

In 1976, Congress enacted the Toxic Substances Control Act. The purpose of this act was to enable some control to be exercised over the large numbers of chemicals before an emergency took place. Under the law, the Environmental Protection Agency (EPA) was given broad powers to regulate the manufacture, use, and disposal of toxic chemicals. The industry will be required to report new chemicals and pretest

certain ones before they can be introduced into manufacturing use or commercial products.

One of the key provisions of the act gives the EPA the authority to collect information on existing industrial chemicals, including chemical names, uses, production figures, by-products, waste disposal, and health effects. Obviously, these data are very useful to emergency response personnel. However, only time will tell how well these requirements will be implemented. At the present time, the only EPA requirement under the law is that manufacturers provide a list of chemicals produced.

In the February 7, 1977, issue of *Chemical and Engineering News*,[4] the following in-

Table 1-2.   Chemicals Manufactured in 1973

Name	Billions of Pounds
Sulfuric acid	63.18
Oxygen, high and low purity	31.87
Ammonia, synthetic anhydrous	30.94
Ethylene	22.41
Sodium hydroxide, 100% liquid	21.36
Chlorine, gas	20.60
Nitrogen, high and low purity	16.38
Sodium carbonate, synthetic and natural	14.99
Nitric acid	14.86
Ammonium nitrate, original solution	13.89
Phosphoric acid, total	13.00
Benzene, all grades	10.65
Propylene	8.76
Ethylene dichloride	7.90
Urea, primary solution	7.12
Methanol, synthetic	7.12
Toluene, all grades	6.78
Ethylbenzene	6.50
Formaldehyde, 37% by weight	6.17
Styrene	6.01
Xylene, all grades	5.90
Vinyl chloride	5.35
Hydrochloric acid, total	4.78
Ammonium sulfate	3.97
Ethylene oxide	3.88

creases in chemical production were announced:

1. Shell Chemical will build units to make 400 million pounds (180 million kilograms) per year of ethylene oxide and 200 million pounds (90 million kilograms) per year of ethylene glycol.

2. Union Carbide will double its manufacture of liquid hydrogen to 17 tons (15,300 kilograms) per day.

3. Farmland Industries started its second 1,000-ton-per-day (900,000 kilograms) agricultural ammonia plant.

4. BASF Wyandotte began operation of a 55-million-pound-per-year (25 million kilograms) plant to manufacture butanediol.

5. Milbrew, Inc., expanded its production of ethanol by 2 million gallons (7.58 million liters) per year.

6. USS Chemicals increased the manufacture of phthalic anhydride to 200 million pounds (90 million kilograms) annually.

And remember, all this was reported in a single week.

A SAMPLING OF INCIDENTS ON RECORD

By looking at some of the previous hazardous materials incidents, several very important lessons can be learned. The case histories that follow are representative of the wide range of possible mishaps in the shipping and handling of hazardous materials. The circumstances and the actions taken in these disasters, which occurred in small as well as in large communities, point up the kinds of problems that must be anticipated in coping with contingencies.

Texas City, Texas, Waterfront Disaster, April 16, 1947[5]

The *S.S. Grandcamp* was docked at Texas City, Texas, to take on a load of ammonium nitrate. Some of the ammonium nitrate had been loaded the day before. At 8:00 A.M. on April 16, 1947, the hatch was opened in preparation for continuing to fill the hold with additional ammonium nitrate.

A stevedore, who had descended into the hold to receive the additional cargo, noticed smoke and found the cargo burning. Several buckets of water were thrown at the fire without having any effect. Next, a soda-acid extinguisher was tried, but this did not work either.

A hose line was ordered, but the order was given not to use it for fear of damaging the cargo. At 8:30 A.M. all personnel were ordered off the ship. The fire department was notified at approximately 8:30 A.M. Two engines were dispatched immediately, with two others responding shortly afterward. The fire department placed some hose lines in service, but the hull was already so hot that the water vaporized as soon as it struck the metal.

An attempt was made to seal the openings and extinguish the fire with the ship's steam-generating equipment. With everything sealed off, the fire spread and gas pressure built up internally. This increased pressure blew off the covers, and large amounts of reddish-brown smoke billowed skyward.

At 9:12 A.M. the *Grandcamp* exploded with great violence, with another explosion occurring approximately five seconds later. The shocks from the explosion were felt in Galveston, 10 miles away, and windows were broken in Baytown, 25 miles away.

A tidal wave more than 15 feet (4.5 meters) high caused water to flow a considerable distance inland. The *Grandcamp* explosions also tore the *S.S. High Flyer* loose from its moorings and into the side of the *S.S. Wilson B. Keene*. The *High Flyer* was loaded with 961 tons (865 metric tons) of ammonium nitrate and 2,000 tons (1,800 metric tons) of sulfur. An attempt was made to separate the two ships and move one away from the dock area. At 1:12 A.M. on April 18, 1947, while tugs were trying to move the *High Flyer*, another explosion occurred. This explosion killed additional personnel, including several aboard the tug.

Damage to property in this disaster was extensive, both in the dock area and the surrounding areas. Approximately 1,000 buildings received major structural damage, portions of the cargo were found over two miles away, and many flammable liquid storage tanks were burning.

There were 468 people killed, with approximately 2,000 suffering injuries. The property loss was put in excess of $40 million.

Investigation indicated that it was the decision to seal off the cargo area and introduce steam to extinguish the original fire that touched off the explosion. The heat and pressure within the sealed hold built up so rapidly that it caused detonation of the ammonium nitrate cargo.

Holland Tunnel Chemical Fire, May 13, 1949[6]

A truck containing 48,536 pounds (21,841 kilograms) of carbon disulfide (in 80, 55-gallon drums) entered the Holland Tunnel at the Jersey City end at approximately 8:45 A.M., on May 13, 1949. (The Holland Tunnel connects New York City with Jersey City, New Jersey, under the Hudson River.) When the truck was about 2,900 feet (870 meters) from the New Jersey entrance [about 6,377 feet (1,913 meters) from the New York exit], it caught fire.

Because of the traffic stalled behind the burning truck, access from the New Jersey side was severely hampered. Traffic in front of the burning vehicle had to proceed out of the tunnel to allow fire equipment in to fight the fire. The high heat damaged the ventilating system, forcing a partial shutdown. This, in turn, caused a build up of toxic fumes in the tunnel.

The fire spread to other trucks, which were abandoned on the roadway by their drivers. These trucks contained meat, bottled bleach, paint supplies, wax, food supplies, paper rolls, and wooden barrels.

With the combined efforts of the New York City and Jersey City Fire Departments, as well as police officers from both jurisdictions, assisted by tunnel emergency crews, they managed to extinguish the fire. Fortunately, there were no fatalities, but there were 66 injured. No long-term study was performed to determine if the injured suffered any residual lung damage from the fumes.

The structural damage to the tunnel exceeded $600,000, while the damage to the trucks, cargos, and telephone and telegraph cables was approximately $400,000.

Problems were encountered with the water supply, breathing apparatus, access, and coordination among the different agencies.

It should be noted that the truck entered the tunnel in violation of the Port Authority regulations with regard to the quantity of chemical carried and the type of containers used. In addition, the truck did not display the required placards.

Brighton, New York, Gas Explosion, September 21, 1951[7]

A regulator vault for the gas distribution system in Brighton, New York, was excavated so that work could be done in connection with a corrosion program. When the work was completed on September 19, 1951, the opening was backfilled and work in preparation for cementing over the sidewalk was begun. On September 21, at about 1:15 P.M., one of the workers set out kerosene warning lanterns to protect the fresh cement. Within a few minutes of placing the lanterns, an explosion rocked the area, with fire erupting at all the broken gas connections.

As a result of damage to the regulating equipment in the vault, gas at 30 psi pressure flowed into the medium- and low-pressure systems. This excess pressure in the low-pressure system caused leaks at the meters, turned pilot lights into torches, came out the bottom of the hot water tanks, or even extinguished the pilot lights and let un-burned gas escape.

The resulting explosions killed 2, injured 30, destroyed 19 houses by internal explosions and fires, badly damaged 25 other dwellings, and caused minor damage to 350 other dwellings. Total loss was estimated at $1,000,000.

Roseburg, Oregon, Explosion, August 7, 1959[8]

A truck driver, driving with a load of 2 tons (1,800 kilograms) of dynamite and 4-½ tons (4,050 kilograms) of blasting agent from the manufacturing plant to Roseburg, Oregon, parked for the night in an alley next to a building material warehouse. Placards labeled "Explosives" were in place on the truck when it was parked.

During the night, the warehouse caught fire. Less than 10 minutes after the fire department arrived on the scene, the truck exploded. The explosion left a crater 52 feet (17 meters) in diameter and 20 feet (6 meters) deep.

In addition to the original building fire, there were now fires in more than 45 buildings in a 7-block area, with damage extending over a 50-block area. Thirteen people, including the assistant fire chief, were killed. Flames from the fire rose over 300 feet (90 meters) in the air. The major initial concern was exposure protection, particularly a 166,000-gallon (629,000-liter) liquefied petroleum gas storage area, less than 400 feet (120 meters) from the blast.

With mutual aid to replace pumpers damaged by the blast and to overcome the rapidly spreading conflagration, the fire was controlled in a little over an hour.

Toledo, Ohio, Gasoline Tank Truck, June 10, 1961[9]

A gasoline tank-semitrailer and a tank-full trailer carrying 7,900 gallons of product lost control on a curve and landed on its side. The center dome opened and released gasoline. Fire ensued immediately, in both the semitrailer and full trailer.

Approximately 7 minutes after the fire began, the tank ruptured, trapping 11 fire fighters. Of these 11, 4 died, and 61 spectators were injured. Four houses were destroyed by the resulting fire.

Cheverly, Maryland, Freight Derailment, May 21, 1964[10]

An 88-car freight train crashed into a switching engine, derailing 19 cars and sending them down a ravine. Three adjacent cars contained liquid oxygen, ethylene oxide, and transformer oil. The drain valve on the liquid oxygen car broke, which allowed oxygen at -297°F to flow out.

Upon impact, the diesel locomotive broke into flame, and the fire was fed by the liquid oxygen. Paper products in some of the other freight cars added fuel to the fire. Fortunately, the ethylene oxide car did not rupture.

To add to the problems, the derailment knocked down a 13,000-volt transmission line. Water supply was the major problem because of the inaccessibility of the area. To get water to the scene required 39 engine companies operating in a relay with 7 miles (11.2 kilometers) of hose.

Philadelphia, Pennsylvania, Refinery Fire, May 23, 1966[11]

A fire at the Gulf Refinery in Philadelphia

began in the still used to manufacture isopropyl benzene. A leak was caused when a 6-inch (15 cm) pipe ruptured and allowed escaping flammable liquid to vaporize. The gas was ignited by an open flame about 150 feet (45 meters) away and flashed back to the still.

The fire ruptured a steam line under 600 psi pressure. The noise created by the escaping steam made communications impossible. The only way to get orders through was to send a runner or to use hand signals. As a result of the great noise, 11 fire fighters were treated for acoustical trauma.

With the use of 22 portable and 4 mounted monitors, the heat was reduced so that attack with foam could be mounted. The foam brought the fire under control and final extinguishment was achieved.

Dunreith, Indiana, Train Derailment, January 1, 1968[12]

A derailment at Dunreith, Indiana, caused several tank cars containing flammable liquids as well as poisons to leak. An ethylene oxide car exploded and severely injured fire fighters and spectators.

The liquid poison was acetone cyanohydrin. The car burned, and some of the liquid leaked onto the ground. From the ground it drained into nearby streams, finally flowing into the river. Cattle along the stream bank were killed in great numbers. Wells were dug to pump out the contaminated ground water. In 11 months over 2 million gallons (7.58 million liters) of water with cyanide were removed for treatment.

The logistics problem associated with an incident of this duration becomes formidable. Reliable water supplies have to be provided for residential as well as industrial use. Law enforcement agencies must assist in maintaining security in the affected area. Temporary assistance must be provided to individuals displaced by the incident.

Orange County, New York, Truck Fire, April 4, 1974[13]

A truck fire on the New York State Thruway appeared to be an ordinary call. The first-in unit found a fire in the undercarriage of the truck. After this fire was extinguished smoke was still coming from the closed cargo compartment.

While the fire officer checked the bills of lading with the driver, the doors were opened. At the rear of the truck were paper bags of material piled on pallets. Rolls of cloth behind these pallets were burning. The bills of lading did not indicate what was in the sacks. The driver explained that the sacks were loaded with Japanese rice.

The label on the sacks was written in an oriental language except for "Tris (2 Hydroxyethyl) Isocyanurate." These sacks were removed and dumped on the ground (some breaking open) so that the fire in the cloth could be extinguished. At this point all personnel were told to don breathing apparatus.

Shortly afterward, one fire fighter began to bleed from the nose. Others began to become disoriented. In all, 8 fire fighters were transported for treatment with 2 in critical condition. With the help of CHEMTREC (the Chemical Transportation Emergency Center, discussed in Chapter 6), the material was identified as a toxic substance that could be absorbed through the skin.

It is important to note that there were no hazardous warning placards on the outside of the vehicle, nor was the bill of lading correctly marked. Fires involving unknown substances or in closed containers must be treated as hazardous substances until identification of the exact contents can be made.

Hutchinson, Kansas, Ammonia Pipeline Break, August 13, 1974[14]

An underground, 8-inch (20 cm) pipeline carrying liquid anhydrous ammonia broke and allowed the product to leak out under pressure. Over 200 residents had to be evacuated, with 4 being treated for inhalation of the ammonia fumes. The fumes killed 11,000 fish and all vegetation in its path.

Because of the distance between control valves, the liquid continued to flow. In this case, 6 miles (9.6 kilometers) of pipe had to be bled before the ammonia was exhausted. During that time all living things in the path of the cloud were in trouble.

Many hazardous materials incidents lead to evacuation of civilians from the area. In this event, emergency response personnel must be prepared to cope with the additional problems of medical assistance, arranging for food, clothing, and shelter, and protection of vacated homes and facilities.

Los Angeles, California, Truck Fire, April 10, 1975[15]

A blowout on a tractor pulling two 27-foot (8-meter) trailers caused the tractor to overturn and begin to burn. One trailer carried 24,000 pounds (10,800 kilograms) of insecticide, methomyl, and the other carried tires. There were no hazardous materials markings on the trailers. Since there was no warning, the fire fighters did not don self-contained breathing apparatus.

When the fire suppression activity in the forward trailer began, it was noted that the boxes were marked poison and flammable. The burning insecticide released toxic vapors that began to affect the fire fighters. The fumes sent 43 fire fighters and 51 civilians to the hospital for treatment.

Milwaukee, Wisconsin, Industrial Leak, May 1, 1975[16]

An industrial firm used both sodium sulfhydrate and sulfuric acid in its processing. Both of these chemicals were stored in tanks that were filled from tank trucks outside the plant via a fill pipe.

Through an error, the operator of a tank truck of sulfuric acid connected it to the sodium sulfhydrate fill pipe. As the acid flowed into the wrong tank, hydrogen sulfide gas was evolved. This poisonous gas dropped workers where they stood because it deadens the sense of smell and then attacks the automatic respiration function of the brain.

Search and rescue efforts were complicated by the fact that an unknown number of people were in the building, others left on their own, and some went directly to the hospital themselves. All told, 3 were killed and 20 were taken to the hospital for treatment, including 6 fire fighters.

As the officer in charge of this incident, how would you find out the nature of the hazard? How would you determine the number of workers? Would the hospital be prepared for a large influx of patients? How much equipment and manpower would be needed at the scene?

Alief, Texas, Chemical Release, June 12, 1976[17]

In this incident, an individual who broke a jar containing a yellowish powder while mowing a vacant lot was covered with the fumes and powder, which caused burning eyes and a swollen face. Fire fighters called to the scene, as well as a neighbor who washed the mower, all experienced the same symptoms. In all, 13 individuals were admitted to the hospital on the first day. By the time removal efforts were completed, 9 more were hospitalized.

Several days later the patients began to show signs of liver and kidney damage, as well as chemical pneumonia. Since the identity of the chemical could not be determined, treatment was very generalized. Each new symptom was treated specifically as it developed. Finally, on June 15, 1976, the chemical was identified as Ortho-Chloro Benzyl Idine Malonitrite. It is used in tear gas cannisters and other antipersonal devices in which it is burned to produce tear gas.

Once the chemical was identified, proper treatment was administered. All those hospitalized recovered and were released.

Could the individual have been rescued without endangering emergency services personnel? Where would you go for an analysis of an unknown chemical? What sources of information on identification are available? Are the medical personnel prepared to help you in this type of situation?

Clifford, Michigan, Train Derailment, October 18, 1976[18]

Because a switch was turned in the wrong direction, a through freight train was suddenly shunted onto an occupied siding. The train crashed into some parked box cars, and derailed them. Two tank cars, one containing butadiene and the other acrylonitrile ruptured and immediately began to burn. The toxic fumes from the burning cars forced the complete evacuation of the town of Clifford. In addition, the town of North Branch, 10 miles (16 kilometers) away, was evacuated.

The fire burned for three days before the chemicals were completely exhausted. Shortly after the fire was extinguished, one of the fire fighters began experiencing symptoms of butadiene poisoning. This individual and 16 other fire fighters were admitted to the hospital. There, oxygen was administered for 36 hours, and all recovered without incident.

Again, a large-scale evacuation was required. Close coordination between all of the emergency forces was necessary to accomplish it quickly and efficiently. Could this be accomplished without ever having practiced it? Would it be possible to think of all of the details without having a written plan?

Montgomery County, Maryland, Pesticide Fire, May 29, 1977[19]

A fire in a storage building of a garden and tree care center sent over 90 personnel, including 74 fire fighters, to area hospitals for treatment of inhalation of pesticides. Persons living in the immediate area were evacuated. Water run-off from the fire fighting attack had to be impounded to prevent pollution of the ground water. Emergency personnel must be prepared for evacuating citizens, large-scale medical emergencies, individuals displaced from their homes, as well as providing heavy equipment to prevent run-off.

Fairbanks, Alaska, Pipeline Break, July 8, 1977[20]

While workers were preparing to put a pump on line at Pump Station 8 of the Alaska Pipeline, a malfunction caused oil from the pipeline to be discharged into the building, where it found a source of ignition. The explosion killed 1 worker and injured 5 others. The oil continued to flow for several minutes before the line was shut down by remote control from Valdez, Alaska, 300 miles (480 kilometers) away. Even after shutdown, however, oil still in the line continued to bleed out. In addition to the structural fire, several brush fires were also set. Fire fighters were hampered by the remoteness of the site, lack of water, and difficult access. Would it be possible to get equipment and water to any place on the pipeline that passes through your area?

Rockwood, Tennessee, Bromine Leak, July 12, 1977[21]

A tank truck containing 1,800 gallons (6,800 liters) of liquid bromine was involved in an accident on an interstate highway near Rockwood, Tennessee. When the tanker rolled on its side it split open, allowing the bromine to escape. The driver was killed and a rider was taken to the hospital with severe respiratory problems from inhaling the fumes. The entire town of 5,259 persons, as well as the 55 patients in a local hospital, had to be evacuated.

What is the status of emergency action plans in your area? Where, for example, could 55 hospital patients be moved so that they could continue to get medical help? These and other questions raised in the foregoing incident reports are typical of those that the emergency response team may be expected to deal with in a hazardous materials incident. The information gathered in this book will help you to prepare for this responsibility.

REGULATION OF HAZARDOUS MATERIALS TRANSPORTATION

There are five major methods for moving hazardous materials from one point to another: train, barge, truck, airplane, pipeline. Within each of the methods are various elements that make up the transportation system:

1. The base on which the system moves

2. The vehicle that does the transport

3. The container for the material

4. The cargo that goes into the container

Each method can interrelate with each element and form many possible combinations, each one presenting a different problem to the emergency response team. A matrix of the methods and elements graphically illustrates this point.

Table 1-3 shows how the seriousness of the hazard is affected by the type of cargo (a solid might just stay in place, while a liquid could cause a running spill fire); by type of container (a box within a freight car might more easily survive a wreck than a tank car full of a flammable gas); by the type of vehicle

Table 1-3 Transportation System Versus Transportation Elements

Transportation System	Base	Vehicle	Container	Cargo
Train	Track	Box car Tank car Hopper car Trailer on flat car (TOFC)	Individual package Whole tank car Whole hopper car	Solid Liquid Gas Cryogenics
Truck	Roadway	Trailer Tank truck Bulk carrier	Individual package Whole tank truck Bulk quantities	Solid Liquid Gas Cryogenics
Vessel	Waterway Inland Ocean Port	Barge Tanker Freighter	Individual package Whole tank Bulk quantities	Solid Liquid Gas Cryogenics
Airplane	Air	Commercial freighter Light plane	Individual package	Solid Liquid Gas
Pipeline	Right-of-way	Transmission line Distribution line	Various size pipe	Liquid Gas

(large-diameter, high-pressure transmission lines take longer to bleed down than smaller distribution mains); and finally, by the type of base (a waterway spill can travel miles downstream, whereas a rail accident can be localized).

Since the Department of Transportation (DOT) has the overall responsibility for regulating the various modes of transportation, an understanding of how it is organized (Figure 1-2) will aid emergency response personnel. The multiple responsibility for overseeing each mode of transportation is evident from looking at the DOT organization chart. Within the Federal Department of Transportation, a different agency has regulatory responsibility for each method of transportation:

Mode	Agency
Train	Federal Railroad Administration
Truck	Federal Highway Administration
Vessels	Coast Guard
Airplane	Federal Aviation Administration
Pipeline	Office of Pipeline Safety

With the regulations concerning the base, vehicle, container, and cargo varying within each agency, it is no wonder that emergency response personnel have difficulty in dealing with hazardous materials transportation incidents. Each agency reacts to pressures on the one side brought by the trade associations and on the other side from the emergency forces, particularly after a disaster. The end result is a confusion of requirements that do not satisfy the needs of emergency response personnel and appear too complex for the shippers. Rules and regulations are ignored or disregarded because the individual shipper is too confused and takes a "who cares" attitude.

In addition to the above-mentioned regulatory agencies, the Materials Transportation Bureau was established within the Department of Transportation in 1975. Within the bureau, the Office of Hazardous Materials Operations was organized. The main functions for this office are:

1. To regulate intermodal transportation (combinations of transportation such as truck and train or train and barge)

DEPARTMENT OF TRANSPORTATION

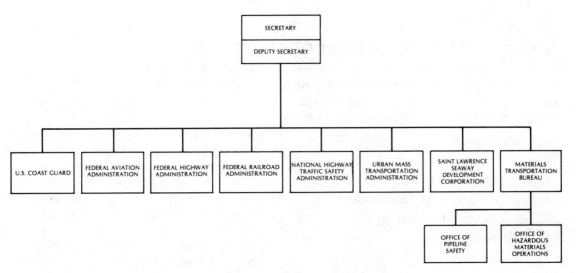

Figure 1-2. Organization chart for the Department of Transportation

2. To collect and analyze hazardous materials transportation statistics

3. To conduct training in handling intermodal transportation

4. To register individuals who transport or ship hazardous materials

The National Transportation Safety Board is an independent agency of the federal government, responsible directly to the executive branch. Their task is to investigate hazardous materials transportation accidents, determine the cause, and make recommendations for improved safety to prevent a recurrence. Their suggestions for improvement are usually made to the regulatory agencies, who must study the recommendations, prepare written changes to the regulations, publish the proposed changes, hold public hearings for comments, study and react to the comments, prepare the final procedures, publish the procedures with an effective date, and finally, enforce the new procedures. Obviously, between the time the safety board makes its recommendation and it is implemented by the regulatory agency, many changes can be made and a great deal of time has elapsed.

While statistics concerning accidents do not tell the whole story, they do indicate trends. In the November 1971 issue of *Fire Journal*,[22] Walter M. Haessler cites the following statistics:

1. Between 1963 and 1968 train accidents increased 66 percent.

2. Highway freight is increasing at four times the population growth.

3. Waterway transportation is increasing at three times the population growth.

The *Philadelphia Bulletin* reports that in 1976 the Southern Railway System recorded a 48 percent increase in accidents over the preceding year.[23]

The transportation hazard is great and will continue to grow larger. As the economic demands push for heavier and larger loads, while at the same time reducing safety factors (road beds, rail maintenance, and waterway lock maintenance), accidents will occur with increasing frequency. Preparation is the key word, if emergency response personnel are to handle an incident in the most efficient manner.

PLACES TO CHECK FOR HAZARDOUS MATERIALS

While emphasis has been placed on incidents arising from the transportation of hazardous materials, this is just one facet of the problem. Hazardous materials must be:

1. Manufactured

2. Stored before transport

3. Transported to a producer

4. Used in manufacturing a product

5. Stored before transport

6. Transported to user

7. Stored by user

8. Transported to waste disposal facility

9. Stored, burned, or buried by a waste disposal facility

The diversity of elements in the chain of hazardous materials handling brings many other federal agencies besides the Department of Transportation into the regulatory business.

The Toxic Substances Control Act, which became effective in 1977, requires the Environmental Protection Agency (EPA) to compile an inventory of all chemicals manufactured between 1974 and 1976. Each chemical is to be studied, with some requiring laboratory testing to assure that they are not hazardous. As new chemicals are introduced, they will also be tested from a safety standpoint.

The EPA is also researching techniques for controlling hazardous spills and has a program for developing and testing devices to be used in case of a spill. Once a prototype device has proved functional, the EPA encourages local jurisdictions to purchase the product for spill control.

The EPA is also responsible for pesticide control. They regulate labeling, packaging, and handling of the many different types of pesticides produced.

The Food and Drug Administration (FDA) also gets involved in the chemical field. While their prime responsibility is the drug industry, the FDA also sets labeling and packaging requirements for drug products. Emergency response personnel must be familiar with labeling techinques in this area.

Each step of the process in the hazardous materials handling chain can present a different type of problem to the emergency response team. The tactics employed in an incident will depend on the stage in the sequence from manufacture to waste disposal in which it occurs.

Manufacturing

"In 1939, the value of chemical shipments exceeded 4 billion dollars; however, by 1966 this figure had risen ninefold to 38.7 billion dollars. Economists estimate that by 1975, the total sales of the chemical industry will reach 70 billion dollars."[24] The magnitude of the chemical manufacturing can be seen in these sales figures. A typical manufacturing plant is shown in Figure 1-3.

Producers

Producers use the basic chemicals in processes for manufactured products. Examples of chemicals in this stage are synthetic

fibers, paints and pigments, and plastics. In addition, the chemicals can be used to provide finished products such as fertilizers, pesticides, paints, drugs, and tires.

Figure 1-3. Chemical manufacturing plant. *Courtesy of the Fertilizer Institute*

Waste Disposal

Hazardous wastes are the particularly dangerous discards of our highly industrialized, science-and-technology-based society. They should be disposed of carefully, but sometimes they are not. Sometimes they are put in open dumps, in landfills designed only for residential refuse, in the ocean, or they are just left around in warehouses. Following are some examples of what can happen:

1. A factory begins depositing chemical wastes in a nearby city dump. Chemicals soon work their way into the soil and down into a spring beneath the dump; from the spring they pass into a creek and thence into one of the nation's major rivers.

2. A commercial laboratory dumps some of its wastes on the open ground within the plant. Soon the soil is contaminated. Then the groundwater is contaminated and becomes unsafe for drinking or for irrigation.

3. In the warehouse of a local weed control agency, someone discovers several drums of a 15-year-old chemical once used to sterilize soil. The drums are taken to a remote area and left there. A rifle shot rings out. A drum explodes. Had the drums been jarred while at the warehouse, several people would have been killed, for the drums of obsolete chemicals had slowly, imperceptibly, turned into time bombs.

4. On the beaches of a southeastern state, glass containers about 5 to 8 inches long are washed ashore from the Atlantic. Investigators determine that the containers were buried at sea shortly after World War II but were broken free of their crates by undersea turbulence. The public is warned not to touch the easily broken containers; they are filled with carbon disulfide, a gas so lethal it could kill if brought in contact with the skin or if inhaled.

Industry generates at least 10 million tons (9 million metric tons) of nonradioactive hazardous wastes a year (this is about 10 percent of all wastes produced by industry), and the amount is growing at a rate of 5 to 10 percent annually. Approximately 90 percent of industrial hazardous wastes are in liquid or semiliquid form. Almost all of the 10 million tons (9 million metric tons) is toxic. Industry's toxic throwaways fall into one or more of these categories: inorganic toxic metals; salts, acids, or bases; synthetic organics; flammables; and explosives.

Still another hazard is presented by the 5,000 tons (4,500 metric tons) of synthetic organic pesticide wastes produced annually as well as the 250 million pesticide containers that though "empty" still contain residue. Hospitals add 170,000 tons (153,000 metric tons) of pathological materials to the hazardous waste load each year.

Radioactive wastes are both high- and low-level residues from nuclear power plants and fuel processing facilities; medical and industrial research labs; and the weapons and research facilities of the U.S. Energy and Research and Development Administration and the Department of Defense. It is estimated that about 24,000 tons (21,600 metric tons) a year of radioactive waste are produced.

With the increasing emphasis on the long-term effects of chemicals on the environment and the human body, waste disposal takes on increasing significance. Because some of these chemicals are slow to degrade, various techniques from burning to burying at sea are used. In each stage of the handling for disposal, an accident can involve the emergency response forces. Everything from landfills to beach areas becomes a locale for a potential hazardous material incident.

Common Locations

When preplanning and studying the hazardous materials situation, many departments completely pass over very common locations. For example, any firm that produces blueprints has a storage of hazardous materials. The original drawing is exposed to ultraviolet light and then developed in ammonia. Other common locations of hazardous materials are a high school with its chemistry laboratory; a construction site

Figure 1-4. Propane distribution

that uses explosives for rock removal; the sporting goods store with its ammunition and fuel for outdoor stoves; and the garden center with fertilizers and pesticides.

Then, too, there are the more obvious places that must be considered. Gasoline stations, chemical distributors, propane distributors (Figure 1-4), bulk storage plants (Figure 1-5), pipeline pumping stations (Figure 1-6), and hospitals are the ones that come to mind immediately.

Each building, manufacturer, storage facility, truck depot, highway, or rail line must be considered a potential location for a hazardous material incident. Visit. Ask questions. Stay alert. These are the watchwords you must follow if you expect to do the job correctly. If you do not, then you will be unprepared for a truck accident involving compressed gas cylinders (Figure 1-7); a train accident involving a flammable liquid (Figure 1-8); or a prank of igniting a 55-gallon drum of flammable liquid in a storm sewer (Figure 1-9).

Figure 1-7. Truck accident involving compressed gas cylinders

Figure 1-5. Bulk storage plant

Figure 1-8. Train accident involving a flammable liquid car

Figure 1-6. Pipeline pumping station

Figure 1-9. A 55-gallon drum of a flammable liquid ignited in a storm sewer

CONTINUING INFORMATION SOURCES

Emergency response personnel must continually stay up-to-date on the latest happenings in the hazardous materials field. In the many periodicals on fire service activities, incidents of various kinds are discussed, both correct and incorrect actions are covered, and results of the techniques used are detailed. These and other sources of information for emergency response personnel are listed in the following pages. For further information write directly to the addresses shown.

Federal Register

All new regulations or changes to existing ones dealing with hazardous materials must be published in the Federal Register. In order to stay abreast of the latest requirements, the *Federal Register* should be read regularly. It is available at most libraries or you or your department can subscribe to it directly. The address is:

Federal Register
Superintendent of Documents
U. S. Government Printing Office
Washington, D. C. 20402

National Technical Information Service

The National Technical Information Service (NTIS) is the publisher of most of the technical documents produced for the federal government by either government agencies or private contractors. The NTIS will provide bibliographies on specific topics, listing research reports on the subject. The individual can then order reports that are of interest. The address for NTIS is:

National Technical Information Service
Department of Commerce
Springfield, Virginia 22151

Periodicals

The periodicals provide up-to-date reports on the tactics used in handling actual hazardous materials incidents. They also publish information on the latest laws and recommended changes to current procedures, as well as technical articles on the latest scientific developments. Addresses for the major publications are:

Fire Engineering
666 Fifth Avenue
New York, New York 10019

Fire Command
National Fire
Protection Association
470 Atlantic Avenue
Boston, Massachusetts 02210

Fire Journal
National Fire
Protection Association
470 Atlantic Avenue
Boston, Massachusetts 02210

Fire Chief
625 North Michigan Avenue
Chicago, Illinois 60611

Firehouse
Firehouse Magazine
Association
33 East 53rd Street
New York, New York 10022

International Fire Chief
International Assocation
of Fire Chiefs
1329 18th Street, N.W.
Washington, D. C. 20036

Federal Agencies

Many Federal Agencies take part in the hazardous materials program. These agencies have specialized information available for use by emergency response personnel. The agencies and their addresses are:

National Transportation Safety Board
800 Independence Avenue, S.W.
Washington, D. C. 20592

Department of Transportation
Railroad Safety Office
Highway Safety Office
Coast Guard
400 Seventh Street, S.W.
Washington, D. C. 20590

United States Fire Administration
National Fire Academy
P. O. Box 19518
Washington, D. C. 20036

Materials Transportation Bureau
Office of Hazardous Materials
2100 Second Street, S.W.
Washington, D. C. 20590

Materials Transportation Bureau
Office of Pipeline Safety Operations
Regional Offices:

EASTERN REGION
Room 6315
2100 Second Street, S.W.
Washington, D.C. 20590
STATES SERVED: Connecticut, Delaware, District of Columbia, Maryland, Massachusetts, New Hampshire, New Jersey, New York, Pennsylvania, Rhode Island, Vermont, Virginia, West Virginia, Puerto Rico

SOUTHERN REGION
1568 Willingham Drive
Atlanta, Georgia 30337
STATES SERVED: Alabama, Florida, Georgia, Kentucky, Mississippi, North Carolina, South Carolina, Tennessee

CENTRAL REGION
Room 1802
911 Walnut Street
Kansas City, Missouri 64106
STATES SERVED: Illinois, Indiana, Iowa, Kansas, Michigan, Minnesota, Missouri, Ohio, Wisconsin, Nebraska

SOUTHWEST REGION
6634 Hornwood Drive
Houston, Texas 77036
STATES SERVED: Arkansas, Louisiana, New Mexico, Oklahoma, Texas

WESTERN REGION
831 Mitten Road
Burlingame, California 94010
STATES SERVED: Arizona, California, Colorado, Idaho, Montana, Nevada, North Dakota, Oregon, South Dakota, Utah, Washington, Wyoming, Alaska, Hawaii

OHM-TADS Project Officer
Division of Oil and Special Materials Control
Environmental Protection Agency (WH-548)
Washington, D.C. 20460

U.S. Environmental Protection Agency
Industrial Environmental Research Laboratory
Oil and Hazardous Materials Spills Branch
Edison, New Jersey 08817

Environmental Protection Agency
Regional Offices:

REGION I
Room 2303
John F. Kennedy Federal Building
Boston, Massachusetts 02203
Telephone: (617) 223-7265

REGION II
Room 908
26 Federal Plaza
New York, New York 10007
Telephone: (201) 548-8730

REGION III
Curtis Building

6th and Walnut Streets
Philadelphia, Pennsylvania 19106
Telephone: (215) 597-9898

REGION IV
345 Courtland Street, N.E.
Atlanta, Georgia 30308
Telephone: (404) 881-4062

REGION V
230 South Dearborn Avenue
Chicago, Illinois 60604
Telephone: (312) 896-7591

REGION VI
Suite 1600
1600 Patterson Street
Dallas, Texas 75201
Telephone: (214) 749-3840

REGION VII
1735 Baltimore Avenue
Kansas City, Missouri 64108
Telephone: (816) 374-3778

REGION VIII
Suite 900
1860 Lincoln Street
Denver, Colorado 80203
Telephone: (303) 837-3880

REGION IX
100 California Street
San Francisco, California 94111
Telephone: (415) 556-6254

REGION X
1200 6th Avenue
Seattle, Washington 98101
Telephone: (206) 442-1200

Energy Research and Development
Administration
(for radiological assistance)
Regional Offices:

REGION 1
Upton, L.I., New York 11973
Telephone: (516) 345-2200
STATES SERVED: Maine, New Hampshire,
Vermont, Massachusetts, Connecticut,
Rhode Island, New York, New Jersey, Pennsylvania, Maryland, Delaware, District of Columbia

REGION 2
P.O. Box E
Oak Ridge, Tennessee 37830
Telephone: (615) 483-8611
Extension 3-4510
STATES SERVED: West Virginia, Virginia, Kentucky, Tennessee, Puerto Rico, Virgin Islands, Mississippi, Louisiana, Arkansas, Missouri

REGION 3
P.O. Box A
Aiken, South Carolina 29801
Telephone: (803) 824-6331
Extension 3333
STATES SERVED: North Carolina, South Carolina, Georgia, Alabama, Florida, Canal Zone

REGION 4
P.O. Box 5400
Albuquerque, New Mexico 87115
Telephone: (505) 264-4667
STATES SERVED: Texas, Oklahoma, Kansas, New Mexico, Arizona

REGION 5
9800 South Cass Avenue
Argonne, Illinois 60439
Telephone: (312) 739-7711
Extension 2111, duty hours
Extension 4451, off hours
STATES SERVED: Ohio, Indiana, Michigan, Illinois, Wisconsin, Iowa, Minnesota, Nebraska, South Dakota, North Dakota

REGION 6
P.O. Box 2108
Idaho Falls, Idaho 83401
Telephone: (208) 526-0111
Extension 1515
STATES SERVED: Colorado, Utah, Wyoming, Montana, Idaho

REGION 7
1333 Broadway
Oakland, California 94704
Telephone: (415) 273-4237
STATES SERVED: Nevada, California, Hawaii

REGION 8
P.O. Box 550
Richland, Washington 99352
Telephone: (509) 942-7381
STATES SERVED: Oregon, Washington, Alaska

Trade Associations

Many of the organizations engaged in hazardous materials work have joined together in numerous trade associations. These associations can provide information on regulations, handling, safety standards, and incidents. Many of them also produce visual aids for use by emergency response personnel in training programs. The various trade associations can be reached at the following addresses:

American Petroleum Institute
2101 L Street, N.W.
Washington, D.C. 20037

American Trucking Association
1616 P Street, N.W.
Washington, D.C. 20006

American Waterways Operator, Inc.
1600 Wilson Boulevard
Suite 1101
Arlington, Virginia 22209

Association of American Railroads
1920 L Street, N.W.
Washington, D.C. 20006

Association of Oil Pipelines
Suite 1208
1725 K Street, N.W.
Washington, D.C. 20006

Bureau of Explosives
1920 L Street, N.W.
Washington, D.C. 20006

Hazardous Materials Advisory Committee
Suite 1107
1100 17th Street, N.W.
Washington, D.C. 20036

Interstate Natural Gas Association
1660 L Street, N.W.
Washington, D.C. 20006

National Tank Truck Carriers, Inc.
1616 P Street, N.W.
Washington, D.C. 20006

Manufacturing Chemists Association
1825 Connecticut Avenue, N.W.
Washington, D.C. 20009

Water Transportation Association
Suite 2007
60 East 42nd Street
New York, New York 10017

The incidents described earlier in this chapter bring home the point that it is important to learn as much as possible about hazardous substances before you become involved in an emergency. In a chemical emergency it is not enough to know the techniques of fire fighting. Without the knowledge of how to deal with them, any accident involving hazardous materials could turn into a major disaster and become a matter of life and death.

The next few chapters cover the fundamentals of the properties and behavior characteristics of various chemicals; the latter part of the book covers the preparation and strategy for handling hazardous materials incidents. This knowledge will not only enable you to make informed decisions at the scene of the incident, but may also be the means of saving lives—among them your own.

QUESTIONS FOR REVIEW AND DISCUSSION

1. Define the term *hazardous materials* and explain why there are so many definitions.
2. For the last year for which data are available, determine how many hazardous materials accidents were reported in your state.
3. For the last year for which data are available, determine the ten top chemicals produced.
4. Select one of the incidents described in the chapter and describe the items which the fire department should have pre-planned.
5. Explain the relationship between the mode of transportation, the base the transportation system runs on, the type of vehicles used, the containers used, and the state of the cargo.
6. Explain the interrelationships and responsibilities of the various divisions of the Department of Transportation.
7. What is the function of the National Transportation Safety Board?
8. For your local community, list ten locations where hazardous materials can be found.
9. What information pertaining to hazardous materials can be obtained from the *Federal Register*?
10. Prepare a form for the communications center containing the reference telephone numbers needed during a hazardous materials incident.

NOTES

1. Associated Press, "U.S. Launches Drive on Stolen Explosives," *Washington Post*, May 20, 1977.
2. American Insurance Association, *Hazard Survey of the Chemical and Allied Industries*, New York, 1968.
3. Ibid., p. 5.
4. *Chemical and Engineering News*, February 7, 1977 (Washington, D.C.: American Chemical Society).
5. National Board of Fire Underwriters, *Texas City, Texas, Disaster, April 16-17, 1947* (New York).
6. National Board of Fire Underwriters, *The Holland Tunnel Chemical Fire, New Jersey–New York, May 13, 1949* (New York).
7. National Board of Fire Underwriters, *The Brighton Gas Fire and Explosion Catastrophe, Town of Brighton (Monroe County), N.Y., September 21, 1951* (New York).
8. National Board of Fire Underwriters, *The Roseburg, Oregon, Fire Explosion and Conflagration, August 7, 1959* (New York).
9. "Large Loss of Life Fires—Transportation," *Quarterly of the NFPA*, January, 1962 (Boston: National Fire Protection Association).
10. "Massive Freight Wreck Tests Maryland Volunteers," *Fire Engineering*, December 1964.
11. James J. McCarey, "Petrochemical Fire Threatens Giant Refinery," *Fire Engineering*, August 1966.
12. Joseph E. Keller, "House Inquiry Looks at Regulations for Shipping Hazardous Materials," *Fire Engineering*, February 1970.
13. Phil Angelo, "Hazardous Cargo," *Fire Command*, August 1974.
14. Thomas E. Wright, "Planning Is First Step in Handling Ammonia Leak From Tanks or Pipelines," *Fire Engineering*, April 1975.
15. Cliff Dektar, "Insecticide Fumes Fell 94 at Truck Fire," *Fire Engineering*, August 1975.
16. R. L. Nailen, "EMT Training Pays Off as Fumes Kill 3, Fell 20 Others in Milwau-

kee," *Fire Engineering*, October 1975.

17. Gary Taylor, "What Is It?" *Houston Post*, June 15, 1976; Dave Precht, "Dangerous Powder Ultimately Pure," *Houston Post*, June 17, 1976.

18. Bob Hiles, "We Got Us a Train Wreck Here," *Firehouse*, January 1977.

19. M. Weil, "90 Examined After Exposure to Toxic Fumes," *Washington Post*, May 30, 1977.

20. "Can Oil Flow Bypass Pump Station 8?" *Fairbanks Daily News Miner*, July 9, 1977, p. 1.

21. "Townspeople Flee Poisonous Fumes," *Anchorage Times*, July 13, 1977, p. 6.

22. Walter M. Haessler, "The Four Problems of Transportation of Goods," *Fire Journal*, November 1971.

23. Stephen M. Aug, "Why Are Wrecks Rising?" *Philadelphia Bulletin*, January 16, 1977.

24. American Insurance Association, *Hazard Survey of the Chemical and Allied Industries*, p. 5.

2 Review of basic chemistry

Thousands of new and exotic chemicals and chemical combinations have been produced to meet a wide range of uses, from sophisticated defense applications to everyday home needs. New ways of combining existing materials are being discovered to introduce new products in the marketplace. Each new chemical combination adds to the already lengthy list of substances, some of which are dangerous, some are not. At the scene of a hazardous materials incident, the questions that need to be asked are:

1. What physical properties does the material have?
2. What will happen when it is heated?
3. Are its vapors toxic?
4. Will it react violently when exposed to or mixed with another chemical?
5. What kind of containers is it shipped in?
6. What kind of extinguishing agents can be safely used on it?
7. What are the long-term effects of exposure to it?

Emergency response personnel must be able to answer these questions in order to make intelligent decisions on the fireground when chemical substances are involved. Normal techniques, such as putting water on the fire or wearing standard protective equipment, may not be the correct procedure. In fact, handling the incident in the regular manner might result in needless injury, death, or extended property damage.

With the thousands upon thousands of possible combinations, it would be impossible to supply specific information on all chemical compounds in one volume. This text endeavors to cover the basic information that emergency response personnel will need for an understanding of how to deal with the broad categories of chemicals that might be encountered. Common sense, coupled with the fundamentals provided here can make an unknown, dangerous situation manageable. Knowledge helps emergency response personnel to become confident in their ability to handle the situation.

ELEMENTARY PRINCIPLES

Emergency service personnel have to become familiar with the basic principles of chemistry in order to interpret the properties and characteristics of the chemical compounds discussed in the following chapters.

Again, the aim is not to make chemists of emergency supervisors, but to cover the fundamentals that will help in understanding how these materials behave.

Chemical Change

Chemicals combine in a variety of ways to produce new substances. Often the actual process of combining will produce excess heat or absorb surrounding heat. For this reason understanding the types of chemical change has particular application to all emergency response personnel.

If two chemicals are combined and the result is a different chemical, the change is called a *combination*. During this type of reaction, heat is usually evolved. Examples of this type of reaction are the heating of a mixture of iron and sulfur to produce ferrous sulfide. The most common reaction that emergency service personnel are familiar with is the combustion of carbon, which is a combination reaction.

When a combination of chemicals is heated and broken down into its constituent parts, it is said to have *decomposed*. An example of the decomposition of a compound occurs when mercuric oxide is heated and broken down into mercury and oxygen. If this occurred under fire conditions, the oxygen would then tend to accelerate the fire.

When a solution of one chemical is combined with the solution of another chemical and an insoluble material is produced, the insoluble material is called a *precipitate*. If a solution of sodium chloride (common salt) is added to a solution of silver nitrate, a precipitate of silver chloride is formed and separates from the solution of sodium nitrate. The silver and sodium have exchanged places to form the new compounds. This is called an *exchange reaction*.

States of Matter

There are three states in which matter can exist: solid, liquid, and gas. Each of the three states has physical properties associated with the particular substance, as shown in Table 2-1.

Almost any pure substance can exist in any of the three states. The variables that influence the changes are the given temperature and pressure. If you say a compound is a liquid, what you mean is that at normal temperatures and pressures it will exhibit the characteristics of a liquid.

Table 2-1   States of Matter and Their Properties

State	Physical Property
Solid	Melting point Solubility or insolubility in liquids Hardness Color Odor Density
Liquid	Boiling point Viscosity Color Odor Density
Gas	Solubility in water Liquefaction Color Odor Density

When a substance goes from one state to another there are specific names assigned to the change. The states of change are shown graphically in Figure 2-1 and explained below:

Vaporization: going from the liquid to the gaseous state or going from the solid to the gaseous state

Condensation: going from the gaseous to the liquid state

Distillation: going from the liquid to gaseous to liquid state

Freezing: going from the liquid to the solid state

Melting: going from the solid to the liquid state

Sublimation: going from the solid to gaseous to solid state

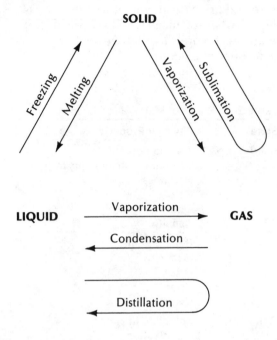

Figure 2-1.   States of matter

Temperature and Heat

Temperature is the condition of an object which determines whether heat will flow to or from another object. Heat will flow from the warmer to the cooler, giving rise to the fact that the object will "feel" hot or cold. Thus, if you touch a piece of metal at a higher temperature than your hand the metal will feel hot because heat will be flowing from the metal (higher) to your hand (lower).

The *quantity of heat*, on the other hand, is the amount of internal energy in a body. It is important to understand that there is a big difference between temperature and a quantity of heat. For example, two burners, both at the same temperature, are held under pans of water: one pan is very small while the other is very large. At the end of the same period of time, the smaller pan will reach a higher tem-perature than the larger pan, even though the heat supplied by the burners was the same.

The quantity of heat is measured in British thermal units (Btu) or gram-calories (gm-cal), depending on the measurement system being used. *One Btu* (British thermal unit) is the amount of heat necessary to raise one pound of water, one degree fahrenheit in temperature from 63°F to 64°F. The *gram-calorie* is the amount of heat necessary to raise one gram of water, one degree centigrade from 14.5°C to 15.5°C. Based on these definitions,

$$1 \text{ Btu} = 252 \text{ gm-cal}$$

Temperatures are measured with thermometers on various scales: the Fahrenheit, Centigrade or Celsius, and Kelvin. The Fahrenheit and Centigrade scales are actually arbitrary values established by scientists based on the boiling point and freezing point of water.

The Fahrenheit scale was set when 32° was chosen as the freezing point and 212° as the boiling point. The difference between the two on the thermometer was then divided into 180 equal parts. Once the number of the divisions was established they were extended above the boiling point and below the freezing point.

The same process was followed for the Centigrade or Celsius (they are used interchangeably) scale, except that 0° and 100° were used as the upper and lower points. Since there are only 100 units between freezing and boiling on the Centigrade scale, the divisions are larger than on the Fahrenheit scale.

It is interesting to note that when the Fahrenheit scale was established, 212° could have been selected as the freezing point. Then we would speak of a lower temperature being warmer. This illustrates the arbitrary way the scale was established.

To convert from either the Centigrade to Fahrenheit scale, or vice versa, the following formulas can be used:

$$°F = \frac{9}{5} °C + 32$$

$$°C = \frac{5}{9} (°F - 32)$$

The point at which all internal energy ceases, meaning that there is a total absence of heat, is called *absolute zero*. Using this as a basis, Lord Kelvin developed a scale at which the zero point was at absolute zero. Then, using the Centigrade interval, the remainder of the scale was established. The Kelvin scale thus runs from 0°K (absolute zero) to 273°K (water freezes) to 373°K (water boils). The relationship between the three temperature scales is shown in Figure 2-2.

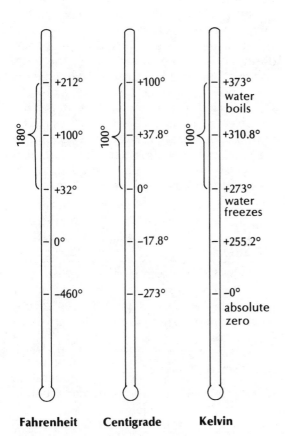

Figure 2-2. Comparison of temperature scales

It is important to remember that heat can be neither created nor destroyed. Heat is really energy, which exists all the time in one form or another. During chemical reactions heat can be either a by-product and output of the process or it can be absorbed during the reaction. Each of these types has a special name. An *exothermic reaction* is one in which the heat is a by-product and it is given off. When heat is absorbed from the surroundings the reaction is called an *endothermic* one. Both types of reactions have an impact on emergency response personnel operating at a hazardous materials incident.

Changes of States of Matter

As shown earlier, the same substance could exist as either a solid, liquid, or a gas, depending on the temperature and pressure. Each time a substance changes state, a quantity of heat is either absorbed or expelled. Both factors affect emergency service personnel at the scene of a hazardous materials incident.

The quantity of heat that must be supplied to a material at its melting point to convert it completely to a liquid at the same temperature is called the *heat of fusion* of the material. To convert ice at 32°F to water as liquid at 32°F requires 144 Btu per pound or 80 calories per gram.

The quantity of heat that must be supplied to a liquid at its boiling point to convert it completely to a gas at the same temperature is called the *heat of vaporization* of the material. To convert water at 212°F to water vapor at 212°F requires 970 Btu per pound or 539 calories per gram.

So if you put water at 60°F on a fire and all of the water were converted to steam, each *pound* of water would absorb:

60° F to 212° F liquid = 152 Btu/pound
212° F liquid to 212° F steam = 970 Btu/pound

1,122 Btu per pound
of water

For each pound of water at 60°F converted to steam, 1,122 Btu of heat would be absorbed.

Ideally, the object would be to throw ice cubes at the fire to gain full heat absorption of the heats of fusion and vaporization. If it were conceivable to do this, the heat absorption would be:

32° F solid to 32° F liquid = 144 Btu/pound
32° F liquid to 212° F liquid = 180 Btu/pound
212° F liquid to 212° F steam = 970 Btu/pound

$$1,294 \text{ Btu per pound}$$
$$\text{of water}$$

Obviously, the key to success in using the water for heat absorption is to convert it to steam. This will provide the maximum cooling capability for the resource. This is especially important when the water is in short supply and must be used to the best possible advantage.

Heat Transfer

One of the big problems at a hazardous materials incident is the large amount of heat produced if a fire develops. The heat being generated can threaten exposures in one of three ways: convection, conduction, and radiation.

Convection is the transfer of heat from one place to another by the actual motion of the hot material. When a pot of water is put on a stove to boil, the hot water at the bottom of the pot heats first, rises, and heats the water above it. This movement of the material actually transfers the heat from the bottom of the pot to the top until eventually it all begins to boil. At a large hazardous materials fire, the heated air can ignite exposures, particularly those at a higher elevation.

Conduction is the transfer of heat through the movement of atoms within a substance. As the atoms on one end of a rod are heated, they increase their vibration, thus colliding with their neighbors a little further out on the rod. Each neighbor then collides with its neighbor, which moves the heat transfer further out. Notice that the energy is passed on with the atoms remaining in their original positions, as opposed to convection, in which

the entire material moves. Fires at a hazardous materials incident which impinge on the metal container of an exposed carrier transmit the heat to the product by conduction. In a tank car exposed to a fire, heat will build up because of conduction.

Thermal conductivity is a measure of the material's ability to conduct heat. The metals as a group have a greater thermal conductivity than the nonmetals. However, other materials do conduct heat.

Heat transmission from *radiation* results from the electromagnetic waves given off when burning occurs. These electromagnetic waves travel at the speed of light. When they strike a nontransparent body they are absorbed and their energy is converted to heat. The radiated energy is made up of many different wavelengths, which change as the temperature increases.

Radiated transfer of heat is very common during a hazardous materials fire. The large volume of fire produces a great many electromagnetic waves which heat nearby objects. The emergency service personnel must cool the objects thus heated. This must be done by putting water directly on the object to absorb the heat. Interposing a water stream between the fire and object will do no good, since the water is transparent and the electromagnetic waves will pass right through.

PRESSURE

When a force is applied over a given area it is called a *pressure*. The common unit for measuring pressure is pounds per square inch. It is important to distinguish between a force and a pressure. If a 400-pound weight is placed on a 4-square-inch platform, the pressure exerted would be:

$$\text{Pressure} = \frac{\text{Force}}{\text{area}} =$$
$$\frac{400 \text{ pounds}}{4 \text{ square inches}} =$$
$$100 \text{ pounds per square inch (psi)}$$

Now, if this weight were placed on an 8-square-inch platform, the pressure exerted would be:

$$\text{Pressure} = \frac{400 \text{ pounds}}{8 \text{ square inches}} =$$
50 pounds per square inch (psi)

The force remained 400 pounds in both examples yet the pressure changed because it is distributed over a broader base.

Atmospheric Pressure

Atmospheric pressure results from the weight of the air elevated above the earth's surface. At sea level the atmosphere exerts a pressure of 14.7 psi. As the altitude increases there is less air and therefore the atmospheric pressure decreases. The converse of this statement is also true. Because there is more air above a point below sea level, atmospheric pressure increases. As a rule of thumb, there is a 0.5 psi change for each 1,000 feet of elevation increase or decrease.

Gauge Pressure

When gauges are installed on any pressure device, they are calibrated so that they read zero psi when the container is empty. Yet, atmospheric pressure is present, so that the gauge is really reading incorrectly. For this reason *gauge pressure* indicates only the pressure in the vessel and is abbreviated psig to show this fact.

Absolute Pressure

Since the gauge pressure does not take into account atmospheric pressure, it is not a true pressure. *Absolute pressure* is the sum of the gauge pressure and atmospheric pressure and is, therefore, a true pressure. To indicate that it is an actual pressure it is abbreviated psia.

If a gauge read 100 psig at sea level, the absolute or true pressure would be 114.7 psia. The formula is:

absolute = gauge + atmospheric
absolute = 100 psig + 14.7 psi
absolute = 114.7 psia

TEMPERATURE–PRESSURE–VOLUME RELATIONSHIPS

Pressure is created in a vessel confining a gas, because the rapidly moving molecules of the gas bang into the walls. The number of impacts over a given time period determines the amount of pressure against the walls.

Pressure–Volume Interaction

Let's say that it takes one second for a molecule to move from wall A to wall B in Figure 2-3. Now, the wall is moved to position C, halfway between A and B. The molecule will have the same velocity, so it will only take ½ second to go between wall A and wall B. Therefore, there will be twice as many impacts. Since pressure is the number of impacts

in a given time period, it has approximately doubled.

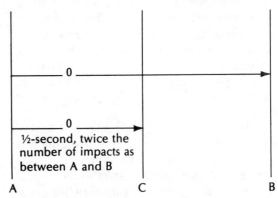

Figure 2-3. Pressure–volume relationship

This behavior of a gas was observed by Robert Boyle in 1660. Known as Boyle's Law, it can be stated as: When the temperature and mass of a gas are kept constant, the product of the pressure and volume is equal to a constant.

Boyle's Law can be expressed mathematically as

$$pV = \text{constant at: constant temperature}$$
$$\text{constant mass}$$

where p = absolute pressure
V = volume

Another way to write this for two different states of the gas at the same temperature is

$$p_1 V_1 = p_2 V_2 \text{ at constant temperature}$$

where p_1 = absolute pressure at point 1
V_1 = volume at point 1
p_2 = absolute pressure at point 2
V_2 = volume at point 2

It must be pointed out, however, that the product pV remains nearly constant at a set temperature; there is some variance at different pressures. Therefore, the equations are good only for something called an *ideal gas*, which actually does not exist. The approximation is still suited for use in developing pressure–volume relationships.

Pressure–Temperature Interaction

As the temperature of a gas increases, there is a corresponding increase in the energy of the molecules. As the velocity of these molecules increases, the number of impacts over a given time period also increases. As was explained above, the number of impacts over the given time period is what creates the pressure. Therefore, as the temperature increases, so does the pressure of a gas in a confined vessel.

This physical fact is known as Charles' Law and can be stated as: If the volume of a gas is kept constant and the temperature is increased, the pressure increases in direct proportion to the increase in absolute temperature.

Using Charles' Law, we can continue to decrease temperature and measure the corresponding decrease in pressure. If you start at 0°C and decrease the temperature to –1°C, keeping the volume constant, the pressure will decrease by 1/273 of its original value. On this basis, decreasing the temperature to –273°C would create a gas with no pressure. This is, in fact, true, because –273°C is known as absolute zero. The molecules of the gas possess no energy and therefore there are no impacts on the walls. The result is a gas with zero pressure.

Vapor Pressure

When a liquid evaporates in a confined area, the vapor that is developed exerts a pressure on its surroundings, just as any gas in a confined vessel causes a pressure based on the temperature and volume. If an excess amount of liquid is used, the space becomes saturated with vapor, and the maximum vapor pressure is created for that particular volume and temperature. The vapor pressures created by water at temperatures between 0°C and 100°C are shown in Figure 2-4.

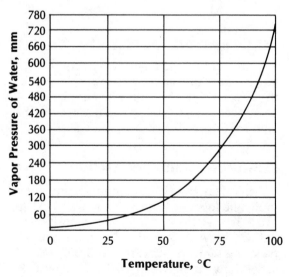

Figure 2-4. Vapor pressure of water between 0°C and 100°C

As the vapor pressure of a substance increases there are increasing problems for emergency response personnel. Obviously, a liquid exposed to a fire at a hazardous materials accident will have an increasing vapor pressure. If the material is flammable, the pressure relief devices will activate and the vapors become ignited. This flame could then act as a blowtorch on other exposures and cause further problems. In addition, if the heat causes a rapid increase in the vapor pressure, pressure relief devices may not be large enough to quickly relieve the overpressurization. This in turn could lead to a violent rupture of the container.

When the vapor pressure created by the liquid is equal to the atmospheric pressure, the liquid will boil. The temperature at which the liquid boils is known as the *boiling point*. As atmospheric pressure changes so does the boiling point. For example, on a mountain top 10,000 feet above sea level where the atmospheric pressure is approximately 9.7 psi (501.5 mm of water), the temperature for boiling water would drop to approximately 89°C (see Figure 2-4).

Loading Outage

When packing liquids in tank cars, sufficient room must be left for the liquid to create vapor without a high vapor pressure. The space that must be left is called *outage*. As the temperature increases so does the vapor pressure. Therefore, the space left must also take into account the temperature to which the storage vessel will be exposed. This is particularly true for train tank cars that sit in the sun at sidings while waiting to be unloaded.

The Code of Federal Regulations, Title 49, paragraph 173.115, requires that flammable liquids having a vapor pressure exceeding 16 psia at 100°F must be loaded so that the minimum outage will be the greatest of one of the following values:

1. Dome capacity

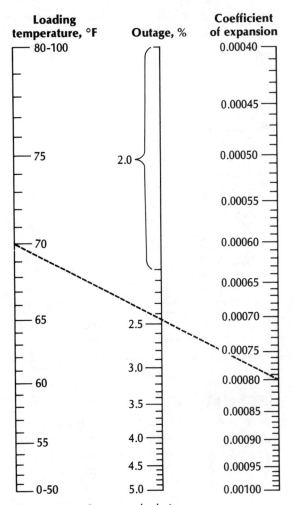

Figure 2-5. Outage calculations

2. Two percent of total capacity of tank and dome

3. Outage as calculated in Figure 2-5.

Suppose gasoline with a coefficient of expansion of 0.00080 is being loaded at 70°F. Using the scales in Figure 2-5, by connecting the 70°F point with the 0.00080 point the required outage for this product is shown to be 2.4 percent.

Some coefficients of expansion for common flammable liquids are shown in Table 2-2.

Table 2-2. Coefficients of Expansion for Common Flammable Liquids

Flammable Liquid	Coefficients of Expansion
Acetone	0.00085
Amyl acetate	0.00068
Benzol (benzene)	0.00071
Carbon bisulfide	0.00070
Ether	0.00098
Ethyl acetate	0.00079
Ethyl (grain) alcohol	0.00062
Methyl (wood) alcohol	0.00072
Toluol (toluene)	0.00063
Gasoline or naphtha:	
50.0–55° A.P.I.[a]	0.00055
55.1–60° A.P.I.[a]	0.00060
60.1–65° A.P.I.[a]	0.00065
65.1–70° A.P.I.[a]	0.00070
70.1–75° A.P.I.[a]	0.00075
75.1–80° A.P.I.[a]	0.00080
80.1–85° A.P.I.[a]	0.00085
85.1–90° A.P.I.[a]	0.00090

[a]A.P.I. (American Petroleum Institute), according to the following formula:

$$°\,A.P.I. = \frac{141.5}{\text{specific gravity}} - 131.5$$

FIRE BEHAVIOR

In order to anticipate what is going to happen at the scene of a hazardous materials incident, emergency response personnel must be familiar with fire behavior.

Fire Tetrahedron

For many years scientists felt that only three ingredients were needed for a fire. This developed into the so-called fire triangle of heat, fuel, and oxygen. However, more recent work in this area has shown that there are really four factors that interact: fuel, temperature, oxidizing agent, and the uninhibited chain reaction. Haessler in his text, *The Extinguishment of Fire,*[1] describes the interaction of these four parts as a tetrahedron, as illustrated in Figure 2-6.

While oxygen is the usual oxidizing agent during the combustion process, there are chemicals that can burn without oxygen's be-

ing present. For example, calcium and aluminum will burn in nitrogen. So, the first side of the tetrahedron is an oxidizing agent that permits the fuel to burn.

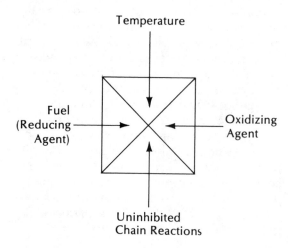

Figure 2-6. Fire tetrahedron

The fuel is the material that is oxidized. Since the fuel becomes chemically charged by the oxidizing process, it is called a *reducing agent*. This makes the second side of the tetrahedron. Fuels can be anything from elements (carbon, hydrogen, magnesium) to compounds (cellulose, wood, paper, gasoline, petroleum compounds).

Some mixtures of reducing agent and oxidizing agent remain stable under certain conditions. However, when there is some "activation energy," as Haessler calls it, a chain reaction is started, which causes combustion. The factor that can trigger this chemical reaction can be as simple as exposing the combination to light. Once the chain reaction begins, extinguishment must take place by interrupting the chain reaction.

Scientists have known for many years that certain chemicals act as excellent extinguishing agents. However, they were at a loss to explain how these chemicals actually accomplished extinguishment, given the triangle of fire model. With the development of the tetrahedron model and the inclusion of the uninhibited chain reaction, a scientifically sound theory could be postulated. With this as a basis, the extinguishing capabilities of the halons and certain dry chemicals were possible.

The final side of the tetrahedron is temperature. The fact that temperature is used instead of heat is deliberate. As explained in an earlier paragraph, temperature is the quantity of the disordered energy, which is what initiates combustion. It is possible to have a high heat as indicated by a large reading of Btu and still not have combustion. The temperature is therefore the key ingredient and the one that influences the action of the tetrahedron.

Classes of Fire

There are four basic categories into which fuels for fire can be broken. Each class of fuels has its own burning characteristics, which require different types of extinguishing agents. The classes of fire are:

1. Class A: Ordinary combustibles, which leave a residue after burning. Items in this class include wood, paper, and textiles.

2. Class B: Flammable liquids and gases, where the burning takes place in the vapor space above the liquid. Products considered in class B are gasoline, kerosene, and propane.

3. Class C: Class A or B fires occurring near energized electrical equipment. This class covers fires in motors, switches, and light fixtures.

4. Class D: Fires occurring in certain combustible metals which are easily oxidized. Examples of combustible metals include sodium, potassium, magnesium, and aluminum.

Properties of Flammable Products

Flammable products can be either solid, liquid, or gas. Each state has characteristic physical properties that enable emergency response personnel to anticipate possible fire behavior.

The *flash point* of a flammable liquid is the temperature at which a liquid will give off sufficient flammable vapor for ignition to occur. This means that once the temperature goes above this point, the liquid is giving off vapors, which, if an ignition source is found, will ignite.

There are two methods for determining the actual flash point, the open cup (OC) and the closed cup (CC) method. In general, the open cup value is about 10 to 15°F higher than the closed cup value.

Just reaching the flash point and having a source of ignition is not sufficient for the fire to occur. There has to be the correct ratio of fuel to oxygen to produce fire. The range of gas or vapor concentrations (percent by volume in air) which will burn or explode if an ignition source is present is called the *flammable limits*. The two limits of the concentration are known as the *lower explosive limit (LEL)* and the *upper explosive limit (UEL).* Below the LEL the mixture is too lean to burn and above the UEL it is too rich to burn.

The *ignition temperature* of a flammable liquid, which is entirely different from the flash point, is the temperature at which the vapor will ignite without a spark or flame being present. Any values for ignition temperatures given in the reference texts are approximate because the value is dependent upon changes in geometry, gas, or vapor concentrations and in the presence of catalysts.

The ratio of the weight of a volume of a liquid to the weight of an equal volume of water is called the *specific gravity* of the liquid; or in other words, the ratio of the density of the liquid to the density of water. In the case of liquids of limited solubility, the specific gravity will predict whether the product will sink or float on water; for example, if the spe-

cific gravity is greater than one, the product will sink, and if the value is less than 1, the product will float.

For a gas, the ratio of the weight of the vapor compared to the weight of an equal volume of air at the same temperature and pressure is known as *vapor density.* Values less than 1 indicate that the vapors will tend to rise, while values greater than 1 indicate that the vapors will tend to settle. Temperature does, however, cause changes in whether the vapors will rise or fall. For example, although methane at 68°F has a vapor density of 0.55, it becomes denser at lower temperatures. At –259°F, its boiling point, the vapor is heavier than air. Vapors from an open container of boiling methane fall rather than rise.

Table 2-3 provides some examples of the values of the physical properties for common products.

Table 2-3. Physical Properties of Common Products

Product	Ignition Temperature (°F)	Flash Point (°F)	Flammable Limits (percent)	Specific Gravity	Vapor Density
Kerosene	410	100	0.7 to 5	<1	—
Gasoline (100 octane)	853	-36	1.4 to 7.4	0.8	—
Heptane	419	25	1.05 to 6.7	0.7	3.5
Ethyl ether	320	-49	1.9 to 36	0.7	2.6
Methane	1004	gas	5.0 to 15.0	—	0.6
Benzine	550	<0	1.1 to 5.9	0.6	2.5
Propane	842	gas	2.2 to 9.5	—	1.6
Ethyl alcohol	689	55	3.3 to 19	0.8	1.6
Acetane	869	0	2.6 to 12.8	0.8	2.0

CHEMICAL BUILDING BLOCKS

Chemistry as a science can be divided into two main categories: organic and inorganic. *Organic chemistry* is concerned with compounds containing the element carbon, and

inorganic chemistry deals with all of the remaining elements. While on the face of this statement, it would seem that inorganic chemistry would be the larger, this is not the case. In actual practice, there are so many compounds containing carbon that the organic branch of chemistry is the larger of the two.

Elements

The building blocks for all words in the English language are the 26 letters of the alphabet. Similarly, the basic building blocks for all chemistry are the 105 *elements*. Of these 105 the first 92, except for 4, occur in nature. The 4 exceptions—technetium, promethium, astatine, and francium—either do not occur in nature or are produced for just an instant by radioactive decay. Elements 93 to 105 have only been produced in the laboratory. Since the *element* is the basic building block, it cannot be decomposed into a simpler substance. The smallest particle of an element that can exist is called an *atom*. The atom is the smallest unit of the element that can enter a chemical reaction.

Compounds

Pure substances composed of two or more elements are called *compounds*. Since they contain at least two elements, compounds are capable of being broken down. Compounds are produced when atoms combine to form molecules or when atoms combine to form large groups of ions.

For example, when two atoms of hydrogen combine with one atom of oxygen, they form a molecule of water. This molecule of water is the smallest unit that the compound can be broken down to. Any further subdivision will break the water molecule down to its two components.

On the other hand, the compound for plain table salt—sodium chloride—is developed by the attraction of a positively charged sodium ion and a negatively charged chloride ion. The compound is held together by the structure of the crystal and is a nonmolecular compound.

Mixtures

Pure elements are very rarely found in nature. Substances are usually found either as a compound or a mixture. A *mixture* is a combination that contains two or more substances that do not lose their individual identities. In addition, there can be variable amounts of each substance within the mixture. A compound, on the other hand, must be made of specific quantities of elements in a chemical combination.

Symbols

As with all branches of science, a shorthand notation for recording information has been developed. Each of the 105 elements is represented by a standard symbol. Derivations for these symbols come from the chemical's present, former, Greek, or Latin name.

When writing the symbol it is important to follow the capital and lowercase letters exactly. There is a big difference between Co, which is cobalt, and CO, which is carbon monoxide.

Table 2-4 is a listing of the symbol and atomic number for all the known elements.

Chemical Formulas

Now, using the symbols for each element, chemical formulas for compounds and chemical reactions can be developed. The expression for water is

$$H_2O$$

This means that a molecule of water contains two atoms of hydrogen and one of oxygen. Sulfuric acid is written as

$$H_2SO_4$$

This means that one molecule contains two atoms of hydrogen, one atom of sulfur, and four atoms of oxygen.

Table 2-4. Symbol and Atomic Number for All Elements.

Name	Symbol	Atomic No.	Name	Symbol	Atomic No.
Actinium	Ac	89	Lawrencium	Lr	103
Aluminum	Al	13	Lead	Pb	82
Americium	Am	95	Lithium	Li	3
Antimony	Sb	51	Lutetium	Lu	71
Argon	Ar	18	Magnesium	Mg	12
Arsenic	As	33	Manganese	Mn	25
Astatine	At	85	Mendelevium	Md	101
Barium	Ba	56	Mercury	Hg	80
Berkelium	Bk	97	Molybdenum	Mo	42
Beryllium	Be	4	Neodymium	Nd	60
Bismuth	Bi	83	Neon	Ne	10
Boron	B	5	Neptunium	Np	93
Bromine	Br	35	Nickel	Ni	28
Cadmium	Cd	48	Niobium	Nb	41
Calcium	Ca	20	Nitrogen	N	7
Californium	Cf	98	Nobelium	No	102
Carbon	C	6	Osmium	Os	76
Cerium	Ce	58	Oxygen	O	8
Cesium	Cs	55	Palladium	Pd	46
Chlorine	Cl	17	Phosphorus	P	15
Chromium	Cr	24	Platinum	Pt	78
Cobalt	Co	27	Plutonium	Pu	94
Copper	Cu	29	Polonium	Po	84
Curium	Cm	96	Potassium	K	19
Dysprosium	Dy	66	Praseodymium	Pr	59
Einsteinium	Es	99	Promethium	Pm	61
Erbium	Er	68	Protactinium	Pa	91
Europium	Eu	63	Radium	Ra	88
Fermium	Fm	100	Radon	Rn	86
Fluorine	F	9	Rhenium	Re	75
Francium	Fr	87	Rhodium	Rh	45
Gadolinium	Gd	64	Rubidium	Rb	37
Gallium	Ga	31	Ruthenium	Ru	44
Germanium	Ge	32	Samarium	Sm	62
Gold	Au	79	Scandium	Sc	21
Hafnium [a]	Hf	72	Selenium	Se	34
Hahnium	Ha	105	Silicon	Si	14
Helium	He	2	Silver	Ag	47
Holmium	Ho	67	Sodium	Na	11
Hydrogen	H	1	Strontium	Sr	38
Indium	In	49	Sulfur	S	16
Iodine	I	53	Tantalum	Ta	73
Iridium	Ir	77	Technetium	Tc	43
Iron	Fe	26	Tellurium	Te	52
Krypton	Kr	36	Terbium	Tb	65
Kurchatovium [a]	Ku	104	Thallium	Tl	81
Lanthanum	La	57	Thorium	Th	90

Thulium	Tm	69	Xenon	Xe	54
Tin	Sn	50	Ytterbium	Yb	70
Titanium	Ti	22	Yttrium	Y	39
Tungsten	W	74	Zinc	Zn	30
Uranium	U	92	Zirconium	Zr	40
Vanadium	V	23			

[a]Unofficial name and symbol

The rules for writing chemical formulas are:

1. The subscript indicates the number of atoms of the element in the compound. When only one atom is present, no subscript is used.
2. When the formula contains more than one group of atoms, parentheses are placed around one of the groups. This can be illustrated in the formula for calcium nitrate:

$$Ca(NO_3)_2$$

Here there are one calcium atom and two nitrate groups. If we look only at the number of atoms there are nine: one calcium, two nitrogen, and six oxygen.

3. The formula does not show the arrangement of the atoms within the compound nor the method of binding the atoms together.

Classification of Elements

Elements can be broken down into three main classifications: metals, nonmetals, metalloids.

A comparison of the physical properties of two classes of elements is shown in Table 2-5. Metalloids exhibit some properties from each of the other classes. Examples of metalloids are boron, silicon, arsenic, and antimony.

Table 2-5. Properties of Metals and Nonmetals

Property	Metal	Nonmetal
State at room temperature	Solid (except for mercury)	Gas or solid, usually
Luster	High	None
Conductor of heat and electricity	Good	Poor
Malleable (may be rolled into sheets)	Yes	No
Ductile (may be drawn into wire)	Yes	No
Melting point	Most have high	Relatively low
Density	Most have high	Relatively low
Examples	Copper, gold, iron, tin, silver, sodium, potassium	Carbon, iodine, phosphorus, sulfur, oxygen, nitrogen, chlorine

METRIC SYSTEM

Reference texts in many of the fields of science use the metric system for measurements. In addition, the United States has begun a gradual shift to the system. For this reason, emergency response personnel must be familiar with these units of measurements.

The official name for the metric system of units is the Systeme Internationale (SI). Under this system there are seven basic units, listed in Table 2-6. Using the units in Table 2-6 as a basis, calculations can be made to determine other measurement values which are needed by scientists. Some of the derived units which are of importance to emergency response personnel are listed in Table 2-7.

Once a basic unit has been determined, a standard prefix can be added to indicate increases or decreases by a factor of 10. The standard prefixes and their abbreviations are:

atto	a	$= 10^{-18}$
femto	f	$= 10^{-15}$
pico	p	$= 10^{-12}$
nano	n	$= 10^{-9}$
micro	μ	$= 10^{-6}$
milli	m	$= 10^{-3}$
centi	c	$= 10^{-2}$
deci	d	$= 10^{-1}$
deka	da	$= 10^{1}$
hecto	h	$= 10^{2}$
kilo	k	$= 10^{3}$
mega	M	$= 10^{6}$
giga	G	$= 10^{9}$
tera	T	$= 10^{12}$

The symbols used for the standard measurements when coupled with the symbols used for prefixes produce a common abbreviation. It is very important to maintain the capital and lowercase letters of these combined abbreviations. Some examples are:

Table 2-6.   Basic Units of the Metric System

Measurement of	Unit	Symbol
Length	meter	m
Mass	kilogram	kg
Time	second	s
Electric current	ampere	A
Temperature	degree Kelvin	°K
	degree Celsius	°C
Amount of matter	mole	mol
Luminous intensity	candela	cd

Table 2-7.   Measurement Formulas Derived from Metric Units

Measurement	Unit	Symbol	Formula
Area	square meter	m^2	m^2
Density	kilogram per cubic meter	—	kg/m^3
Heat quantity	joule	J	N x m
Force	newton	N	$kg \times m/s^2$
Power	watt	W	J/s
Pressure	pascal	Pa	N/m^2
Velocity	meter per second	m/s	m/s
Volume	cubic meter	m^3	m^3
	liter	l	dm^3

millimeter	mm	= 0.001 meter
centimeter	cm	= 0.01 meter
kilometer	km	= 1,000 meters
cubic centimeter	cm³	= 1cm × 1cm × 1cm
milligram	mg	= 0.001 gram
centigram	cg	= 0.01 gram
kilogram	kg	= 1000 grams

Conversion values for English system to metric, or from metric system to English, are given in Table 2-8.

Table 2-8. Conversion Values

English to metric system	Metric to English system
1 inch = 25.4 mm	1 mm = 0.0394 in.
= 2.54 cm	1 cm = 0.394 in.
1 foot = 304.8 mm	= 0.033 ft
= 30.48 cm	1 m = 39.37 in.
= 0.305 m	= 3.28 ft
1 mile = 1609.34 m	1 km = 3280.83 ft
= 1.609 km	= 0.621 mi
1 in.^{2} = 645.16 mm^2	1 cm^2 = 0.155 in.^{2}
= 6.45 cm^2	= 0.0011 ft^2
1 ft^2 = 929.03 cm^2	1 m^2 = 10.764 ft^2
1 in.^{3} = 16.38 cm^3	1 cm^3 = 0.061 in.^{3}
= 0.016 *l*	1 *l* = 61.02 in.^{3}
1 ft^3 = 28.32 *l*	= 0.035 ft^3
1 gal = 3.79 *l*	= 0.264 gal
	1 m^3 = 35.31 ft^3
	= 264.19 gal
1 oz = 0.278 N	1 g = 0.0022 lb
1 lb = 0.4536 kg	1 kg = 2.205 lb
= 4.448 N	1 N = 0.225 lb

QUESTIONS FOR REVIEW AND DISCUSSION

1. Why is a knowledge of basic chemistry necessary for emergency response personnel?

2. By what methods do chemicals combine and what hazards to emergency response personnel do each present?

3. Explain the processes that take place when a substance changes states (gas, liquid, or solid).

4. Convert 86°F to the Celsius scale and convert 28°C to the Fahrenheit scale.

5. Explain the various methods of heat transfer.

6. Explain the temperature, pressure, volume relationships, and how this affects gases.

7. Explain how vapor pressure is created.

8. Why should emergency response personnel learn about outage?

9. Explain how the fire tetrahedron brings together a reducing agent, an oxidizing agent, temperature, and an uninhibited chain reaction to produce a fire. How does this affect emergency response personnel?

10. Define flash point, flammable limits, ignition temperature, specific gravity, and vapor density.

NOTES

1. Walter M. Haessler, *The Extinguishment of Fire* (Boston: National Fire Protection Agency, 1974), pp. 2–7.

3
Health hazards

Trying to predict the behavior of a hazardous commodity is a very risky business. Reliance must be placed on the placard and label for information, the reference sources to provide technical data, and the technical experts who are on call to recommend handling procedures. However, the basic responsibility falls back on the emergency response team.

This chapter discusses the basic health hazards inherent in chemicals used as agricultural and garden pesticides, corrosive materials, and poisons. It is important to understand the nature and extent of the hazards, and the necessity of extreme caution in approaching and dealing with these materials. However, the reader must keep in mind that only broad categories are covered. Emergency response personnel must adapt the generalized information presented here to fit a particular situation.

AGRICULTURAL AND GARDEN CHEMICALS

Over the years, humans have waged a great war against insects. With over 5,000 different kinds of insects in the United States, and the large economic losses caused by their damage, the war continues. Insects ruin crops, cause cattle to grow ill, spoil plants, and cause injury and death to humans.

As the arsenal of weapons increases, so do the problems associated with applying the chemicals. Insects become immune to the particular compound that was designed to eradicate them. Furthermore, side effects on the environment, wildlife, fish, and humans all must be weighed against efforts to control insects.

The economic loss to our society caused by insects makes the development and application of agricultural and garden chemicals big business. As greater quantities are produced, new combinations introduced, and discarded wastes increase, emergency response personnel are faced with greater probabilities of encountering these materials in hazardous circumstances.

Definition

The major agricultural and garden chemicals are used to control insects or pests. These groupings of chemicals, called *pesticides*, are defined as chemical agents used to destroy, prevent growth of, or control any living plant or animal considered to be a pest. Within the general category of pesticides, chemicals are broken down into four classes:

1. Insecticides (to control insects)
2. Fungicides (to control plant disease organisms)
3. Herbicides (to control weeds)
4. Rodenticides (to control pest animals)

Scope of the Problem

There are about 1,400 basic chemicals that are combined in various forms with each other as well as with other materials to produce about 35,000 pesticide products. These are the ones that made it to the marketplace. In addition to these are the ones that were tried, tested, and then discarded for various reasons. These pesticides are produced by 4,600 companies at 7,200 plants for an annual volume of 1.6 billion pounds worth $3 billion as of 1975.

As an example of what an insecticide goes through, let's follow the development of the insecticide Orthene, a product of the Chevron Chemical Company, as described in "The Long Road to Market."[1]

Statistically, about 16,000 chemical compounds must be synthesized to find one commercially successful pesticide. Using normal work rates, it would take three chemists about 30 years to find one successful product. New products are produced by modifying known compounds or inventing totally new chemical structures.

Once the chemical is produced in the laboratory, it receives preliminary testing in the bio-lab. Here its effectiveness on plants and insects is determined. Once the evalua-tion is completed (80 percent never get past this step), further experiments are run on plants, insects, and animals. At this stage Orthene was found to be excellent against sucking and chewing insects, yet was relatively ineffective to mammals, birds, and fish.

Following satisfactory laboratory test results, an economic study of manufacturing costs versus marketability is made. If the new product appears to be economically justified, the first "real world" test is conducted. At this point, two to five pounds of the chemical are applied to crops during a growing season in various test areas around the country. This is known as stage 1 testing.

Stage 2 testing is conducted for those pesticides that show further promise. Here the product is tested by universities, government agencies, and foreign marketing companies. At the conclusion of the growing season all crops are destroyed.

Now comes the major decision. Full-scale development is dependent on an analysis of the technical test data as well as the marketing and economic conditions. Once the go-ahead is given, registration applications must be drafted, production plans established, and long-range toxicology tests run. Production plans are complex because the need to manufacture six million pounds of Orthene is quite different from the small test amounts that have been made up to this point.

Long-range toxicology tests document:

1. Whether the chemical builds up in the food chain.
2. How much remains on the plant as residue.
3. How quickly it breaks down in the environment.
4. How toxic it is.
5. Whether it creates any toxic by-products.
6. Whether it is absorbed through the skin and lungs.

After the long-range tests have been

completed, governmental approval and registration for full-scale marketing are necessary. For U.S. registration, Chevron Chemical submitted almost 40,000 pages of data on Orthene. The new product is currently registered for over 100 insect species and about 180 plants.

What does this mean to emergency response personnel? Let's look at the various development stages and see how each would affect you.

1. On-going research in the laboratory means that there are many unknown compounds in various stages of development. An incident in such a facility will mean that little information on hazards will be available.

2. For testing at the bio-lab, the product must be transported and stored before actually being applied. An incident at this stage will mean that little information about the product is available. Certainly CHEMTREC (the Chemical Transportation Emergency Center, discussed in Chapter 6) will not yet have any information. Emergency response personnel will have to rely on the manufacturer's chemists, if they can be reached.

3. "Real world" tests spread the product over a wider geographical area. Communities with large state argicultural colleges as well as state agricultural extension agents could have some of the experimental chemicals in their area. Again, the product is transported, stored, and applied. Are there any such experimental plots in your community?

4. Once full-scale manufacturing is begun, emergency response personnel face the possibility of an incident in the manufacturing plant; during transport of the product; in storage of the product in distribution centers, lawn and garden centers, or farm buildings; and during the application of the chemical.

Of course, all pesticides must be registered with the Environmental Protection Agency (EPA). Emergency response personnel could then assume that at least the chemical has met certain standards. However, the EPA has the ability to exempt the restrictions under certain "emergency conditions." For example, in the *Federal Register* on June 2, 1977, an exemption was granted to the state of Idaho. The introduction to this exemption states:

> The Environmental Protection Agency (EPA) has granted a specific exemption to the Idaho State Department of Agriculture (hereafter referred to as the "Applicant") to use dinoseb for the control of various broadleaf weeds infesting 30,000 acres to be planted in lentils. This exemption was granted in accordance with, and is subject to, the provisions of 40 CFR Part 166, which prescribes requirements for exemption of Federal and State agencies for use of pesticides under emergency conditions.[2]

The question that must be answered is, did the emergency response personnel know that dinoseb was being used in their state? In addition, since it had not yet been registered, where would these personnel receive information on hazards and treatment? Again, we are back to reading the *Federal Register* to find out what hazardous materials are being used in the local areas.

As you can see, the scope of the problem for emergency response personnel is enormous. You can come across pesticides in the most unexpected places. Be prepared to take positive action when an incident occurs.

Identification

Required labeling and placarding for the identification of pesticides are described in Chapter 6. The following discussion will help to ensure an understanding of the basic information supplied by the labels.

Each pesticide label contains a signal word that indicates the degree of danger. The signal words are:

1. Danger (with skull and crossbones)—high hazard.
2. Warning—moderate hazard
3. Caution—low hazard

The product name will be clearly shown on the front panel of the label. As with the other chemical hazards, make sure that the exact spelling of the name is used when requesting information from any of the reference sources. In addition to the product name, the label will have the active ingredients by chemical name and it may also show the common name.

It is important to note that the inert ingredients (those which do not act as the pesticide) only have to be shown as a percentage. This means that a flammable liquid could be used as a solvent base and only identified on the label as "inert ingredients, 35%." Emergency response personnel would therefore not know initially that they had a flammable hazard as well as a health hazard.

One other piece of helpful information on the label is the EPA registration number. This number is issued to all pesticides before they can be marketed in the United States. Using this number, reference sources such as CHEMTREC can provide the backup data needed for handling the incident.

Pesticide Poisoning

Among the great dangers at a pesticide incident are the many ways emergency response personnel can become exposed to toxic products. The poisoning can occur if you are exposed to the chemicals themselves; to the smoke from burning pesticides; to contaminated water runoff from fire fighting operations; as well as during spill control operations.

The amount of exposure to a pesticide can have a direct relationship to the possible hazard. However, even short-term exposure to a highly toxic one can be fatal. The more concentrated the pesticide the shorter the exposure time need be before health problems will develop.

Short-term exposure over a long period can develop a chronic poisoning. Recent cases are on record of several chemical workers who have experienced such effects as sterility and unusual types of cancer. On the other hand, acute poisoning usually results from a single exposure to a highly toxic material.

There are a variety of ways for the toxic fumes to enter the body. If the pesticide is in contact with the eyes or the skin it is easily absorbed. Of course, inhalation brings the toxic products into the lungs and from there it is distributed to other parts of the body. Finally, the pesticide can be ingested through the mouth. Before you begin to discount eating the pesticide as a possibility, think about such things as stopping for a cigarette, chewing on the tip of a pen, or putting the stem of your eyeglasses in your mouth. These things can and do happen at the scene of an incident without anybody's really thinking about it.

The absorption of the pesticide through the skin is the most common method of getting it into the body. The amazing part of this method is that there may be no sensation of irritation of the skin at the point of entry. Small amounts absorbed through the skin can cause sickness, injury, and even death.

The fastest way for the pesticide to enter the body is through the eyes. With the eyes so closely linked to the central nervous system, any material that enters this way can do substantial damage very quickly. While one way for the pesticide to enter the eyes is from being splashed with the product or contaminated water runoff, the most common way is for toxic smoke to enter the eyes.

Breathing in toxic fumes at either a large spill or at a fire can very quickly cause poisoning. This is particularly true of those products that affect the central nervous system. Some fumigants are so powerful that a few breaths will cause death.

It should be obvious that certain critical actions are necessary at a pesticide incident. Since some can be absorbed through the skin (can you tell which ones?), complete and full protective clothing is absolutely necessary. Absorption through the eyes and lungs can be prevented if fully self-contained breathing apparatus is worn. In addition, a positive pressure mask will ensure that even small face piece leakage will be prevented. Gloves are essential to prevent skin absorption as well as the possibility of transferring the product to your mouth after you leave the contaminated area.

One of the hardest things for emergency response personnel to do is to wear full protective gear including breathing apparatus when outside fighting fire. If it is at a pesticide incident their failure to take full precautions can increase the chance of injury and may even cost a life. Do not worry about looking foolish, but be concerned about sufficient protection for personnel.

Spills and Fires

Fires or spills involving pesticides and/or fertilizers can be disastrous. Special hazards that are created under these conditions must be recognized by emergency response personnel. As explained in the preceding paragraphs, you must be alert to:

1. Toxic smoke
2. Water runoff
3. Pesticide residue
4. Pressurized containers
5. Materials that can explode when heated

The toxicity of the pesticide is determined by the following:

1. Kind of pesticide
2. Concentration of pesticide
3. Amount of exposure
4. Type of exposure

Emergency response personnel will have no control over the kind or concentration of the pesticide they are encountering. The only thing they can do is to regulate the amount and type of exposure. This means full protective clothing and self-contained breathing apparatus for both spills and fires. It also means wearing the gear even if the incident is not in an enclosed area.

The primary hazards faced by the emergency response personnel at the scene of a pesticide emergency—the most dramatic aspects such as the threat or outbreak of fire, the exposure to toxicity, usually receiving wide press coverage—are not the only dangers to be taken into account.

Often, secondary effects, such as water and soil contamination as well as air pollution, may well involve emergency response personnel in future emergency incidents. Long-term effects of the residue, when everyone has long since forgotten about the pesticides, can take their toll and present hazards that emergency response personnel will be expected to deal with at some future time.

The key to conducting a successful attack and achieving control of any hazardous materials incident is *preplanning*. How to go about preparing a preplan and the important advantages to be gained by it are discussed in detail in Chapter 7.

Following is a suggested procedure to be followed at the scene of a chemical spill or fire. Some of the steps are featured in Figure 3-1.

Procedure on Arrival at the Scene

1. Attempt to identify the material or materials involved in fire. If you have not obtained this in advance from your preplan, contact the owner, manager, or a plant worker to give you the information. Check the label, if available, for physical

FIRE FIGHTING TACTICS

FOR FIRE DEPARTMENTS

1 **Contact facility operator** Determine type, quantity and hazards of products. Determine if fire should be fought after weighing fire fighting & postfire hazards vs. possible salvage.

2 **Notify physician to stand by** Physicians may obtain poison control information by contacting Chevron Chemical Co. See Front Cover of Guide. Many other manufacturers will provide similar information.

3 **Contact Chevron Chemical Company** Maintain liaison for specialized information, particularly during a large fire. See Front Cover of Guide. Many other manufacturers will provide similar information.

4 **Evacuate downwind & isolate area** Patrol area to keep out spectators.

5 **Wear personal protective equipment** Wear rubber or neoprene gloves, boots, turn-outs & hat. If contact cannot be avoided (such as entering an unventilated building for rescue) also wear self-contained breathing apparatus (Air Paks).

6 **Attack fire from upwind & from a safe distance** Bottles, drums, metal & aerosol cans are not vented and may explode.

7 **Contain fire & protect surroundings** Prevent spread of fire by cooling nearby containers to prevent rupture (move vehicles & rail cars if possible). Burning chemicals cannot be salvaged.

8 **Use as little water as possible & contain run-off** Contaminated run-off can be the most serious problem. Water spreads contamination over a wide area. Construct dikes to prevent flow to lakes, streams, sewers, etc. Cooling effect of water retards high-temperature decomposition of the chemicals to less toxic compounds.

9 **Use water fog spray not straight stream** Fog spray is more effective for control. Avoid breaking bottles and bags: adds fuel and contamination. Straight streams spread fire and contamination.

10 **Poisoning—Avoid product, smoke, mist and run-off** In case of contact or suspected poisoning, leave site immediately, follow instructions on Back of Guide. Any feeling of discomfort or illness may be a symptom of poisoning. Symptoms may be delayed up to 12 hours. Chemicals may poison by ingestion, absorption through unbroken skin, or inhalation. Wash face and hands before eating, smoking or using toilet. Do not put fingers to mouth or rub eyes.

Figure 3-1. Tactics for handling a pesticide spill or fire. *Reproduced with permission of the Chevron Chemical Co.*

or chemical hazard statements. Any special flammability, combustion, or explosive characteristics check on the label and on the side panel of carriers.

2. Initiate calls to the reference sources which are discussed in Chapter 6. The incident commander should try to make these calls. It is also important to establish contact with your local medical authorities to alert them for a call for help if it becomes necessary.

3. Evacuate downwind and isolate the spill or fire area from spectators.

4. Ensure that all emergency response personnel (fire, police, medical) wear full protective equipment.

Attack Strategy

1. Begin the attack from a safe distance and upwind. Use as little water as possible. Consider the use of high-expansion foam in the initial fire fighting stages. Remember, most pesticides decompose into less toxic forms at high temperatures. Also, use fog streams instead of straight streams to avoid breaking bottles and bags.

2. Build a dike to prevent contaminated water from escaping into sewers, ponds, or streams. The dikes can be made of earth by using heavy equipment to build it. (Do you have resources for getting this equipment?) If necessary, earth for the dike will have to be brought in by truck.

3. Cut ventilation holes into the roof, which may allow enough oxygen to enter to increase the temperature and cause the pesticide to decompose into less harmful products.

4. Plug up drains to storm sewers or basements. A special foam just being developed can be used as a plug for sewers or to seal manhole covers in streets. Large plastic sheets can be used to line a hole or trench to prevent the water runoff from being absorbed into the soil.

5. Protect exposures from radiated heat.

6. If water runoff does escape, make sure that the proper authorities are notified. These include the local, state, and federal EPA, the local and state Department of Health, and the various police agencies.

7. Ensure that a guard is left at the containment dike until a professional company can be called to remove it. There are many such specialized companies that do this and you should know how to reach them.

Decontamination Procedures.

Decontamination procedures must be accomplished under very careful conditions. The wash water must be contained and then disposed of with the other contaminated material. The steps to be followed are:

1. Remove protective clothing upon leaving site and impound with contaminated response equipment.

2. Upon return to quarters, shower and shampoo thoroughly with soap and water, change into clean clothing, and wash inner clothing with detergent.

3. Watch for signs and symptoms of pesticide poisoning. (These are discussed under the next heading.)

4. Put on coveralls and rubber or neoprene gloves and decontaminate protective clothing and equipment using a strong detergent solution. Decontaminate in an isolated area.

5. Contaminated cotton-jacketed hose may have to be destroyed; most are weakened by strong detergents.

6. Seek assistance of the Pesticide Safety Team Network, activated through CHEMTREC (Chapter 6), to get specific advice on the product to use for decontamination.

Symptoms and First Aid for Pesticide Poisoning

It is important for all emergency response personnel to be able to recognize the symptoms of pesticide poisoning. Prompt treatment is extremely important. Any unusual appearance or feeling of discomfort or illness can be a sign or symptom of pesticide poisoning. These symptoms may occur immediately or up to 12 hours later.

The major signs that indicate poisoning are:

1. Any unusual behavior or appearance—staggering, drooling, slurred speech, pinpoint pupils, profuse sweating, convulsions, or coma.
2. Feelings of discomfort—muscle cramps, stomach cramps, dizziness, skin or eye irritation, localized pain, or difficulty breathing.
3. Illness—headache, nausea, or vomiting.

A good source of information on the signs, symptoms, handling, and first aid treatment is the label on the product, if available. There may be such important information as precautionary handling advice, instructions on what to do if exposed, and a "note to the physician" giving antidotal or treatment suggestions.

As a further guide, Tables 3-1 through 3-9 list the most widely used pesticide compounds, how absorbed, levels of toxicity, and symptoms of poisoning.

Table 3-1. Chlorinated Hydrocarbon Insecticides

Compound	How Absorbed	Toxicity	Symptoms
Aldrin Benzene Hexachloride (BHC) Chlordane DDT Dicofol (Kelthane) Dieldrin Endrin Kepone Lindane (Isomer of BHC) Mirex Thiodan Toxaphene	Ingestion Inhalation Skin absorption	Mild to high	Onset 20 minutes to 4 hours Nausea Vomiting Restlessness Tremor Apprehension Convulsions Coma Respiratory failure Death

Table 3-2. Organophosphous Compound Insecticides

Compound	How Absorbed	Toxicity	Symptoms
Abate DDVP (Vapona) Diazinon Dicapthon Dimethoate (Cygon) Dursban Ethion Fenthion (Baytex) Gardona Malathion Naled (Dibrom)	Ingestion Inhalation Skin absorption	Moderate to extreme	1. *Mild:* anorexia (loss of appetite), headache, dizziness, weakness, anxiety, tremors of tongue and eyelids, excessive contraction of pupils, impairment of visual acuity. 2. *Moderate:* nausea, salivation, lacrimation, abdominal cramps, vomiting, sweating, slow pulse, muscular tremors. 3. *Severe:* diarrhea, pinpoint and nonreactive pupils, respiratory difficulty, pulmonary edema, cyanosis (turning blue), loss of sphincter control, convulsions, coma, and heart block.

Table 3-3. Carbamate Insecticides

Compound	How Absorbed	Toxicity	Symptoms
Baygon Carbaryl (Sevin) Thiram Vapam Zectran	Ingestion Inhalation Skin absorption	Slight (Carbaryl) to moderate (Baygon)	Constriction of pupils Salivation Profuse sweating Lassitude Muscle incoordination Nausea Vomiting Diarrhea Epigastric pain Tightness in chest

Table 3-4. Arsenical Insecticides

Compound	How Absorbed	Toxicity	Symptoms
Cacodylic (Dimethyl arsinic acid) DSMA, MSMA (Sodium methanearsonates) Paris Green Sodium Arsenite	Ingestion Inhalation	Mild to high	*Onset:* 30 minutes to many hours Vomiting, profuse painful diarrhea—bloody later. Colicky pains in esophagus, stomach, and bowel. Dehydration, thirst, muscular cramps. Cyanosis, feeble pulse, and cold extremities. Headache. Dizziness, vertigo. Delirium or stupor. Skin eruption. Convulsions. *Three terminal signs:* coma; general paralysis; death. *Ingestion:* chief initial symptoms are those of a violent gastroenteritis, burning, esophageal pain, vomiting, watery or bloody diarrhea containing much mucous, later collapse, shock, marked weakness. Death generally due to circulatory failure. *Inhalation:* may cause pulmonary edema, restlessness, dyspnea, cyanosis, and foamy sputum.

Table 3-5. Halogen Fumigants — Insecticides

Compound	How Absorbed	Toxicity	Symptoms
Methyl bromide Sulfuryl fluoride (Vikane)	Ingestion Inhalation Skin absorption	Mild to high	*Onset:* 4 to 12 hours following inhalation. *Symptoms* include dizziness, headache, anorexia, nausea, vomiting, and abdominal pain. Lassitude, weakness, slurring speech and staggering gait. Mental confusion, mania, tremors and epileptiform convulsions.

Bromides cause rapid respiration, pulmonary edema, cyanosis, collapse, and death.
Coma, areflexia, and death due to respiratory or circulatory failure. Late manifestations may include bronchopneumonia, pulmonary edema, and respiratory failure.
Methyl bromide may produce cutaneous blisters and kill via dermal exposure.

Table 3-6. Cyanide Fumigants — Insecticides

Compound	How Absorbed	Toxicity	Symptoms
Hydrocyanic acid (Prussic) Hydrogen cyanide Organic Bound Cyanides (e.g. Acrylonitrile) Cyanogas	Ingestion Inhalation	Extreme	One of fastest acting known poisons. *Massive dose*—unconsciousness and death without warning. *Smaller doses*—illness may last one or more hours. Following ingestion, bitter, acrid, burning taste followed by constriction of membrane in throat. Salivation and nausea without vomiting. Anxiety, confusion, and dizziness. Variable respirations—inspiration short, expiration prolonged. Odor of bitter almonds in breath and vomitus. Initial increase in blood pressure and slowing of heart followed by rapid and irregular pulse, palpitation, and constriction of chest. Unconsciousness, convulsions, and death from respiratory failure.

Table 3-7. Phosphine Fumigants—Insecticides

Compounds	How Absorbed	Toxicity	Symptoms
Celphos Delicia Phostoxin (Aluminum phosphide)	Inhalation	Extreme	Nausea, vomiting, diarrhea, great thirst, headache, vertigo, tinnitus, pressure in chest, back pains, dyspnea, a feeling of coldness, and stupor or attacks of fainting. May develop hemolytic icterus and cough with sputum of a green fluorescent color. Chronic poisoning may be characterized by anemia, bronchitis, gastrointestinal disturbances, dental necroses, and disturbances of vision, speech, and motor functions.

Table 3-8. Coumarins, Indandiones—Rodenticides

Compound	How Absorbed	Toxicity	Symptoms
Diphacin Fumarin Pival Pivalyn PMP (Valone) Warfarin	Ingestion	Slight (single dose) to to high (multiple doses)	After repeated ingestion for several days: Bleeding from nose, gums, and into conjunctiva, urine, and stool. Late symptoms: Massive ecchymoses or hematoma of skin, joints, brain hemorrhage. Shock and death.

Table 3-9. Herbicides

Compound	How Absorbed	Toxicity	Symptoms

Organic Acids and Derivatives
Dichlorophenoxyacetic acid
  (2, 4-D)
Silvex: 2, 4, 5-TP (proprionic acid derivative)
Trichlorophenoxyacetic acid
  (2, 4, 5-T)
	Ingestion	Mild to high	Weakness, and perhaps lethargy. Anorexia, diarrhea. Muscle weakness—may involve the muscles of mastication and swallowing. Ventricular fibrillation and/or cardiac arrest and death.

Urea Betasan Bromacil Hyvar-X Karmex (Diuron) Telvar (Monuron) Urab (Fenuron-TCA) Urox	Ingestion	Mild to moderate	During handling may cause irritation of eyes, nose, throat, and skin. *Ingestion:* May cause gastroenteritis.
Miscellaneous Diquat Endothal (Dicarboxylic acid derivative) Paraquat (Quarternary ammonia derivatives)	Ingestion	Mild to moderate	Lethargy Convulsions Coma

Treating the Poison Victim

The first thing to do in a case of suspected poisoning is to get the exposed victim to competent medical help as quickly as possible. Contact a Poison Control Center if there is one in your community, or a toxicologist, for advice about what to do until you can reach a physician. On the scene and during transport of the victim, emergency medical personnel should employ the following general first aid measures:

1. Establish an airway.

2. Remove secretions from the throat.

3. Administer oxygen.

4. Decontaminate patient with detergent and lots of water if there is any chance that skin or hair are contaminated.

5. Monitor vital signs.

6. Begin cardiopulmonary resuscitation if necessary.

7. Emergency medical personnel must avoid contact with any excretions from the patient. They should wear rubber gloves while working with the patient to prevent skin absorption.

8. For any toxic fumes in the eye, flush with

clear water *immediately* and continue for 15 minutes.

9. Provide as much information on the suspected poisoning agents as possible for medical authorities.

Pesticides may appear in the least expected places. You must be prepared to find them at almost any call. In a recent incident a 2½-year-old boy was hospitalized in a rural community. His illness was diagnosed as organophosphate poisoning. The boy had been playing in a pile of 55-gallon (209 liters) drums about 50 feet (15 meters) from the front door of his home. The city had obtained the drums from an aerial insecticide applicator and planned to use them as trash containers. In fact, the city had urged residents to come by and pick up a drum.

Investigation showed that the drums contained enough pesticide residues to harm anyone, child or adult, who touched them.

CORROSIVES

Both acids and bases are considered to be in the general category of chemicals called corrosives. Both groups will cause severe damage to living tissue when brought in contact with it. The Department of Transportation definition of a corrosive is a liquid or solid that causes visible destruction or irreversible alternations in human skin tissue at the site of contact, or in the case of leakage from its packaging, a liquid that has a severe corrosion rate on steel. There are other substances that will meet this definition of corrosives, but they are classified by the Department of Transportation under another hazard. An example of this is hydrogen peroxide, which is a corrosive but is classified and shipped as an oxidizer.

Acids and bases are completely opposite in character, yet they both meet the definition of a corrosive. In fact, acids and bases are so different that they can be used to neutralize each other.

Guy-Lussac, a chemist, concluded in 1814 that it was necessary to define bases and acids in terms of each other. This basic concept has formed the foundation for our current definition of these two chemical classes. A very basic definition of an *acid* states that it is a compound that will yield hydrogen ions (H^+). (To be chemically exact, an acid yields hydronium ions (H_3O^+) when dissolved in water.) On the other hand, *bases* are compounds that will yield hydroxide ions (OH).

The process of neutralization of an acid by a base, or vice versa, occurs when the hydrogen ions and the hydroxide ions form neutral water:

$$H^+ + OH^- \longrightarrow H_2O$$

It is important to remember that while the reaction will create a neutral substance, there can be a great deal of heat given off by the reaction. Sometimes, the reaction can take place so fast that it appears to be explosive. In any case, splattering can occur if the conditions for mixing are not carefully controlled.

Emergency response personnel must also be aware of the fact that the term *neutralization* does not necessarily mean *safe*. The degree of neutralization is dependent upon concentration of the two reacting chemicals. *Concentration* is defined as the percentage of the substance in water. Commercial concentrations can vary from 38% hydrochloric acid to 98% sulfuric acid.

When reading the label or shipping papers for a corrosive, emergency response personnel must not confuse concentration with strength. The strength of an acid is dependent upon the number of H+ ions the compound contains. Those with more H+ ions are strong, while those with less are considered weak. It is therefore possible to have a weak acid in a high concentration (an 85 percent concentration of boric acid).

Neutralization of an acid or a base by the addition of its opposite can thus be divided into three general categories:

1. A strong acid plus a strong base gives a neutral solution.

2. A strong acid plus a weak base will produce an acidic solution.

3. A weak acid plus a strong base gives a basic solution.

pH Measurements

As discussed in the preceding paragraphs, the acidity of an aqueous solution is dependent upon the concentration of H^+ ions. One way of expressing this concentration is to use a numerical scale known as the pH scale, which bases its values on mathematical conversions. Table 3-10 contains the pH scale and Table 3-11 lists some examples of the pH values of some common compounds.

Table 3-10. pH Scale

Mole/Liter (H^+)	pH Value	Acid or Base	Degree of Acidity
1×10^{-14}	14		
1×10^{-13}	13		
1×10^{-12}	12		
1×10^{-11}	11	Basic	
1×10^{-10}	10		
1×10^{-9}	9		
1×10^{-8}	8		Increasing
1×10^{-7}	7	Neutral	Acidity
1×10^{-6}	6		
1×10^{-5}	5		
1×10^{-4}	4		
1×10^{-3}	3	Acidic	
1×10^{-2}	2		
1×10^{-1}	1		
1×10^{0}	0		

Table 3-11. pH Values of Common Compounds

Solution	pH Value
Blood	7.4
Milk	6.6
Carbonated Water	3.0
Vinegar	2.8
Lemon Juice	2.3
0.1 Mole HCl	1.0

When using Table 3-10, remember that a concentration of $2.3 \times 10^{-9} H^+$ ions means this value falls between a pH of 8 and 9. The actual pH value would be about 8.9.

Acids

The derivation of the word *acid* comes from the Latin word *acidus*, which means sour. The early scientists noted a common tart taste to the known acids and so named the group. One of the earliest acids known was vinegar, which in Latin is known as *acetum*. However, the so-called sour taste is not common for all acids.

When an acid is dissolved in water the free hydrogen ions (positively charged) will migrate toward a negative electrode if electricity is passed through the solution. In this way, the acid solution can become an efficient conductor of electricity. The familiar lead-acid battery in an automobile uses sulfuric acid, H_2SO_4, as an electrolyte (carrier of electricity).

There are two basic groups into which acids can be classified: organic and inorganic. The basic definition of an *organic acid* is one that contains carbon within the compound. Conversely then, an *inorganic acid* is one without any carbon.

The most important class of organic acids are known as carboxylic acids. This name is derived by a contraction of the two components of the chemical group that make up the acid, *carb*onyl and hydr*oxyl*. In the shorthand notation of the chemist, the carboxyl group is written:

$$C \overset{\displaystyle O}{\underset{\displaystyle OH}{\big<}}$$

The symbol is also abbreviated as COOH. The carboxyl group is then combined with either carbon and hydrogen in a chain or carbon and hydrogen in a ring.

In general, organic acids are flammable because of their carbon content, while inorganic acids are not flammable. However, inorganic acids can act as an oxidizing agent that can ignite other combustible materials in case of a spill.

A list of organic and inorganic acids and their relative strengths are shown in Table 3-12. Note that, in general, organic acids are weaker than the inorganic ones. The table also contains the physical properties of some common acids.

Bases

A base was defined in an earlier paragraph as a compound that yields a hydroxide ion (OH⁻) in solution. Other names for bases are *caustics* and *alkalines* and these terms can be used interchangeably.

Bases have the same general properties as the acids. They cause severe damage to human skin tissue, can react violently and release heat, and, of course, react with acids, as described in the preceding paragraphs.

Table 3-13 contains the physical properties of the two most common bases, sodium hydroxide and potassium hydroxide.

Hazards of Corrosives

As described in the preceding paragraphs, corrosives present a number of hazards to emergency response personnel. The officer in charge of an incident involving a corrosive must ensure that proper protection for all personnel is maintained. At times, the vapor given off from a leak is so penetrating that special acid suits and breathing apparatus must be worn for the protection of the wearer. To determine the correct methods of attack at a spill or fire, the first step is to identify the product.

The following general summary of the possible hazards of corrosive materials will aid emergency response personnel in making tactical decisions:

Table 3-12.   Organic and Inorganic Acids

Name of Acid	Chemical Symbol (I) Inorganic (O) Organic		Flash Point		Specific Gravity	Vapor Density	Boiling Point		Freezing Point	
			°F	°C			°F	°C	°F	°C
Perchloric	HClO₄	(I)	—	—	1.70	—	397	202.78	0	−17.78
Sulfuric	H₂SO₄	(I)	—	—	1.84	—	640	337.78	50	10.00
Hydrochloric	HCl	(I)	—	—	1.20	1.3	−121	−85.00	−175	−115.00
Nitric	HNO₃	(I)	—	—	1.50	—	187	86.11	−44	−42.22
Phosphoric	H₃PO₄	(I)	—	—	1.69	—	500	260.00	108	42.22
Hydrofluoric	HF	(I)	—	—	1.00	0.7	68	20.00	−117	−87.78
Formic	HCOOH	(O)	156	68.89	1.22	1.6	213	100.56	47	8.33
Acetic	CH₃COOH	(O)	109	42.78	1.05	2.1	244	117.78	61	16.11
Hydrocyanic	HCN	(O)	0	−17.78	0.7	0.9	79	26.11	—	—
Butyric	CH₃(CH₂)COOH	(O)	161	71.67	0.96	3.0	327	163.89	—	—
Propionic	CH₃CH₂COOH	(O)	130	54.44	0.99	2.6	286	141.11	—	—

(increasing acid strength ↑)

Table 3-13.   Bases

| Name of Base | Chemical Symbol | Boiling Point | | Melting Point | | Specific Gravity |
		°F	°C	°F	°C	
Sodium hydroxide	NaOH	2500	1371.11	600	315.56	2.1
Potassium hydroxide	KOH	2400	1315.56	680	360.00	2.0

1. The most obvious problem with a corrosive is that it attacks the skin as well as metal. This means that emergency response personnel must be completely encased in nonpenetrable clothing, with self-contained breathing apparatus inside. All exposed skin must be completely covered.

2. Acids such as nitric acid act as an oxidizing agent. When brought into contact with products containing cellulose, there is a possibility of spontaneous ignition. An explosion can also occur if hot nitric acid mixes with common organic materials.

3. When water is added to sulfuric acid a large amount of heat is produced. Therefore, on spills of this acid, large quantities of water can produce heat that results in boiling and spattering of the acid, as well as the possibility of igniting nearby combustibles. Be sure of the exact compound before deciding to dilute an acid spill with water.

4. Besides being corrosive, some acids are extremely poisonous. Hydrocyanic acid, a fairly weak acid, is a strong poison. To show you how dangerous it is, one part of hydrogen cyanide in 3,700 parts of air (270 parts per million) can be immediately fatal. The threshold limit of exposure is 10 parts per million. Obviously, this chemical must be handled very carefully.

5. As described in Table 3-3, the organic acids are flammable. So, in addition to the concern for the health aspects of the incident, emergency response personnel must ensure that sources of ignition are removed. In addition, as explained, inorganic acids can oxidize and cause ignition of nearby combustibles.

6. Some acids can act as an explosive. Peracetic acid (CH_3COOOH) has an NFPA 704 reactivity rating of 4. It is extremely dangerous when exposed to heat and becomes heat and shock sensitive. Picric acid also has a reactivity rating of 4, and is classed as a high explosive.

Emergency response personnel must determine the name of the corrosive before beginning to take corrective action. Just determining that the product is a corrosive or seeing the word *acid* on the label does not mean that the only hazard is to health. Until the product is properly identified, emergency response personnel must exercise extreme care because there could also be danger as a result of oxidation, flammability, reactivity, or reaction to extinguishing agent.

Case History of a Silicon Tetrachloride Leak

On April 26, 1974, in a bulk storage tank farm on the south side of Chicago, a storage tank of silicon tetrachloride developed a leak. About four hours after the leak first started, the spill had worsened to the point where evacuation was needed. The surrounding residential area housed 20,000 to 30,000 persons. The Governor of Illinois alerted three National Guard battalions to assist. An hour later, as evening approached, the decision was made to evacuate the residents.

Still, the leak continued. Experts from 15 different chemical companies could not come up with a solution. The corrosive material remained in the diked area, but transfer of the remaining product still had not begun.

One suggestion for reducing the fumes rising from the diked area was to use high-expansion foam. Early the next morning, some 16 hours into the incident, the foam was applied. This, however, only intensified the fuming. A decision was made to add the foam after covering the liquid with number 6 fuel oil. Four hours later the fuel oil was added along with eight truckloads of lime. This dramatically reduced the fumes within one hour.

Later that morning, 22 hours after the start of the incident, the transfer pumps were started. They began pumping at the rate of 12,000 gallons (48 cubic meters) per hour. Because of the corrosive effects of the liquid on the pump one broke down, leaving only one to handle the load. It was estimated at this time that it would take 30 to 40 hours to transfer the 750,000 gallons (2.8 million liters) of product remaining.

Because the covering had reduced the hazard from the fumes, the residents were allowed to return home. In addition, fire and police personnel were reduced.

At approximately 8:00 A.M. on the third day, 44 hours after the leak started a heavy rain fell. There were now 80,000 gallons (303,000 liters) in the moat. The rain caused a return of severe fuming. As a result, the electric lines were corroded and power was cut. An hour and a half later, all pumping stopped because of corrosion to the pumps. At this point 100,000 gallons (400 cubic meters) had leaked and 80,000 gallons (320 cubic meters) had been transferred.

By the afternoon of the third day the main pipe leading to the tanks appeared to be weakening from the acid corrosion. If this pipe broke, the remaining 600,000 gallons (2.27 million liters) would be turned loose.

At this point communications problems were developing. Cooperation of the telephone and electric utilities was necessary to ensure that the command post could function.

Finally, early on the morning of the fourth day, the leak was patched with 150 cubic yards (120 cubic meters) of quick-setting cement. A total of 280,000 gallons (1.06 million liters) of the product had leaked into the diked area. It took five more days for the product that had leaked to be pumped back into a storage tank.

At the height of the incident there were representatives of 15 organizations trying to coordinate with each other and implement decisions. Coordination at a time like this becomes extremely important. Would your department be ready to handle an incident of this magnitude?

POISONS

A poison is any material that can harm living tissue. The degree of toxicity varies with the particular chemical. Some can kill with great speed even in minute quantities.

Emergency response personnel must be concerned with poisons entering the body and causing damage. There are three basic ways in which poisons attack:

1. By *asphyxiation*. The asphyxiants work by excluding oxygen. Poisons can do this by simply displacing the oxygen in the air and causing the percentage of oxygen available to fall below that which is necessary to support life. Other asphyxiants are absorbed by the bloodstream and prevent oxygen from being transported throughout the body. Finally, some poisons are transported by the bloodstream to the tissues, where they prevent the tissues from absorbing oxygen.

2. By *ingestion*. Some poisons are caustic and can damage and destroy human tissue when brought in contact with it. Internal organs and tissue are damaged to the point at which they cease to function.

3. Through *bodily functions*. Some poisons directly attack the body's vital organs such as the liver or heart. Others affect the nerves and the ability of the nerves to transmit impulses. Finally, muscle tissue can be damaged that will affect an individual's ability to move.

Health Hazards of Poisons

Fire fighters must exercise extreme care when working around poisons. Poisons can enter the body through inhalation, absorption through the skin, through the mouth, or through an outside opening in the body (existing or accidental).

Full protective equipment is necessary. This means that special suits and breathing apparatus should be worn. All skin openings must be covered.

Before attempting a rescue, emergency response personnel must be sure that the type of poison is known and that adequate protection for the rescuers is available. It will only compound the problem if the rescuers become injured and, in turn, need assistance. A few seconds to evaluate the situation can save many minutes during the rescue.

Handling a Poison Incident

A poison incident must be approached with extreme care. A slight miscalculation by emergency response personnel can result in loss of life. Following are suggested procedures when called to the scene of an accident where poisonous materials are present:

1. Evacuate personnel from surrounding areas. Start with those downwind, but do not assume that the wind direction will stay the same. All civilians should be removed from the immediate vicinity.

2. Try to identify the product involved. Check labels and bills of lading for information; consult reference sources to determine the hazards.

3. Position personnel and apparatus upwind from the spill or leak. Make sure there is a path for escape for both personnel and apparatus.

4. Use the type of protective equipment and breathing apparatus recommended by the reference sources.

5. If possible, contain the leak with natural or artificial barriers. Try to divert flow from exposures, mixing with other chemicals, or from entering the sewage system.

6. An attempt should be made to close or stop the leak. It may be necessary to call on specialized personnel from the chemical manufacturer to assist. Personnel performing this task must be afforded all possible protection in case an accident occurs and they need rescuing.

7. If the poison is also flammable and burning, special additional precautions are necessary. The instructions provided by the reference sources should be followed.

8. Decontamination of equipment, protective clothing, personal clothing, and skin are necessary after exposure. Water run-off from the decontamination must also be handled carefully.

Physical Properties

The physical properties of some of the more common poisons are listed in Table 3-14.

Montgomery County, Maryland, Poison Spill,[3] April 4, 1977

A firm that stores thousands of chemicals for use in research by various government agencies is located in Montgomery County, Maryland. An employee was moving a vial

Table 3-14.   Physical Properties of Selected Poisons

Product	Color	Odor	State	Usual Shipping Container	Life Hazard	Threshold Limit Value, ppm
Acetaldehyde	Colorless	Fruity	Liquid below 69° F (20.5° C)	5- to 55-gallon metal drums	Eye, skin, and lung irritant. Narcotic effect from long inhalation.	200
Acrolein	Colorless	Disagreeable	Liquid	5- to 55-gallon metal drums	Small amounts are highly poisonous.	0.5
Bromine	Reddish brown	—	Liquid or gas	Glass bottles, 10-gallon drums, tank cars	Irritating to eyes and respiratory tract at 10 ppm level.	0.1
Ethylene Dichloride	Clear	Chloroform	Liquid	1-, 5-, and 55-gallon drums and tank cars	Inhalation, skin contact, or oral intake. Eye irritant.	100
Fluorine	Yellow	Pungent	Gas	Special steel cylinders	Injury to eyes, skin, or respiratory tract.	0.1
Hydrogen Fluoride (Hydrofluoric Acid)	Colorless	—	Liquid below 67° F (19.4° C)	Carboy, bottles, or cylinders	Irritating to eyes, skin, and respiratory tract in both liquid or gas states.	3.0
Phenol	Colorless or white; reddish when impurities are present.	—	Solid	Bottles, cans, drums, and tank cars	Causes severe skin burn. Absorbed through the skin or via inhalation of dust.	5.0
Silver Nitrate	Colorless to white	—	Solid	Amber or black bottles	Skin, eyes, or respiratory irritant.	—
Hydrogen Sulfide	Colorless	Rotten eggs	Gas	Steel cylinders	Eye, skin, and respiratory irritant.	—
Hydrogen Cyanide	Colorless	Bitter almonds	Liquid below 79° F (26° C)	Steel cylinders; tank cars	Few breaths can cause death. Can be absorbed through skin.	—
Dimethyl Sulfate	Colorless	Onionlike	Liquid	Glass bottles, drums, tank cars	Inhalation, skin contact, orally. Eye irritation and severe burns. Damage may not show up for several hours after exposure.	—
Cyanogan	Colorless	Pungent	Gas	Metal cylinders of less than 125 pounds (56.25 kilograms) capacity	Inhalation or skin contact. Small amount can cause death.	—
Chloropicrin	Colorless	Penetrating	Liquid	Metal drums of less than 30 pounds (13.5 kilograms) capacity	Skin or lung irritant. Short exposure can cause fatal lung damage.	—
Benzotrifluoride	Clear to white	Aromatic	Liquid	55-gallon drums (20.9 liters)	Skin, lung, and eye irritant. Forms an acid in contact with moisture.	—

(*continued* on page 60)

Table 3-14. *(Cont.)*

Product	Color	Odor	State	Usual Shipping Container	Life Hazard	Threshold Limit Value, ppm
Chlorine Trifluoride	Greenish-yellow liquid Colorless gas	Chlorine	Liquid below 530° F (276.7° C)	One-ton (1 metric ton) containers and tank cars	Skin, lung, and eye irritant. Pulmonary edema from inhalation.	—

from the refrigerator when it fell to the floor and broke.

Employees of the company opened a window to ventilate their office. The chemical was so potent that the building ventilation system picked up the fumes and carried them to other tenants in the same building. Another problem was that rags used to soak up the spill were placed in the regular trash. The trash collector had great difficulty eliminating the fumes from trash pickup equipment.

Personnel in nearby offices complained of headaches, dizziness, and nosebleed. Others complained of trouble breathing. Some had symptoms of the poisoning five days after the incident.

Workers in nearby offices had to be kept from their offices for more than a week before the fumes were finally removed. The odor was so strong that it was very evident more than two blocks away.

The ultimate problem in incidents like this is that the long-term effects of the exposure are unknown. No one can say what complications or disease may show up in 20 years. The quantities of poisonous material released in the original exposure need not be large. Small amounts can be spread through a ventilating system to extend the problem way beyond the initial area of the incident — as for example, in a high-rise building.

QUESTIONS FOR REVIEW AND DISCUSSION

1. What are the four major categories of pesticides and how do they act to control the problem?

2. What information can be obtained from a pesticide label?

3. How can pesticides enter the body?

4. What determines the toxicity of the pesticide?

5. Outline the steps for handling a pesticide spill or fire.

6. What are the symptoms and first aid for pesticide poisoning?

7. List 15 chemicals that fall in the category of a corrosive.

8. What is the difference between the strength and concentration of an acid?

9. What does a pH measurement indicate?

10. Summarize the hazards that emergency response personnel face when handling a corrosive incident.

11. What three ways do poisons act on the body?

12. Outline the steps emergency response personnel should take when handling a poison incident.

NOTES

1. "The Long Road to Market," *Chevron World*, Spring 1977.

2. "Issuance of a Specific Exemption to Use Dinoseb to Control Various Broadleaf Weeds in Lentils," *Federal Register* 42, no. 106 (June 2, 1977): 28172.

3. Robert Meyers, "Chemical Spill Fumes Make Workers Ill," *Washington Post*, April 9, 1977.

4
Explosives, oxidizers, and radioactive materials

Many of the chemical substances in use today present explosive or radioactive hazards. Because of the great number of products that fall into these categories, this chapter presents only broad, generalized information on the major hazards. The materials are described on the basis of their common properties and potential for problems, and techniques are given for handling an incident.

Understanding the behavior characteristics of a chemical is the first step toward knowing how to bring it under control in an emergency. The problems are complicated by the fact that a product may have hazardous characteristics that fall into more than one category. The labels do not tell the whole story, because chemicals are labeled according to their primary hazardous classification. The basic information given here is only a starting point. To be able to deal with the specific hazards of a given material under the special conditions of an emergency incident, it is necessary for emergency response personnel to consult with reference sources listed in this text.

EXPLOSIVES

The greatest hazard in an accident or fire in which explosives and blasting agents are present is the threat of an explosion. An emergency incident is much more likely to occur during transport than in stationary storage. A vehicle carrying explosives could have an accident that might trigger a fire or explosion; a storage magazine is usually well protected from accidental damage.

During 1976 there were 3 billion pounds (1,350,000,000 kilograms) of explosives produced in the United States, according to an article in the May 20, 1977, issue of the *Washington Post*.[1] Of this amount, 200,000 pounds (90,000 kilograms) of blasting agents and 29,000 blasting caps were reported stolen. The article further reports that explosive blasts caused 69 deaths, 239 injuries, and more than $12 million in property damage in 1976. It may be assumed that stolen material was involved in some of these incidents.

In addition to dealing with the potential

hazards of explosives legally stored and transported, emergency response personnel must be on the lookout for stolen materials. A routine fire incident could turn into a disaster with the detonation of stolen explosives. There is no way to predict when and where they might show up.

The Department of Transportation regulates the markings necessary for transporting explosives, but the sole use and storage are regulated by the Department of the Treasury, Bureau of Alcohol, Tobacco, and Firearms (ATF) under Title XI of Public Law 91-452. On August 3, 1977, this agency proposed sweeping changes to this law.[2] Emergency response personnel are therefore affected by still another government agency.

Definition of Explosive

An *explosive* can be defined as a mixture of solids or solids and liquids which upon rapid and violent decomposition produces large volumes of gas. Explosives can be categorized in two classes on the basis of the speed of reaction time: high explosives react extremely quickly—on the order of millionths of a second—while low explosives react slightly slower. As far as emergency response personnel are concerned, the difference in timing is immaterial. Both types are dangerous and you cannot relax because of the use of the word *low* in the description.

A high explosive is said to explode or detonate; *detonation* is the creation and rapid release of the gas generated by the explosion. On the other hand, a low explosive burns very rapidly (the time difference in creating the gas); this is termed *deflagration*.

Black Powder (Low Explosive)

The most common low explosive is black powder. It is composed of sulfur, charcoal, and either potassium nitrate or sodium nitrate. Explosions from black powder have a shearing and heaving action, while burning with a very hot flame. Most of the black powder manufactured today is used inside safety fuses, which are used in setting off high explosives. Black powder may also be used in an explosive train (described under the next heading), sporting firearms, military devices, and fireworks. Some of the properties of black powder are listed in Table 4-1.

Table 4-1.   Properties of Black Powder

Property	Description
Sensitivity	Relatively insensitive to shock Relatively insensitive to friction Relatively insensitive to static electricity Ignites from spark or heat that reaches ignition temperature
Ignition	572° F (282° C) from any source (spark, fire) Blasting accessories (blasting caps, detonating cord)
Velocity	Open burning, five seconds per foot In steel pipe (coarse granulations) 168 meters (560 feet) per second In steel pipe (fine granulations) 620 meters (2,070 feet) per second
Fumes produced	Carbon monoxide and hydrogen sulfide

High Explosives

In the safe detonation of a high explosive, a series of controlled explosions takes place in sequence. This sequence of events, called an *explosive train*, is shown in Figure 4-1.

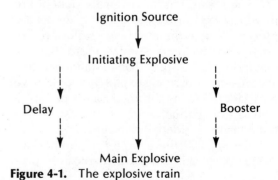

Figure 4-1.   The explosive train

To initiate the explosive train there is a *primary* or *initiating* explosive. This, in turn, causes detonation of the main charge. Of course, there has to be an ignition source to detonate the initiating explosive. In addition, various devices can be inserted between the initiating explosive and the main explosive to perform several jobs. Some of these extra devices are used to ensure complete detonation. Others are used to slightly delay detonation of the main charge, which allows sequencing of the detonation. Each device used within the explosive train is an explosive and must be treated and handled as such.

Initiating Explosives

The Du Pont *Blasters' Handbook*[3] defines the uses of initiating explosives as designed to: "(1) initiate charges of explosives, (2) supply or transmit flame to start an explosion, or (3) carry a detonation wave from one point to another or from one charge of explosives to another."

Electric Blasting Caps. The most common initiating device is the electric blasting cap, shown in Figure 4-2. These caps use an electric current, which can be initiated from a remote location. The caps consist of a metal shell into which are loaded several powder charges and and electric ignition element to which are attached the electrical wires (leg wires), as shown in Figure 4-3. A rubber plug, crimped to the metal, seals the explosives in the cap.

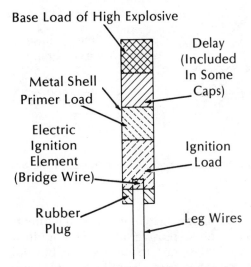

Figure 4-3. Cross section of an electric blasting cap

Some caps contain a delay charge to give a predetermined delay interval between the application of the electric current and the detonation of the cap. The delay can vary from a few thousandths of a second to 12 seconds. The delay enables a complete round of explosions to be fired from a single application of current.

Blasting caps come in eight sizes, designated 1 through 8, with a number 8 being the most powerful. These numbers indicate the relative strength of the cap. In addition to this numbered designation, delay caps are numbered on the basis of their delay period. For example, Du Pont (MS) Delay E. B. Caps have 19 different delay intervals and are numbered as follows:

Period 1	25 millisecond delay
Period 2	50 millisecond delay
through	through
Period 19	1 second delay

Nonelectric Blasting Caps. Blasting caps can be attached to safety fuse on one end (Figure 4-4). The other end of the fuse can then be lit

Figure 4-2. Electric blasting cap

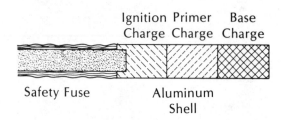

Ignition Primer Base
Charge Charge Charge

Safety Fuse   Aluminum
Shell

Figure 4-4. Cross section of a nonelectric blasting cap

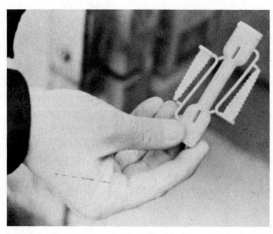

Figure 4-5. Connector delay

with an ordinary match or a special igniter to initiate the explosive train. The ignition charge is designed to ensure that burning from the safety fuse is continued. This flame front then ignites the primer charge, which in turn detonates the base charge. The shell of the cap is made of aluminum or copper.

Safety Fuse. Safety fuse is used to allow personnel to detonate the main charge from a safe distance. It is made of black powder formulated with potassium nitrate. The exterior is covered to protect against abrasion and penetration from water.

Safety fuse will burn at the rate of 90 to 120 seconds per yard (0.9 meter). Within the United States, the 120 seconds per yard (0.9 meter) material is standard.

If a series of fuses have to be lit in sequence, igniter cord is used. This cord burns with a short, hot flame and ensures quick ignition. Burning rates for igniter cord vary from 8 to 20 seconds per foot (0.3 meter).

Detonating Cord. Detonating cord is a flexible cord, which contains a high explosive in the core. Emergency response personnel must exercise extreme care that this cord is not confused with the safety fuse. Detonating cord is a high explosive.

The core of detonating cord is usually pentaerythritol tetranitrate (PETN). When PETN is used, detonating cord will explode at the rate of 21,000 feet (6,300 meters) per sec-

ond. Obviously this is a great deal faster than the 90 seconds per yard (0.9 meter) of safety fuse. Initiation of the explosive train for this type of explosive takes place with a blasting cap.

Connector Delay. When using detonating cord, it is sometimes necessary to delay the explosion of the main charge so sequencing of the explosion can occur. To accomplish the required delay a connector (Figure 4-5) is inserted in the line. These delays can vary from 5 milliseconds to 25 milliseconds.

Boosters. A booster (Figure 4-6) consists of a metal casing about the size of a small flashlight battery, which contains a high explosive inside. Boosters are designed to ensure detonation of the primary charge by increasing the intensity of the initiating charge.

Main Charge Explosives

The major high explosives used as main charges are dynamite and water gel explosives. The destructive power of an explosive is measured in comparison to standard dynamite. The *strength* of the dynamite, which can be defined as the energy content of the explosive, is based on the nitroglycerin content of straight dynamite.

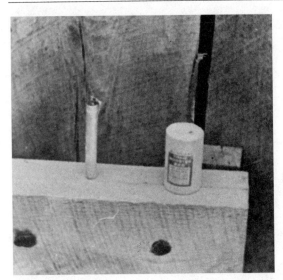

Figure 4-6. Booster—two cylinders on a wooden shelf

Therefore, a 40 percent straight dynamite will contain 40 percent nitroglycerin and a 50 percent stick will contain 50 percent nitroglycerin. However, this relationship is not a straight line. A 60 percent dynamite is not twice as strong as a 30 percent one, but somewhat less than twice as much.

Straight Dynamite. Straight dynamite contains nitroglycerin as its only ingredient. It is manufactured in strengths varying between 30 and 60 percent. When detonated this dynamite produces toxic fumes. In addition, it is very sensitive to shock and friction.

Dynamite Mixtures. Dynamite with ammonium nitrate added is a common mixture. The strength of this type of explosive varies from 20 to 55 percent. Figure 4-7 shows some typical mixtures.

Gelatin. Gelatin dynamite uses a nitrocotton-nitroglycerin gel. The gel is a very thick, rubberlike liquid, which is insoluble in water. The strength of the gelatin can vary from 20 to 90 percent. The fumes produced by the explo-

sion usually do not present a problem. A special mixture of the chemicals is manufactured to ensure the retention of sensitiveness and high velocity even after long storage periods.

The addition of ammonium nitrate to some gelatin mixtures tends to lower the velocity somewhat. These mixtures are not as water-resistant as the straight gels. They are used primarily because they are more economical.

Water Gels. Water gel explosives are a relatively recent development. They have several advantages over the other types of explosives that have been discussed so far:

1. High density
2. High degree of water resistance
3. Lower cost
4. Ease of handling
5. Reliability

Water gel explosives generally consist of a gelled ammonium nitrate–sodium nitrate solution with a sensitizer. Originally, the water gels were used only to replace the large-diameter dynamite [10 centimeters (approximately 4 inches) in diameter]. However, recent technology advances have enabled the development of smaller diameter gels, which allow complete replacement of dynamite.

The greatest advantage of the water gel explosives is its safety as compared to dynamite. Under DOT regulations the water gels are shipped as Explosives B.

Storage

Explosives should be stored in magazines (Figure 4-8), which will protect against accidental detonation or theft. The door to the magazine is usually secured with two locks, which are covered in such a way as to prevent them from being cut. The surface of the magazine should be painted with a reflective color to prevent heat absorption, which will cause deterioration of the contents.

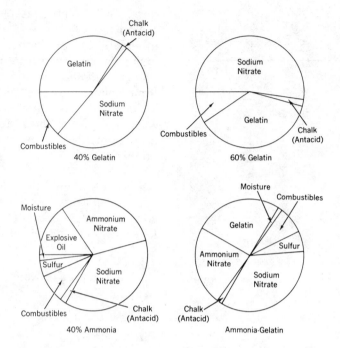

Figure 4-7.   Dynamite mixtures. From James Meidl, *Explosive and Toxic Hazardous Materials,* Glencoe, 1970.

Figure 4-8.   Storage magazine

Safety Precautions

In an earlier paragraph the meaning of the strength of an explosive was discussed. However, the destructive force of the explosive is dependent upon several variables:

1. How tightly the explosive is packed in the container.
2. How easy the explosive is to ignite.
3. How shock-sensitive is the explosive.

Packing an Explosive. The more tightly the explosive is packed within the container, the more destructive is the resulting explosion. This is a result of the increased amount of explosives as well as the fact that the gas created has no place to dissipate.

Explosive Ignition. Heat can cause deterioration of dynamite over a period of time. In addition, exposure to heat can cause dynamite to become shock-sensitive. Since explosives will detonate when a certain temperature is reached, exposure during a fire becomes a very dangerous situation. Table 4-2 gives the approximate temperature at which some of the more common explosives will detonate.

Table 4-2. Approximate Detonation Temperatures for Explosives

	Temperature	
Explosive	°F	°C
TNT	880	471
Lead Azide	640	338
Nitroglycerin	420	215
PETN	405	189
Mercury Fulminate	350	176

Accidental explosive ignition can also occur if electric blasting caps are used in the vicinity of radio transmitters. A field of electrical energy created in the air can be enough to trigger the cap and detonate the charge. The minimum distances from transmitter of varying power are shown in Table 4-3.

Table 4-3. Minimum Distance from Radio Transmitter for Electric Blasting Caps

	Minimum Distance	
Transmitter Power, watts	Meters	Feet
5–25	30	100
25–50	45	150
50–100	66	220
100–250	105	350
250–500	135	450
500–1,000	195	650
1,000–2,500	300	1,000
2,500–5,000	450	1,500
5,000–10,000	660	2,200
10,000–25,000	1,050	3,500
25,000–50,000	1,500	5,000
50,000–100,000	2,100	7,000

Heat and Shock Sensitivity. During exposure to heat, a gradual decomposition of the explosive can take place. The most common sign of decomposition is the emission of nitrogen oxide gas. When this distinctive brownish-orange gas becomes visible, the area must be immediately evacuated.

An explosive may also be subject to detonation from a shock wave when a nearby explosive detonates. Whether the exposed explosive does detonate is dependent upon the terrain between the blast and the exposure, the type of intervening material (rock, air, soil, water), and the strength of the original blast.

Handling an Explosives Incident

An incident involving explosives can occur at any of three points:

1. During transportation.
2. During storage.
3. During use.

The most likely possibility for an emergency call will be during transportation. If there is an accident with no fire, then the best advice is to eliminate ignition sources. Any spilled flammable liquids should be washed from the area. Before moving any of the wrecked vehicles, remove the explosives at least 200 feet (60 meters) away, if possible.

DON'T FIGHT EXPLOSIVE FIRES is the advice of the Institute of Makers of Explosives. The procedures to follow are outlined in their emergency procedure card, illustrated in Figure 4-9. In an accident involving a transport vehicle carrying explosives in which there is a cargo fire, traffic should be stopped and the area cleared for 2,000 feet (600 meters) in all directions. If there is a tire fire, plenty of water should be used to extinguish the fire. Dry chemical and dirt can be tried in place of water. However, the fire can easily reignite, so the best possible method is to remove the tire. For an engine or cab fire, extinguish the fire

EMERGENCY INSTRUCTIONS
IF ANY OF THESE SIGNS APPEAR ON TRUCK, TRACTOR OR TRAILER

DO THIS: **1** Identify the Cargo!
Look for these signs . . .

2 Act Fast!
Take the right action . . .

CARGO FIRE
- Don't fight cargo fire! (Cargo May Explode).
- Stop all traffic and clear the area for 2000 feet in all directions.
- When tractor-trailer is involved, separate tractor from trailer if possible.

TIRE FIRE
- When tractor-trailer is involved, separate tractor from trailer if possible.
- Use plenty of water — douse it. If water is not available use dry chemical fire extinguisher or dirt.
- **CAUTION:** Fire may start again. Stand by with extinguisher ready.
- Conserve dry chemical — use in short bursts.
- Get tire off and away from vehicle.

ENGINE OR CAB FIRE
- When tractor-trailer is involved, separate tractor from trailer if possible.
- Use dry chemical fire extinguisher, water or foam.
- Disconnect one battery cable when possible.

BODY FIRE
- Clear area before fire reaches cargo.
- When tractor-trailer is involved, separate tractor from trailer if possible.
- Use dry chemical extinguisher, water or foam.
- Do not fight fire when it reaches cargo.

In event of fire, accident, leakage, loss of cargo, delay or other emergency, contact your company immediately.

After Completing Emergency Procedures **CALL THIS NUMBER** and Report Incident

Figure 4-9. Emergency procedure card for explosive fires. *Courtesy of Institute of Makers of Explosives*

with the materials available (water, dry chemical, a foam). In all fires in a tractor-trailer rig, if time permits and emergency response personnel know the correct techniques, the tractor should be separated from the trailer. However, in doing this, care must be exercised to ensure that the cargo will not be damaged. (Refer to Chapter 12 for further discussion of procedures.) Remember, explosives exposed to heat can become unstable. After the fire is extinguished, do not touch, handle, or move any of the cargo.

Marshalls Creek, Pennsylvania, Explosion, June 26, 1964

On June 26, 1964, a truck carrying explosives detonated during transit. The disaster killed three fire fighters and three bystanders and injured ten. In addition, over a million dollars' worth of property was destroyed.

The truck, carrying 26,000 pounds (11,700 kilograms) of nitro carbonitrate (a blasting agent), 4,000 pounds (1,800 kilograms) of 60 percent standard dynamite, and 99 blasting caps, pulled off to the side of the road with a tire fire. The time was about 3:15 A.M. The driver disconnected the tractor from the trailer and drove off in search of a telephone. After the driver had gone, another truck driver came by and, upon seeing the fire, drove to a nearby telephone and called the fire department. The time was now about 4:00 A.M.

The fire fighters from the Marshalls Creek Volunteer Fire Department arrived on the scene about 4:10 A.M. They set up foam lines to fight the tire fire, which had, by now, engulfed the entire trailer. Fire fighters arriving on the scene reported that the trailer did not contain any placards or warnings. About 30 seconds after the fire fighters arrived on the scene the explosives detonated.

In addition to killing six and injuring ten, the blast released hundreds of venomous snakes from a tourist reptile exhibit nearby. The blast also damaged several nearby houses

and destroyed three fire trucks.

Truck fires, particularly those involving cargo, must be approached with extreme caution. Until the cargo is known, emergency response personnel must assume that it is dangerous. The absence of placards does not indicate that the load is nonhazardous. Be careful and use common sense.

OXIDIZERS

The term *oxidation* was originally used to describe a reaction of an element or compound with oxygen. The opposite reaction is termed *reduction*, and occurs when a compound gives up oxygen in a reaction.

As more and more research into combustion was done, it was learned that many reactions do not actually involve oxygen. For example, carbon burns in fluorine in a reaction that resembles the reaction of carbon and oxygen. Another example is the burning of hydrogen in either fluorine or chlorine as well as oxygen. On this basis, an *oxidizer* is an element or compound that permits a substance to burn.

Oxidizers, under certain circumstances, can be as dangerous as explosives. The fine line that divides oxidizers from explosives can be crossed very easily. Emergency response personnel must be aware of the fact that oxidizers can be extremely unstable.

Handling an Oxidizer Incident

Emergency response personnel must approach an oxidizer incident in the same manner as they would treat an explosive. Caution must be exercised at all times. The following procedures are suggested:

1. Evacuate personnel from surrounding areas.
2. Try to identify the product involved. Use reference sources to determine hazard.
3. Position personnel and apparatus upwind from the spill or leak. Make sure there is a

path for escape for both personnel and apparatus.

4. Use full protective equipment and breathing apparatus.

5. If possible, contain the leak with the use of natural or contrived barriers. Try to divert flow from exposures, mixing with other chemicals, or from entering the sewage system.

6. Attempt to close or stop the leak. Personnel making this attempt should be protected from the fumes and from possible ignition of the vapors.

7. If the oxidizer is burning, use the extinguishing method suggested by the reference sources. In most cases water is effective in extinguishing the fire.

8. Most oxidizers are soluble in water, so solutions of the material can be absorbed in many places. These include wood floors, merchandise, and other combustibles. Then, as the material dries out, it could possibly ignite spontaneously. Overhaul, therefore, is extremely important after extinguishment is accomplished.

Blasting Agents

The Du Pont text defines a *commercial blasting agent* as a "cap insensitive chemical composition or mixture which contains no explosive ingredient and which can be made to detonate when initiated with a high strength explosive primer."[5] One type of blasting agent is handled as an oxidizer because it does not contain any high explosives. The general group of these oxidizers is known as nitro carbonitrates (NCN). Some various trade names for it are Nitramon, Nitramite, and ANFO-P.

Special primers are necessary to set off NCN blasts. Another way is to set off a dynamite charge with a conventional cap. This, in turn, will ignite the NCN.

Ammonium Nitrate

The American Insurance Association has issued a special interest bulletin on ammonium nitrate. The following description of ammonium nitrate comes from that bulletin:

Ammonium nitrate has been an important ingredient in most commercial formulations of dynamite and industrial blasting agents since 1933. In recent years the manufacture of blasting agents, and the field compounding these materials at or near the site of use, has found wide acceptance. These preparations have partly or wholly displaced black powder and dynamite explosives in many conventional uses. About 60 percent of blasting done in the country is accomplished with such agents, consuming 5 to 10 percent of the nitrate supply for this purpose.

There is need for making some distinction between the danger potentials of fertilizer grade ammonium nitrate (currently coated with small percentages of organic surfactants and inorganic noncaking agents) and the commercial blasting agents compounded with oil and, in shipping classification, termed "nitro-carbo-nitrate". While both of these materials are designated for Federal regulatory purposes in normal transportation as "oxidizing materials" (yellow label), they pose different degrees of hazard relative to decomposition, sensitivity to explosion or detonation when exposed to shock, intensive heat or when involved in major fires.[6]

Health Hazard. When ammonium nitrate thermally decomposes, extremely toxic gases are given off. These oxides of nitrogen can cause severe injury and death upon short exposures. Ventilation is extremely important, not only to disperse the toxic gases, but to prevent pressure build up.

Physical Properties. In its pure form, ammonium nitrate is a colorless to white crystalline powder. In commercial grade, anticaking agents are added, which makes the color light to dark brown. There are a variety of forms in which ammonium nitrate can be found:

1. Porous pellets or prills
2. Flakes
3. Grains
4. Fertilizer grade
5. Dynamite grade
6. Technical grade
7. Nitrous oxide grade

The American Insurance Association Bulletin 311 describes the action when ammonium nitrate is exposed to heat:

> Ammonium nitrate melts at 337°F (151°C), although on gradual heating, the salt begins to slowly decompose into ammonia and nitric acid at around 175°–200°F (79°–93°C). From 350° to 390° (176°–199°C), the breakdown is more rapid, with production of whitish fumes due to release of nitrous oxide ("laughing gas," a potent supporter of combustion), and water vapor. This gas production increases, with puffs of flame at a steady rate up to 500°F (260°C). At some point near 575°F (301°C), the reaction may accelerate with sudden rushes of light brown to orange-copper colored fumes, indicating the formation to toxic higher oxides-of-nitrogen with the possibility that the decomposition may become vigorous enough to produce a detonation.[7]

Fire Fighting Incidents. Fires in ammonium nitrate storage areas should be fought with large amounts of water. Maximum cooling should be attempted.

If the fire involves ammonium nitrate with the possible contamination from other chemicals or building materials stored nearby, there is danger of the material reacting explosively. In this case, withdrawal and evacuation by emergency forces is warranted. Unattended master stream devices should be used. Residues must be completely cleaned up. Areas must be scrubbed clean of any material.

Ammonium Nitrate, Fertilizer Grade

Ammonium nitrate is produced synthetically by chemically combining ammonia with nitric acid. Another technique is to use ammonia in a water solution and when the reaction is complete, evaporating the water. The resultant product is then granulated.

Physical Properties. The problems in heating fertilizer grade ammonium nitrate are summarized in a publication of The Fertilizer Institute.

> Ammonium nitrate melts at about 169°C (336°F). If the melt is heated to about 210°C (410°F) in an open container, it will begin to decompose into nitrous oxide (the "laughing gas" of commerce) and water. This reaction liberates heat, and it would, therefore, ordinarily be expected that a mass of decomposing ammonium nitrate would become progressively hotter and decompose at an accelerating rate until finally the decomposition would become explosive. However, a process which absorbs heat is going on at the same time as the heat-producing decomposition which gives nitrous oxide and water. This process includes the simultaneous vaporization and dissociation of ammonium nitrate into gaseous ammonia and nitric acid.
>
> The heat-absorbing process tends to offset the heat-producing reaction. No runaway temperature and pressure rise will occur, so long as the water, ammonia, nitric acid, and nitrous oxide can escape freely, or in other words, so long as the pressure above the decomposing cape freely, or in other words, so long as the pressure above the decomposing ammonium nitrate is not too high.
>
> In the manufacture of nitrous oxide ammonium nitrate is decomposed safety at temperatures up to nearly 300° C

(570°F) when the gaseous products of decomposition are allowed to escape freely.

The burning rate of paper or polyethylene bags and any combustible dunnage such as paper, wood, etc., will be increased by the presence of ammonium nitrate. So long as the products of reaction escape freely and the back pressure of the gaseous products does not rise excessively, burning combustible material in contact with ammonium nitrate will not lead to an explosion.

When ammonium nitrate is heated to a high temperature in a bomb or other tight, strong, closed container which retains the products of decomposition and the heat of reaction, the temperature and pressure may suddenly rise dangerously high and may result in an explosion.

If a dynamite charge or other high explosive primer is detonated in ammonium nitrate or in mixtures containing large proportions of ammonium nitrate, an explosive rate of decomposition may result.[8]

Storage. Ammonium nitrate in the fertilizer grade is shipped and stored in bags or bulk. The material the bags are made of is moisture resistant and multilayer. In bulk quantities, ammonium nitrate is shipped in trucks, box cars, or closed hopper cars and stored in buildings in bulk quantities.

Organic Peroxides

Organic peroxides are used in the processing of reinforced plastics, plastic films, and synthetic rubber. They are also used in the bleaching of oils, fats, waxes, syrups, gums, and flours. In addition, heavy use is made of organic peroxides in the textile, printing, and pharmaceutical industries.

The hazard of organic peroxides is very clearly defined by the American Insurance Association:

These active oxygen compounds are vigorously reactive; they present explosion potentials by virtue of their thermal and shock sensitivity and their tendency to undergo autoaccelerative combustion and, in some instances, even violent decompositions to a point of detonation. Therefore, the problem of safety must, of necessity, be approached earnestly and intelligently by all concerned if these materials are to continue to find safe commercial acceptance.[9]

Handling Organic Peroxide Incidents. The organic peroxides contain a large amount of active oxygen, which directly supports the combustion and explosion processes, even with air excluded. Liquid and paste peroxides will decompose and burn intensely and sometimes violently when exposed to heat. The generalized procedures for handling oxidizers also apply to organic peroxide.

Health Hazard. The peroxides will cause irritation to the upper respiratory tract and mucous membranes when encountered as a dust or mist. Contact of the skin can cause burns.

Traskwood, Arkansas, Train Derailment, December, 1960

A 50-car train derailment in Traskwood, Arkansas, involved two 50-ton cars of fertilizer grade ammonium nitrate.[10] One of the box cars contained bagged material, while the other was a hopper car with the material in bulk.

In addition to the carloads of ammonium nitrate, there were several tank cars of aviation gas and other petroleum products, as well as a tank car of nitric acid.

There was a detonation of the ammonium nitrate, probably only the material from the bagged car. Fortunately, the derailment was in such a remote area that injuries and property damage were limited.

RADIOACTIVE MATERIALS

Since the end of World War II, there has been a greatly increased need for radioactive materials. This need results from peaceful as well as defense requirements.

When the word *radioactive* is mentioned, most people imagine the great mushroom cloud created from an exploding atomic bomb. However, there are many peacetime uses of radioactive material that have a great beneficial effect on the public. Among these uses are:

1. Generation of electric power.
2. Treatment of disease.
3. Identification in criminology.
4. Food preservation.
5. Age determination.
6. Finishing and hardening of materials.
7. Making of plastics.
8. Electricity in out-of-the-way places.
9. Nondestructive testing of materials and welds.
10. Monitoring of chemical processes.

As with all of the other hazardous materials discussed, radioactive substances are produced, transported, used, and finally dumped as wastes. There is a risk of an accident at each of these stages, which endangers those in contact with the substances in each stage, as well as emergency response personnel who are called on for assistance.

Transportation

The Department of Transportation estimates that approximately 2½ million packages of radioactive material are shipped each year in the United States.[11] However, the General Accounting Office reported that in 1974 federal inspectors found more than 2,000 safety violations, with more than half of these being serious enough to cause radiation exposure. Despite these findings, the DOT reports that:

> Thus far, based upon best available information, there have been no known deaths or serious injuries to the public or to the transportation industry personnel as a result of the radioactive nature of any radioactive material shipment.[12]

A summary of the regulations for transporting radioactive materials is provided by the Department of Transportation:

> Under the Department of Transportation Act of 1966, the U.S. Department of Transportation (DOT) has regulatory responsibility for safety in the transportation of radioactive materials by all modes of transport in interstate or foreign commerce (rail, road, air, water), and by all means (truck, bus, auto, ocean vessel, airplane, river barge, railcar, etc.), except postal shipments. Postal shipments come under the jurisdiction of the U.S. Postal Service, formerly known as the Post Office Department. Shipments not in interstate or foreign commerce are subject to control by a state agency in most cases.

> The Interstate Commerce Commission (ICC) formerly had the jurisdiction over both the safety and economic aspects of the transport of radioactive materials by land, but the jurisdiction over safety was transferred to the Department of Transportation when DOT was formed in April 1967. The ICC (for land shipments) and the Civil Aeronautics Board (for air shipments) still exercise jurisdiction over the economic aspects of radioactive materials transport through the issuance of operating authorities to carriers, and control of shipping costs (freight rates).

> Under the Atomic Energy Act of 1954, as amended, the U.S. Nuclear Reg-

ulatory Commission (NRC) also has responsibility for safety in the possession and use, including transport, of byproduct, source and special nuclear materials. Except for certain small quantities and specific products for which the possession and use are exempted, a license is required from the NRC for possession and use of such materials. The NRC has established, in 10 CFR Part 71, requirements which must be met for licensees to deliver licensed material to a carrier for transport if fissile material or quantities exceeding Type A are involved. The NRC also assists and advises DOT in establishment of national safety standards and in review and evaluation of packaging designs.

Several states have entered into formal agreements with the NRC whereby the regulatory authority over byproduct, source and less than critical quantities of special nuclear material has been transferred to the states from the NRC. These "Agreement States" have adopted uniform regulations pertaining to intrastate transportation of radioactive materials which require the shipper to conform to the packaging, labeling, and marking requirements of the U.S. Department of Transportation to the same extent as if the transportation were subject to the rules and regulations of that agency.[13]

Most of the radioactive material transported by truck travels on the interstate highways. Interstate 80, which stretches from San Francisco to New York City, is one of the most heavily used routes for carrying radioactive materials. Bruce Ingersoll, writing in *The Evening Bulletin*, reports that:

> Within 25 miles (40 kilometers) of I-80, 10 nuclear power reactors are generating electricity and 14 more are under construction or planned. In addition, the government has three small research reactors at Argonne National Laboratory.

> Only a few miles to the south of I-80 are the destinations for much of the nuclear waste moving in interstate commerce - General Electric's Morris (Ill.) storage facility and Nuclear Engineering's Sheffield burial ground.[14]

The other major interstate highway that carries a great deal of radioactive materials truck traffic is Interstate 95, covering the states between Maine and Florida. Within 25 miles of this road are 14 reactors with another 12 in the planning stages.

Radioactive materials are not permitted to be transported through New York City under their Health Code. This prohibition means that all radioactive materials destined for Long Island cannot come by land routes as the only access is through the city. This regulation is being challenged by the Associated Universities, Inc., who operate a laboratory that utilizes radioactive materials in research.[15] A ruling by the Materials Transportation Bureau on the restrictiveness of the New York City law has not been issued at the time of publication of this text. However, this conflict does point out the problems experienced by local government in trying to protect the health and safety of its citizens and the need for free interstate commerce as supervised by the federal government.

The transportation factor brings the possibility of an incident to all corners of the country.

Radiation

Persons on earth are continually being exposed to low levels of radiation. If a very sensitive meter were used, it would measure radioactivity everywhere you went. The causes of this background radiation are:

1. Cosmic rays from space.
2. Natural radioactive materials in the earth and water.

Research in the various scientific fields has enabled production of materials with much higher levels of radiation. With the availability of these new chemicals and processes, there is a corresponding increase in the probability of an incident. This would then cause a possible release of the radioactive material.

Those elements that emit nuclear radiation spontaneously are said to be *naturally* radioactive. However, scientists have discovered that they can make stable elements radioactive also. These are said to be *artificially* radioactive elements and are referred to as *radioisotopes*. Radium is a naturally radioactive element, while cobalt is not radioactive until it has been bombarded by neutrons in a nuclear reactor.

There are over 40 kinds of radioactive elements occurring in nature, each emitting one or a combination of the three kinds of radiation in a fixed pattern uniquely characteristic of that element. Some of the materials that will be encountered give off only one of the three kinds, but others can give off a combination of all three. Since protection for each of the three differs, a description of them is necessary. The three different types of radiation are:

1. Alpha rays
2. Beta rays
3. Gamma rays

A ray is defined as a stream of the particles.

Alpha Rays. Alpha rays are made up of very large particles, which are the same as the nucleus of the helium atom. They do not travel very far in air, about 3 inches (7.6 centimeters) maximum, and can be stopped by something as thin as a sheet of paper as demonstrated in Figure 4-10.

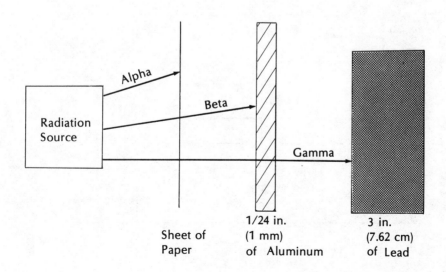

Figure 4-10.   Penetration power of radiation particles

Normally, alpha particles will not penetrate the skin. However, if they do enter the body, they can attack the organs or be absorbed by the bones. Therefore, care must be exercised to prevent breathing or swallowing radioactive substances.

Beta Rays. Beta radiation is made up of electrons, which are a portion of the atomic structure. In comparison with alpha particles, they are very much smaller (about 2,000 times) and travel about 8 times faster. A piece of aluminum about 1/24 inch (about 1 millimeter) thick will be able to stop a beta ray (Figure 4-10). Even plexiglass will stop a stream of beta particles. Because beta particles are so much smaller than the alpha particles, they are less able to cause cell damage. Unbroken skin keeps the beta particle from entering, although in large quantities it can produce serious damage to the skin and tissue directly be-

neath the skin. Like alpha particles, if radioactive materials emitting beta particles enter the body, damage to the organs can occur.

Gamma Rays. Gamma rays are very similar to X-rays, except that they are much shorter. Just as light is a form of electromagnetic radiation, so are gamma rays, except that the wavelength of the gamma rays is much shorter.

Gamma rays have over 100 times the penetrating power of beta rays and about 10,000 times the penetrating power of alpha particles. With their great penetrating power, gamma rays can pass through the skin and damage internal body structures. In addition, when striking the skin, they can cause serious localized damage.

Barriers for blocking gamma rays include earth, water, and the dense materials such as lead (Figure 4-10). The thicknesses required for reducing gamma radiation in half are shown in Figure 4-11.

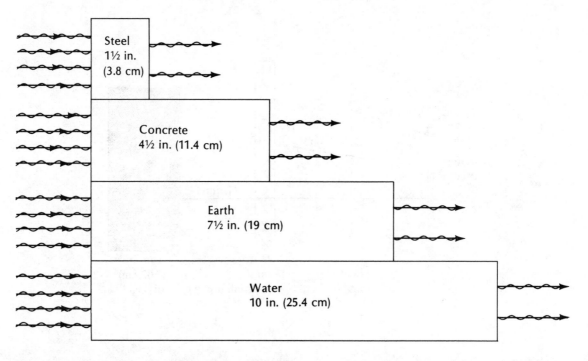

Figure 4-11. Comparative thickness required to reduce the intensity of gamma radiation by one-half. *Courtesy of United States Atomic Energy Commission*

Half-Life. The length of time it takes for one-half of a given amount of radioactive element to change into the next element is called its *half-life.* The half-life of known radioactive elements ranges from very short times, expressed in fractions of a second, to others that are very long, expressed in billions of years.

For example, the half-life of radium is about 1,600 years. If one ounce (28 grams) of radium had been set aside 1,600 years ago, today it would be ½ ounce (14 grams) of radium and ½ ounce (14 grams) of stable lead. It would take another 1,600 years for the ½ ounce (14 grams) of radium to be reduced to ¼ ounce (7 grams) of the original 1 ounce (28 grams). The decay of a radioactive material is shown in Figure 4-12. The half-life of several common radioactive elements is shown in Table 4-4.

Health Effects of Radiation

All of the radiations discussed above are capable of affecting the chemical composition of the living cells that compose our body tissues. If, for example, an alpha particle penetrates the walls of a particular cell, the rather complex chemical processes and delicate electrical balance of this cell might be upset to such an extent as to create a deteriorating situation that may eventually result in the death of the cell. When a cell is thus penetrated and the radiation gives up its energy within the cell we say the radiation has been absorbed. How dangerous would the death of a particular body cell be to the body's functions?

The body is made up of millions upon millions of cells. All cells are not of equal importance to our bodily functions. Muscle cells are not as important from a functional basis as, say, nerve cells that control reflexes. The body has recuperative capabilities, which may vary according to the health of the person affected. From this it is obvious that the answer to the preceding question must be a qualifying one. If there are enough cells destroyed of the important kind, and/or fast

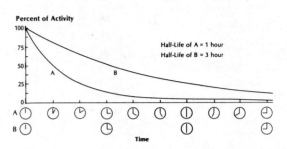

Figure 4-12. Decay of two radioactive substances with half-life of one and three hours, respectively. *Courtesy of United States Atomic Energy Commission*

enough so that the body does not have time to replace them, then a deteriorating situation prevails. This could result in the death of the person. On the other hand, if the death rate of individual cells is very slow, or takes place over a long period of time, or the exposure is over only such parts of the body as arms or legs, the recuperative ability of the body would have a chance to replace dead cells. Under these conditions permanent damage might not result.

An individual can be harmed by radiation in one of three ways:

1. Radiation sickness
2. Radiation injury
3. Radioactive poisoning

The sickness is caused by internal changes to the body chemistry, whereas injury is actual damage to the body organs such as the skin, kidneys, or liver. Many radioactive materials have poisonous qualities in addition to their radioactive property. They act just like the poisons outlined in Chapter 3.

Table 4-4. Half-Life of Common Radioactive Elements

Element	Half-Life
Uranium 235	7 hundred million years
Radium 226	1622 years
Strontium 90	25 years
Cobalt 60	5 years
Iodine 131	8 days
Sodium 24	15 hours
Polonium 212	Less than 1/1,000,000 second

The dangers from radioactive sources can take two forms:

1. The closed radioactive source contained in a vessel that prevents any radioactive substances from escaping becomes a hazard when accidentally opened. Then, these sources can cause external radiation of the human body.

2. The open radioactive source, particles from which can pass into the environment (air, water, and soil) and irradiate people both externally and internally.

Units of Measurements. The amount of radioactivity as well as the strength of the radiation all have an effect on the health of the individual exposed. Emergency response personnel must therefore have an understanding of the units of measurement and the relative amounts that can be dangerous.

The unit of radioactivity is the curie. The *curie* is defined as the amount of radioactive material that will give 37 billion disintegrations per second. One gram of radium will produce about this amount of disintegrations, so it has a radioactivity of one curie.

The roentgen (abbreviated R) is the unit that expresses the ionization effects of gamma radiation (and X-rays) in air. One roentgen is the amount of gamma radiation that will cause the formation of slightly over 2 billion

ion pairs in dry air. Since the measurement is taken in air, the figure is not always an accurate indicator of what the exposure will do to the body. To overcome this shortcoming another measure has been established, the roentgen equivalent man (REM).

The REM is the quantity of radiation of any type that produces the same biological effects in a person as the absorption of one roentgen of gamma radiation. Under many conditions, within the accuracy required for radiation protection, the terms roentgen and REM can be used interchangeably.

Certain limits have been established, for health reasons, to which an individual could be exposed during the year. The limits are:

1. General public   170 milliREM
2. Once-in-a-life-time exposure (emergency) industry worker   25 REM
3. Over 18 years old   5 REM

Individuals can receive exposure from natural and medical sources. Figure 4-13 shows the estimated yearly radiation to the general population if everyone's exposure were the same.

Exposure Effects of Radioactive Materials. The effects of exposure to radioactive materials varies with the amount of exposure. The possible harm from exposure is summarized in Table 4-5.

Table 4-5. Exposure Effects From Radioactive Materials

| Single dose over whole body | | |
Roentgens	Milliroentgens	Effect
Less than 20	Less than 20,000	Clinically not detectable
25	25,000	Maximum limit during one emergency exposure
50	50,000	Blood changes, but no illness expected
100	100,000	Blood changes, nausea, fatigue
200	200,000	10 percent fatal, genetic complications
300	300,000	20 percent fatal, 100 percent illness
450	450,000	50 percent fatal, high percentage of sterility
650	650,000	95 percent fatal
1,000	1,000,000	Survival improbable

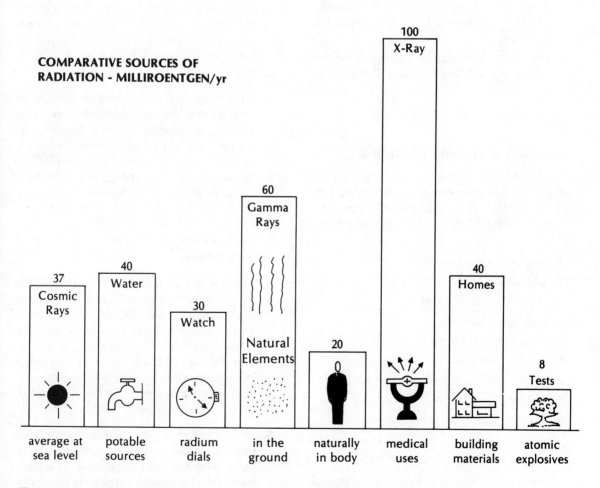

COMPARATIVE SOURCES OF RADIATION - MILLIROENTGEN/yr

Figure 4-13. Sources of radiation exposure. *Courtesy of United States Atomic Energy Commission*

Protection from External Radiation. Emergency response personnel can secure protection from external radiation in three basic ways:

1. Time: The less time an individual is exposed, the smaller the dosage. This is based on the formula
   dose rate × exposure time = total dose
   Obviously the total dose goes down as the exposure time is reduced.
2. Distance: As the distance increases, the amount of exposure to the radiation decreases. The decrease is in accordance with the inverse square law. This means that as the distance increases, the amount of radiation decreases as the square of the distance.
3. Shielding: The relative efficiency of the different shielding methods is shown in Figure 4-11.

Handling a Radiation Incident

A radiation incident can occur during any one of the handling processes. Emergency response personnel should be alert to a possible radioactive problem when responding to a call at:

1. Universities and colleges
2. Industrial plants
3. Industrial laboratories
4. Biological firms
5. Hospitals and clinics
6. Doctor's offices
7. Atomic power plants on land and sea
8. Military installations
9. Transportation by land, air, and sea

Following is a suggested sequence of events for handling a radiation incident:

1. Keep away from the wreckage, material, container, or other material involved, *except* to rescue people.

2. Isolate the scene of the incident and surrounding area. Detain any persons that may have been contaminated, exposed to radiation, or required to handle contaminated personnel and items. Isolate any contaminated vehicles or facilities. If vehicles or equipment must leave the emergency scene, know where they can be located and instruct the departing personnel to practice isolation of themselves and equipment until the radiation team can check them out.

3. Request radiological help. Be familiar with the local radiological agencies that can assist you: Civil Defense, Department of Energy, university laboratories, private laboratories, and individual personnel located in your area who have had such experiences.

4. Protect emergency response personnel from contaminants. Always use full protective equipment. When properly protected to handle radioactive incidents, whether it is a fire or an injured person, do not hesitate to respond. The radiological team will aid you after the emergency is secured. In the meantime, attempt to be aware of the location of the radioactive materials and try to prevent their added involvement.

5. When the emergency is terminated, do the minimum of overhauling and retreat to the isolation area. Do not return equipment back into service unless absolutely necessary. When the Emergency Radiation Assistance Team arrives, thoroughly inform them as to what type of radioactive materials have been observed and the events that have taken place.

6. The radiation team will attempt to check out personnel and equipment that have been involved and then seek to determine what possible exposures have occurred to personnel, general public, and the surrounding environment.

If a military weapon is involved the following additional procedures should be followed:

7. Report the accident as soon as possible to the nearest military authority or ERDA (Energy Research and Development Agency) office.

8. Keep sightseers away from the accident area. Get other witnesses to help in doing this. In the open, an exclusion distance of at least 500 yards (450 meters) should be established to minimize the chance of fatal injuries from the direct blast effects of conventional high explosives. Even at this range, however, there is danger of injury from secondary missiles that might be flung in the air by an explosion.

9. If there is a fire, stay out of the smoke except for the purpose of rescuing people. Always approach the accident scene from upwind and from uphill, if possible. Control water runoff, if possible.

10. Do not try to fight a fire, especially if it is believed that explosives may be present.

11. Do not permit anyone to touch anything unnecessarily or retain as souvenirs any objects found in the accident area.

After rescuing an individual exposed to radiation, there are several facts that must be considered. There are three types of radiation accident patients. One type is the individual who has received whole or partial body external radiation and may have received a lethal dose of radiation, but is no hazard to attendants, other patients, or the environment. This patient is no different from the radiation therapy or diagnostic X-ray patient.

Another type is the individual who has received internal contamination by inhalation or ingestion. This patient also is no hazard to attendants, other patients, or the environment. Following cleansing of minor amounts of contaminated material deposited on the body surface during airborne exposure, treatment is similar to the chemical poisoning case, such as lead. Body wastes should be collected and saved for measurements of the amount of nuclides to assist in determination of appropriate therapy.

A third type is one who has received external contamination of body surface and/or clothing by liquids or by dirt particles, with problems similar to vermin infestation. Surgical isolation techniques to protect attendants and cleansing to protect other patients and the hospital environment must take place to confine and remove a potential hazard.

When external contamination is complicated by a wound, care must be taken not to cross-contaminate surrounding surfaces from the wound and vice versa. The wound and surrounding surfaces are cleansed separately and sealed off when clean. When crushed dirty tissue is involved, early preliminary wet debridement following wound irrigation may be indicated. Further debridement and more definitive therapy can await sophisticated measurement and consultant guidance.

If contamination of emergency response personnel is suspected, the following decontamination procedures should be followed:

1. Remove outer clothing and place in plastic bag or wrap in a blanket.

2. Avoid shaking clothing in the air to reduce contamination.

3. Wash face first, with a washcloth. Then wash the hands. Place washcloth in plastic bag.

4. Do not leave area in which decontamination is taking place, as shoes may spread the material to other areas.

5. The transport vehicle must be checked and decontaminated also.

Shipping Containers

Transmission of radiation from radioactive materials can be greatly reduced or stopped entirely by means of shielding. Hence, radioactive materials are shipped in containers that are likely to be rather bulky and heavy. Radioactive materials to be shipped by land, air, or water transport are packaged, identified, and marked in accordance with definitely prescribed regulations for the protection of the public health and safety.

Packaging Definitions. When considering packaging, certain terms must be defined. All numerical references in the following refer to Title 49 of the Code of Federal Regulations:

Transport Group. Any one of seven groups into which normal form radionuclides are classified according to their radiotoxicity and their relative potential hazard in transportation. (173.389(h)). The Code of Federal Regulations provides a transport group for all radioactive materials.

Transport Index. A number placed on a package of "Yellow Label" radioactive materials by the shipper to denote the degree of control to be exercised by the carrier; i.e., the number of yellow-labeled packages that may be placed in a single vehicle or storage location. The transport index is actually the measured dose rate of radiation at three feet (0.9144 meter) from the surface of the package (173.389(i)).

Type "A" Packaging. Packaging designed in accordance with the general packaging requirements of parts 173.24 and 173.393 and which is adequate to prevent the loss or dispersal of the radioactive contents and to retain the efficiency of its radiation shielding properties if the package is subjected to the test prescribed in part 173.398 (b) (Normal conditions of transport) (173.389(j)).

Type "A" Quantity Radioactive Material. Material that may be transported in type "A" packaging (173.389(l)).

Type "B" Packaging. Packaging that meets the standards for type "A" packaging, and in addition, meets the standards for the

hypothetical accident conditions of transport as prescribed in part 173.398(c) (173.389(k)).

Type "B" Quantity Radioactive Material. Material that may be transported in type "B" packaging (173.398(l)).

Types A and B Packaging. Table 4-6 lists the transport groups and the maximum quantity in each group that can be shipped in either type A or B packaging. If the quantity in the particular transport group exceeds that listed for type B, it is considered a "large quantity" under the DOT definition. Typical packaging for type A material is shown in Figure 4-14. Type A packaging must withstand normal conditions of transport only without loss or dispersal of the radioactive control contents. Figure 4-15 illustrates typical type B packaging which must stand both normal and accident test conditions without loss of contents.

Package must withstand normal conditions (173.398[B]) of transport only without loss or dispersal of the radioactive control contents.

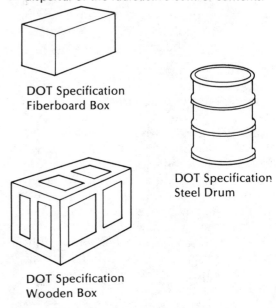

DOT Specification
Fiberboard Box

DOT Specification
Steel Drum

DOT Specification
Wooden Box

Figure 4-14. Typical Type A packaging. *Courtesy of Department of Transportation*

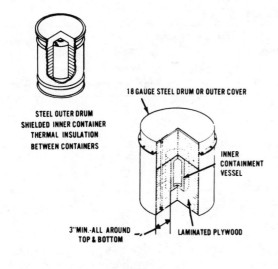

18 GAUGE STEEL DRUM OR OUTER COVER

STEEL OUTER DRUM
SHIELDED INNER CONTAINER
THERMAL INSULATION
BETWEEN CONTAINERS

INNER
CONTAINMENT
VESSEL

3"MIN.-ALL AROUND
TOP & BOTTOM

LAMINATED PLYWOOD

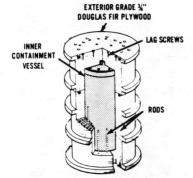

EXTERIOR GRADE ¼"
DOUGLAS FIR PLYWOOD

INNER
CONTAINMENT
VESSEL

LAG SCREWS

RODS

Figure 4-15. Typical Type B packaging. *Courtesy of Department of Transportation*

Label Requirements. The label that must be applied to the package is dependent upon the external radiation level. Table 4-7 shows the limits that require a specific label.

Transport Index. The regulations prescribe that the maximum permissible dose rate at contact with the accessible exterior surface of any package of radioactive materials offered for transport shall not exceed 200 mR/hr, or 10 mR/hr at three feet (0.9 meter). The latter value is equal to the maximum transport index. Higher dose rates are also prescribed, which are allowable provided that "sole use" of the transport vehicle is assured by the shipper.

To control the radiation level of accumulations of multiple numbers of packages once in the transportation environment, the regulations require that the carrier shall maintain certain prescribed separation distances between radioactive materials packages and other areas that are continuously occupied by persons and/or photographic film. These separation distances relate the storage time against the transport index, which was defined earlier as the dose rate in mREM/hr at three feet (0.9 meter) from the accessible exterior surface of the package.

No package offered for transport (in other than sole-use vehicles) may have a transport index exceeding 10. The establish-

Table 4-6. Type A and Type B Package Quantity Limits

Transport Group	Type A Quantity, curies	Type B Quantity,[a] curies
I	0.001	20
II	0.05	20
III	3	200
IV	20	200
V	20	5,000
VI, VII	1,000	50,000
Special Form	20 [b]	5,000

[a] Quantities exceeding Type B are "large quantity" (large radioactive source).
[b] Except for Californium-252, for which the limit is 2 curies.

Table 4-7.   Radioactive Material Package *Label Criteria*

Label	Dose Rate Limits	
	At Any Point on Accessible Surface of Package, mR/hr	At 3 Feet (0.9 meter) from External Surface of Package (Transport Index), mR/hr
Radioactive—White I	0.5	0
Radioactive—Yellow II	50	1.0
Radioactive—Yellow III [a]	200	10

[a] Requires Vehicle Placarding (This label mandatory for any fissile Class III (173.389A) or large quantity package (173.389B), regardless of dose rate levels.)

ment of this maximum transport index of 10 is based on considerations for prevention of fogging of "fast" photographic film. The total transport index of any aggregate number of packages in any single transport vehicle (other than a sole-use vehicle) or storage location may not exceed 50. The regulations provide graded tables of stowage distances vs time for stowage in accordance with the cumulative transport index.

However, the packaging is not measuring up to all it was supposed to be designed for. Syndicated columnist Jack Anderson reports as follows that government testing of containers for radioactive materials showed that they would not hold up in a crash:

> Last April, the internal reports show that a supposedly "fail safe" cask was badly damaged during a routine test. Hasty repairs were made, but the cask also failed a second round of examinations.[16]

Civil Defense V-715 Monitor

Many types of devices are available for providing a quantitative readout of the amount of radioactivity released in an incident. Emergency response personnel must be familiar with the use of the various types of instruments available to them, several of which are described in the following paragraphs.

Dosimeter. A dosimeter, illustrated in Figure 4-16, is roughly the size of a pen. It is designed for measuring accumulated exposure doses of gamma radiation received by emergency response personnel. If possible, personnel should obtain a dosimeter, set it to zero, and wear it during the incident. In this way exposure to radiation levels can be accurately determined.

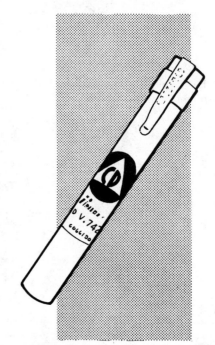

Figure 4-16.   Dosimeter. *Courtesy of Office of Civil Defense*

Dosimeter Charger. A dosimeter charger (Figure 4-17) is used to charge (zero) and to read the dosimeter. The reading on the scale shows how much radiation the wearer has accumulated since setting the unit to zero.

Civil Defense V-700 Monitor. The civil defense V-700 monitor is a low-range instrument (Figure 4-18) that measures from 0 to 50 milliroentgens per hour (mR/hr). In its low range, it measures gamma dose rates and detects the presence of beta radiation. It is important to note that it only detects beta radiation and does not provide a quantitative amount of this type of radiation. A shield in the probe can be opened so that the instrument can detect both types of radiation. In addition, this unit can be modified to provide a probe that will detect alpha particles also.

Civil Defense Monitor V-715. The civil defense V-715 monitor is a high-range instrument that can measure from 0 to 500 roentgens per hour. As discussed earlier, 500 roentgens is an extremely dangerous level. The V-715 monitor is designed to measure a large release, such as after a nuclear incident. It does not have the capability for beta measurement.

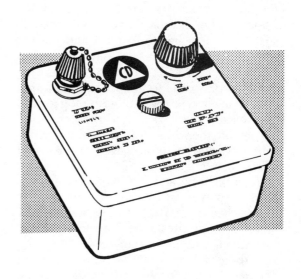

Figure 4-17. Dosimeter charger. *Courtesy of Office of Civil Defense*

Fairfax, Virginia, Radioactive Truck Transport Crash,[17] January 22, 1975

A truck carrying 80 drums and 10 eight-cubic-foot boxes of nuclear waste to a burial ground in South Carolina was involved in an accident. There was no release as a result of the incident, and the waste material was delivered to burial ground.

Athens, Alabama, Nuclear Reactor Fire,[18] March 22, 1975

The Tennessee Valley Authority (TVA) operates the Browns Ferry nuclear power station. At about 12:15 P.M. on March 22, 1975, a worker used a candle to test for air leakage at a point where some cables penetrated a wall.

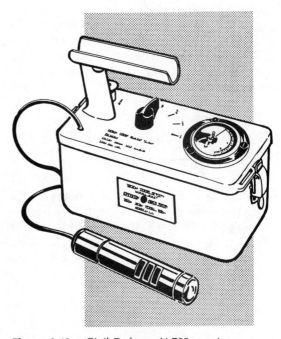

Figure 4-18. Civil Defense V-700 monitor. *Courtesy of Office of Civil Defense*

Since there was air leakage, the flame from the candle was drawn into the cable tray and some flexible polyurethane foam, used to seal the space, began to burn and the fire moved along the cables toward the reactor building.

The fire spread to the polyvinyl chloride (PVC) that covered the cables running toward the reactor building. After the insulation burned off, the wires shorted together, which shut down control to the valves, blowers, and pumps.

Fire fighting activities were undertaken by plant personnel. Because of the electric wiring they were afraid to use water. They therefore used carbon dioxide and dry chemical initially with little results.

About an hour after the fire started, plant personnel were ordered out of the building and fire fighting efforts were temporarily discontinued. This was necessitated by the dense PVC smoke and the fact that plant personnel did not have breathing apparatus. Except for some sporadic efforts, fire fighting did not resume for over 3 hours.

One hour after the fire started, the Athens, Alabama, Fire Department was called, and they arrived on the scene about 20 minutes later. However, the plant manager did not allow the use of water, so the fire fighters assisted in using the CO_2 system in the cable spreading room. Almost 4½ hours after the start of the fire, fire fighting efforts in the reactor building started again. Finally, 6½ hours after the start of the fire, permission to use water was received. At this point, fire fighters determined that the threads on the facilities' hose did not match the fire department's thread. This further hindered fire fighting operations. The fire was declared out 7½ hours after it began and an hour after water was used.

The greatest danger from this fire was during the first hour when the ability to remotely open valves to the reactor to pressurize it were lost. Then, during the last four hours, when control of the relief valve was lost, so was the ability to add water to the reactor. Both of these could have led to a nuclear incident.

If you have a nuclear facility in your area, has an emergency plan been developed to help if a nuclear incident occurs? Are the emergency response personnel aware of how many individuals would have to be moved and in what period of time? Preplanning is essential to handle an emergency of this nature.

College Park, Maryland, Radioactive Leak,[19] May 17, 1977

A routine examination of some laboratories at the University of Maryland disclosed the presence of a radioactive material. The material was found to be tritium. One of the key factors in this incident is that the material used in an experiment was not supposed to be radioactive.

The student had been using the metal in various experiments for about nine months. Since the radiation was beta rays, and of a low level, there did not appear to be any health damage. However, the long-term effects are unknown.

In addition, a fairly high level of radiation was found in the student's closet in his home. Smaller amounts were also found in his car and on the roof of the laboratory building. Clean-up operations took five days.

A school official is quoted as saying: "We now think it might have been a mistake in labeling on the part of the people down at the Savannah River plant."

Would you be prepared to detect, monitor, and decontaminate from such an incident? From whom could you request assistance? How long would it take to get assistance to the incident?

QUESTIONS FOR REVIEW AND DISCUSSION

1. What is the definition of an explosive?
2. What is the difference between a high and low explosive?
3. What are the parts of the explosive train?
4. What techniques can be used to initiate an explosive train?
5. Why are connector delays and boosters used in the explosive train?
6. What are the requirements for storing explosives?
7. What steps should be followed by emergency response personnel when handling an explosive incident?
8. Why can an oxidizer be considered as dangerous as an explosive?
9. What steps should be followed by emergency response personnel when handling an oxidizer incident?
10. What hazards must be prepared for when ammonium nitrate is burning?
11. Explain the health hazards associated with an organic peroxide incident.
12. In your local area, where can radioactive materials be found?
13. Explain the different types of rays that radioactive materials emit. How can they be stopped?
14. Explain the term *half-life* when referring to a radioactive substance.
15. Explain the health affects of exposure to a radioactive substance.
16. What units are used to measure the amount of radioactivity of a substance?
17. What steps should emergency response personnel take when handling a radiation incident?
18. Define the transport group and transport index when referring to packaging for radioactive materials.
19. What are the criteria for the three types of radioactive labels?
20. Explain the different devices used for radiation measurement.

NOTES

1. "U.S. Launches Drive on Stolen Explosives," *Washington Post*, May 20, 1977, p. A6.
2. "Explosive Materials Regulations," *Federal Register*, Superintendent of Documents, August 3, 1977, pp. 39316–39327.
3. *Blasters' Handbook*, 15th ed. (Wilmington, Del.: Du Pont Company, 1969, p. 89.
4. "Big Blast at Marshalls Creek," *Volunteer Firefighter*, September 1964, p. 13.
5. *Blasters' Handbook*, p. 47.
6. American Insurance Association, "Ammonium Nitrate-Fire-Explosion-Health Hazards," Bulletin No. 311 (New York, 1966).
7. Ibid.
8. Fertilizer Institute, "Fertilizer Grade Ammonium Nitrate" (Washington, D.C.), p. 21. Reprinted with permission.
9. American Insurance Association, "Fire, Explosion, and Health Hazards of Organic Peroxides" (New York, 1966), p. iii.
10. American Insurance Association, "Ammonium Nitrate-Fire-Explosion-Health Hazards."
11. Materials Transportation Bureau, Office of Hazardous Materials, "A Review of the Department of Transportation (DOT) Regulations for Transportation of Radioactive Materials" (Washington, D.C.: Department of Transportation, 1976), pp. 2–4.
12. Ibid.
13. Ibid.

14. Bruce Ingersoll, "Nuclear Cargoes Hit Road," *Evening Bulletin*, November 17, 1977, p. 22.

15. "Receipt of Application for Inconsistency Ruling," *Federal Register*, Superintendent of Documents, August 15, 1977, pp. 41204–41205.

16. Jack Anderson, "Nuclear Container Research Hits," *Washington Post*, November 11, 1977, p. D17.

17. Megan Rosenfeld, "Truck with A-Waste in Crash," *Washington Post*, January 23, 1975, p. B2.

18. Gary Kaplan, "The Browns Ferry Incident," *IEEE Spectrum*, October 1976, pp. 55–61.

19. David A. Maraniss, "Radioactive Gas Leak Found at University of Maryland Lab," *Washington Post*, May 18, 1977, p. C1.

5
Cryogenics, compressed gases, and flammables

Continuing with the descriptions of the categories of hazards, this chapter discusses the properties and potential hazards of cryogenic fluids and compressed gases, as well as the four classes of flammable products defined by the Department of Transportation: flammable gases, flammable liquids, combustible liquids, and flammable solids. These families of chemicals have been the cause of some of our most disastrous hazardous materials incidents. It is important for emergency response personnel to become familiar with them and learn as much as possible about how to deal with them.

As in the preceding chapters, coverage is very broad. This is just the starting point from which tactics can be planned. For specific information about the particular chemical involved and the circumstances of a particular incident, reference must be made to the various sources given in this text for the safest way in which to effect control.

CRYOGENICS

Originally, scientists used the word *cryogenics* to describe the science that deals with the equipment, processes, and end products of low-temperature refrigeration. However, in recent years, the word has evolved to cover the field of science that deals with low temperatures.

In the context of this book, the broadened definition of cryogenics will be used. Since the temperature is the criterion for a cryogenic, the upper and lower limits must be established. The lower limit is established as absolute zero in theory, while in practice it is the lowest level that scientists can achieve. The upper limit is set arbitrarily at $-101°C$ ($-150°F$) or about $123°K$.

The value of $-150°C$ ($-238°F$) was chosen because to achieve temperatures below this requires special refrigeration equipment and special insulation. The science of cryogenics deals with liquids and gases at temperatures between $-150°C$ ($-238°F$) and absolute zero.

Refrigeration Process

Most refrigeration processes involve a complete cycle. This means that a substance is used to extract the heat from one reservoir, give this heat up to another reservoir, and then the cycle is repeated all over again.

The refrigeration process is diagrammed in Figure 5-1. The refrigerant is kept in the liquid storage compartment, which is part of a closed system. The refrigerant feeds through the throttling valve very slowly. As it goes through the valve it expands, the temperature is sharply reduced, and some of the liquid vaporizes. In this state it can easily absorb heat from the cold reservoir, which causes a complete evaporation of the refrigerant. This change takes place in the evaporator.

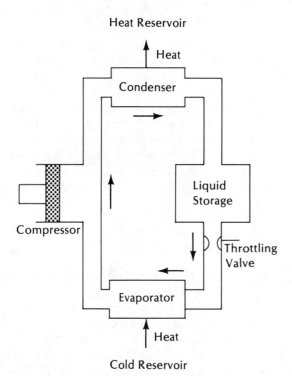

Figure 5-1.   Refrigeration process

The refrigerant is now hot and passes through to the compressor. Here, its temperature is raised even higher. At the condenser the heat is given up to the heat reservoir and the gas is converted back to a liquid. As it moves from the condenser it is stored, ready to begin the process all over again.

Gas Liquefaction

The liquefaction of any gas can be accomplished by removing sufficient thermal energy to allow the attractive forces of the molecules or atoms of the substance to become effective. Once they become attracted and close together the gas becomes a liquid. Those particles that have the least attraction require the lowest temperatures to liquefy. The limiting factor, therefore, in converting a gas to a liquid is the ability to develop a refrigeration process that will be able to extract heat at sufficiently low temperatures.

Another factor that aids in converting a liquid to a gas is increased pressure. Therefore, the combination of pressure and temperature is the determining factor for changing the gas into the liquid.

Above a certain temperature, called the *critical temperature*, no matter what pressure is applied the gas will not liquefy. The pressure required to liquefy a gas at its critical temperature is called the *critical pressure*. Some examples of the critical temperature and pressure for some common substances are shown in Table 5-1.

Table 5-1.   Examples of Critical Temperatures and Pressures

Substance	Critical Temperature, $^\circ K$	Critical Pressure, atm
Hydrogen	33.24	12.8
Nitrogen	126.0	33.5
Oxygen	154.3	49.7
Water	647.1	217.7
Ammonia	405.5	111.5
Carbon Dioxide	304.2	73.0

Above the critical temperature, the energy of the molecules is enough to overcome whatever inherent attraction they might have. As the temperature is decreased, the molecules are able to get closer together and form a liquid easily. Pressure, as it increases, brings the molecules closer together so that again, liquefaction can occur. However, once the temperature is above the critical level, no amount of pressure is enough to bring the molecules together to form a liquid. It takes the proper combination of pressure and temperature to bring about the liquefaction of a gas.

Liquefied Gas Properties

Liquefied gases, once they are in a liquid state, can be used as refrigerants. When used as a cooling agent, the big advantage is that the complicated refrigerant equipment is not necessary. This means that cryogenics will become more and more common. Table 5-2 lists the physical properties of some common cryogens.

Liquid Oxygen. When oxygen is liquefied, it turns into a light blue liquid. Usually at low temperatures, reactions take place slowly. However, when dealing with liquid oxygen, a small amount of heat added in the right place can cause an explosion in a system containing oxygen and another substance with which it combines chemically. A substance such as a drop of oil is sufficient to cause the explosion with just the heat from friction or mechanical impact. For these reasons, all piping and equipment for handling the liquid oxygen have to be extremely clean. All storage tanks must be completely free of any foreign particles.

Liquid Hydrogen. Liquid hydrogen presents a great problem to emergency response personnel because of its extreme flammability hazard. The key to successfully handling the liquid is to prevent its uniting with oxygen.

Insulation

The most common way of insulating cryogenic containers is to use double-walled construction with a vacuum between the two layers, as shown in Figure 5-2. The vacuum space prevents heat transfer by convection since there are no molecules to conduct the heat.

To keep the two walls of the container separated, structural supports are necessary. However, by placing a solid object touching the two walls, heat is absorbed through conduction. For this reason, the structural supports are kept to the minimum possible and the material used is chosen for minimum heat conduction.

Table 5-2. Common Cryogen Physical Properties

Substance	Normal Boiling Temperature, °K	Normal Melting Temperature, °K	Critical Point Pressure, atm	Critical Point Temperature, °K
Helium[4]	4.2	Does not solidify	2.3	5.2
Neon	27.2	24.5	26.9	44.5
Nitrogen	77.3	63.3	33.5	126.3
Carbon Monoxide	81.7	68.0	34.5	133.0
Argon	87.3	83.9	48.0	150.7
Oxygen	90.2	54.9	50.1	154.8
Ammonia	239.7	195.0	111.5	405.6

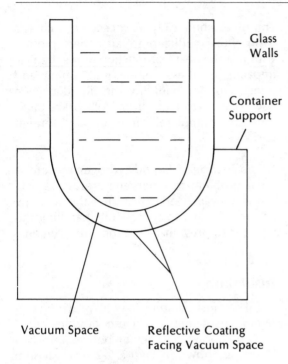

Glass Walls

Container Support

Vacuum Space

Reflective Coating Facing Vacuum Space

Figure 5-2. Cryogenic container insulation

Any substance whose temperature is above absolute zero continuously gives off energy in the form of electromagnetic radiation. The rate at which energy is radiated from a surface is proportional to the fourth power of its absolute temperature. As a result, the heat energy which is exchanged is equal to the difference of the fourth power of the absolute temperatures of each of the facing surfaces. On this basis, heat exchange from radiation takes place from the outer container to the inner container. To inhibit radiant energy transfer, the facing surfaces of the container are coated. The coating on the inside of the warm container keeps the radiation at a low level, while the coating on the cold container, reflects the radiated energy back into the vacuum space.

Refinements in design of insulation for cryogenic storage have enabled larger and larger quantities to be stored. A multilayer insulation has been developed which uses 75 to 150 layers of aluminum foil and glass-fiber paper per inch (2.54 cm). Vessels using this type of insulation have a rate loss of less than 5 percent per year.

In addition, a liquid nitrogen shield, evacuated foams, evacuated perlite, and evacuated multilayer superinsulations are used.

Handling and Storage of Liquid Hydrogen

Once cryogenic hydrogen has been liquefied and then purified to the required level, a transfer and storage system is required. The primary objectives of any cryogenic storage and transfer system are to minimize cryogenic fluid losses and maintain the desired transfer rates. The transfer system includes transportation by trucks, trains, and ships.

Losses during handling and transfer are significant. It is not unusual for large quantities of liquid to be lost when improper handling techniques are used. At a liquid hydrogen plant in Powersville, Ohio, 85 percent of the liquid produced is shipped and 15 percent is lost during handling at the plant.[1] The same reference reports that between 10 percent and 20 percent of the product is lost during transfer from the transport tank to the stationary tank. Overall, during the entire process, between 30 percent and 70 percent of the liquid hydrogen can be lost (Figure 5-3). These losses present a serious potential hazard.

A study conducted for the Office of Naval Research indicates the safety considerations in dealing with liquid hydrogen. They report that:

The growing applications for liquid hydrogen in low temperature research and propulsion systems have made the control of liquid-hydrogen hazards a matter of major concern. This control problem exists in small laboratories transferring a few liters of liquid-

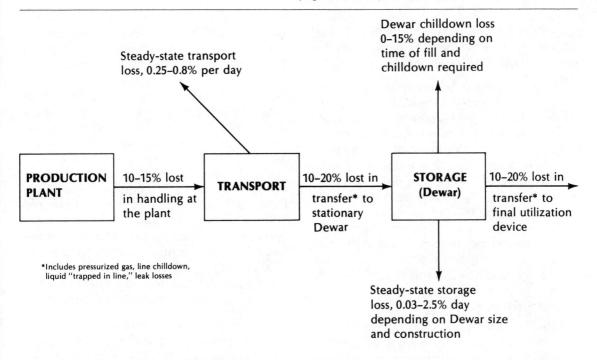

Figure 5-3.   Liquid hydrogen losses during transport, transfer, and storage[3]

hydrogen as well as large facilities handling thousands of gallons per day.[2]

Liquid hydrogen rapidly expands its volume approximately 850 times when released to the air. The gas has a wide flammability range, 4.1 percent to 74.8 percent, and the low ignition energy required combine to create a severe fire hazard. To give you an idea of what can happen in a release, hydrogen gas, when mixed with sufficient air to yield 4 percent of the lower flammable limit, forms a combustible hydrogen-air mixture with a volume 21,000 times that of the original liquid. This indicates the potential destructive capacity of a small liquid hydrogen spill.

Of a total of 96 accidents involving liquid hydrogen experienced by NASA, 80 involved a release of the product.[4] The 80 accidents can be broken down to 66 that involved a release to the atmosphere, 20 to an enclosure, and 6

to both locations. When hydrogen was released to the atmosphere, ignition occurred 62 percent of the time; in release to enclosures, ignition occured every time (100 percent). This emphasizes the danger when hydrogen is released inside an enclosure. Not only is the ignition dangerous in a confined space, but the explosion potential is also greater.

Two important concerns associated with liquid hydrogen spills are the duration of the hazardous period following such a spill and the extent to which the vapor cloud in the vicinity of the spill forms a combustible mixture downwind of the spill. In one study,[5] the spill tests indicated that vapor cloud ignition produces a hot fireball which ignites combustible material within the confines of the fireball. The major emphasis of the study was to provide data for the prediction of evaporation rates from the ground and to determine the distance downwind that a hazardous condition will exist. The results indicated:

1. That, initially, all heat supplied to evaporate the liquid comes from the ground. In later stages of evaporation (i.e., after approximately 3 min) some heat contribution is made by condensation of air into the hydrogen pool. The evaporation rate has an initial value of the order of 5 to 7 in./min (12 to 24 centimeters/min) decreasing rapidly to a steady-state value of about 1 ½ in./ min (4 centimeters/min). It was also found that ignition of the vapor did not significantly affect the rate of evaporation, but that use of a pebble bed of crushed rock greatly increased the evaporation rate.

2. That in spill tests vapor clouds were formed extending up to 200 ft (60 meters) *downwind*. Upon ignition at the pool the flame traveled *downwind* for over 100 ft (30 meters).

3. That in discharges of liquid hydrogen from a pipeline at rates varying from 30 to 300 gpm, a vapor cloud is formed, which persists near ground level for 500 to 700 ft (150 to 210 meters) downwind. Ignition of the cloud was only accomplished within 100 ft (30 meters) of the vent. (The data were too preliminary to conclude whether or not under certain conditions, the vapor could not be ignited at greater distances.)

Liquid hydrogen should not be exposed to the air. The low temperature of liquid hydrogen can solidify air. Vents from liquid hydrogen storage vessels may be plugged by accumulations of frozen moisture condensed from the air. The resulting pressure buildup can be sufficient to rupture the container, releasing hydrogen to the air. Thus, openings in liquid hydrogen containers should be examined periodically to make sure that they do not become plugged with frozen mois-

ture. Also, if air or oxygen is allowed to condense and solidify in liquid hydrogen, a potential explosion hazard can result.

Avoid contact with hydrogen liquid or gas, or with uninsulated pipes or vessels containing liquid hydrogen. Due to its extremely low temperature, liquid hydrogen can produce an effect on the skin similar to a burn; these "burns" can also be produced by the very cold gas formed by evaporation of the liquid. Particularly hazardous is contact of these cold gases with delicate tissues, e.g., the eyes. Splashing of liquid hydrogen is a common hazard. Wear full protective clothing when dealing with liquid hydrogen.

A typical liquid hydrogen container (Figure 5-4) contains several safety features:

1. An inner vessel, ¼-inch (6.3 millimeters) safety valve (4, Figure 5-4) set at 10 psig

2. An inner vessel, ½-inch (12.7 millimeters) safety head (5, Figure 5-4) with a rupture pressure of 48 psig

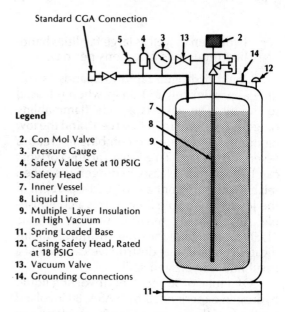

Standard CGA Connection

Legend

2. Con Mol Valve
3. Pressure Gauge
4. Safety Value Set at 10 PSIG
5. Safety Head
7. Inner Vessel
8. Liquid Line
9. Multiple Layer Insulation In High Vacuum
11. Spring Loaded Base
12. Casing Safety Head, Rated at 18 PSIG
13. Vacuum Valve
14. Grounding Connections

Figure 5-4. Cutaway view of liquid hydrogen container. *Courtesy of Air Reduction Company*

3. An outer casing, 1-inch (25.4 millimeters) safety head (12, Figure 5-4) with a rupture pressure of 18 psig

To protect the safety heads from the weather, polyethylene cap plugs are installed.

Handling and Storage of Liquid Helium

Liquid helium has unique properties. Its extreme cold makes it invaluable for cryogenic investigation and cryogenic industrial applications. At one atmosphere pressure, liquid helium has a boiling point of –452.1°F (–268.95°C), a mere 7.6°F (–13.56°C) above absolute zero. Helium is the only known substance that remains liquid under ordinary pressure at temperatures close to absolute zero.

Liquid helium is a colorless, odorless, nontoxic fluid having a density one-eighth that of water. It is chemically inert under all temperature and pressure conditions.

To vaporize one volume of liquid helium requires only one-sixtieth the quantity of heat needed to vaporize an equal volume of liquid nitrogen. Therefore, containers used to store liquid helium must be exceptionally well insulated to prevent heat from getting to the liquid.

There are three basic types of containers for shipping and storing liquid helium: the liquid nitrogen-shielded type; the cold helium gas-shielded type; and the vacuum type.

The nitrogen-shielded container has four concentric compartments (Figure 5-5). The inner compartment, containing the liquid helium, is protected from the ambient temperature by an inner vacuum jacket, then a liquid nitrogen radiation shield, and a second vacuum jacket next to the outer casing. Encompassing the four compartments is a stainless steel casing that provides moderate protection during handling. These containers have a dual neck. The central neck passage is for filling, liquid withdrawal, and for normal venting

at 0.5 psig. The outer passage surrounding the central neck tube is for secondary venting through a safety relief valve set at 7 psig.

Radiation shielding for the gas-shielded containers (Figure 5-6) is provided by a copper shell located in the vacuum space that surrounds the liquid helium reservoir. This shell is cooled by the normal vent line as it carries cold helium gas from the lower end of the central fill tube to the atmosphere. The normal vent valve is set at 15.2 psia.

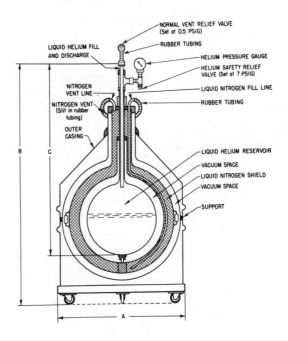

Figure 5-5. Nitrogen-shielded container for liquid helium storage. *Courtesy of Air Reduction Company*

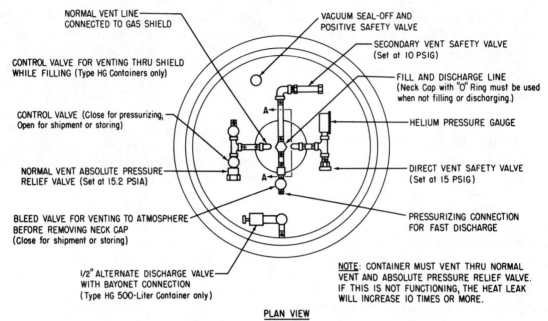

NORMAL VENT LINE—
CONNECTED TO GAS SHIELD

VACUUM SEAL-OFF AND
POSITIVE SAFETY VALVE

SECONDARY VENT SAFETY VALVE
(Set at 10 PSIG)

CONTROL VALVE FOR VENTING THRU SHIELD
WHILE FILLING (Type HG Containers only)

FILL AND DISCHARGE LINE
(Neck Cap with "O" Ring must be used
when not filling or discharging.)

CONTROL VALVE (Close for pressurizing,
Open for shipment or storing)

HELIUM PRESSURE GAUGE

NORMAL VENT ABSOLUTE PRESSURE
RELIEF VALVE (Set at 15.2 PSIA)

DIRECT VENT SAFETY VALVE
(Set at 15 PSIG)

BLEED VALVE FOR VENTING TO ATMOSPHERE
BEFORE REMOVING NECK CAP
(Close for shipment or storing)

PRESSURIZING CONNECTION
FOR FAST DISCHARGE

1/2" ALTERNATE DISCHARGE VALVE
WITH BAYONET CONNECTION
(Type HG 500-Liter Container only)

NOTE: CONTAINER MUST VENT THRU NORMAL
VENT AND ABSOLUTE PRESSURE RELIEF VALVE.
IF THIS IS NOT FUNCTIONING, THE HEAT LEAK
WILL INCREASE 10 TIMES OR MORE.

PLAN VIEW

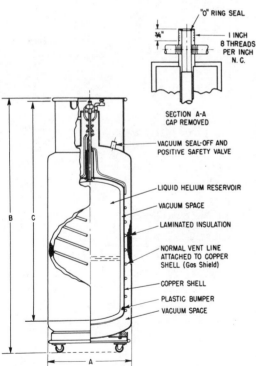

Figure 5-6. Gas-shielded container for liquid helium storage. *Courtesy of Air Reduction Company*

A secondary vent line from the upper end of the central fill tube leads to a safety valve set at 10 psig. A third line leads to a safety valve set at 15 psig. This third line provides an independent passage directly from the helium chamber to the atmosphere. All three valves are marked with their operating pressures.

The normal vent line has a shutoff valve for the purpose of building up pressure when withdrawing liquid. This valve should be returned to the open position after making a liquid withdrawal.

The vacuum type container (Figure 5-7) is equipped with an 18 psia absolute pressure relief valve, a 10 psig relief valve, a 37 psig bursting disk, and an 18 psig casing bursting disk. In addition, there is a secondary relief path bursting disk of 100 psig and a secondary path relief valve set for 50 psig.

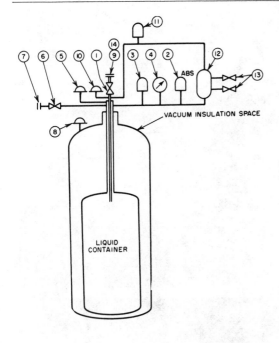

VACUUM INSULATION SPACE

LIQUID CONTAINER

LEGEND

1. LIQUID LINE ACCESS VALVE
2. 18 PSIA ABSOLUTE PRESS. RELIEF VALVE
3. 10 PSIG RELIEF VALVE
4. 0-30 PSIG PRESSURE GAUGE
5. 37 PSIG BURSTING DISK
6. GAS PHASE VALVE
7. GAS PHASE CONNECTION
8. 18 PSIG CASING BURSTING DISK
9. LIQUID LINE ACCESS PORT
10. SECONDARY RELIEF PATH BURSTING DISK 100 PSI
11. SECONDARY PATH RELIEF VALVE, 50 PSI
12. THERMAL OSCILLATION DAMPER
13. THERMAL OSCILLATION DAMPER PURGE VALVES
14. LIQUID LINE ACCESS PORT ADAPTER

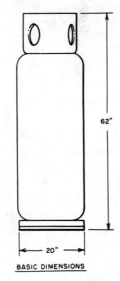

62"

20"

BASIC DIMENSIONS

Figure 5-7. Vacuum-type container for liquid helium. *Courtesy of Air Reduction Company*

Transportation of liquid helium also occurs by tank truck, as shown in Figure 5-8. These tanks also have relief valves and casing disks for safety purposes.

A potential hazard in liquid helium containers is the formation of plugs in the narrow passages, which will prevent venting or liquid removal and allow the pressure to build up to the burst point. A plug is formed as air and water vapor condense and are frozen by the cold helium vapor escaping from the container. If a helium container is open to the atmosphere through mishandling, a steady stream of air will flow down the sides of the tube, condense, freeze, and form a solid plug. Pressure variations due to change of altitude can cause excessive boiling, harden relief valve seats so that they leak, and allow air to enter the open lines. The tendency for air and moisture to enter openings in the container and cause plugging is increased when shipments are made by air.

Handling and Storage of Liquid Oxygen, Nitrogen, and Argon

Liquid nitrogen, liquid oxygen, and liquid argon are extremely cold liquids. At atmospheric pressure liquid oxygen exists at –297.3°F (–182.9°C), liquid nitrogen at –320.5°F (–195.8°C), and liquid argon at –302.6°F (–185.9°C). Liquid nitrogen and argon are colorless; liquid oxygen has a pale blue color. These liquids possess a viscosity and density approximating that of water.

The heats of vaporization of liquid oxygen, nitrogen, and argon are low. In addition, a small quantity of liquid produces a large volume of gas at atmospheric pressure. One cubic foot of liquid oxygen, for example, will produce 860 cubic feet of oxygen gas. A small amount of heat flow from the atmosphere into the liquid will, therefore, produce an appreciable volume of gas. For this reason, all storage vessels are provided with pressure relief devices unless the container is vented

Figure 5-8. Liquefied helium truck transport. *Courtesy of Air Reduction Company*

properly to provide escape of evaporating gases. All lines and vessels in which the liquid may be trapped between closed valves are equipped with pressure relief valves. If there is any likelihood that the relief valve may freeze, as, for instance, from ice formed from dripping water or condensed moisture, such vessels and lines must be equipped with rupture disks. Both pressure relief valves and rupture disks are placed and protected so that water cannot splash or condense upon them. In addition, it is necessary to vent relief valves and rupture disks to the outside atmosphere.

Liquid oxygen, nitrogen, or argon are transported in insulated containers that pro-

vide means for the escape of gas as liquid evaporates. Never cork or plug the outlet to such containers. These chemicals are also stored on-site in large fixed containers, as illustrated in Figure 5-9.

Because of their extremely low temperatures, liquid oxygen, nitrogen, or argon will "burn" the skin like hot liquids. Therefore, it is imperative that emergency response personnel not allow these liquids to come in contact with the skin or to soak into clothing. Liquid oxygen must never be poured upon clothing, fabrics, rags, waste, or other readily combustible materials because a spark can start a serious fire and cause personal injury.

LIQUID STORAGE TANK   CONTROL HOUSE AND COLD BOX   COMPRESSION EQUIPMENT

Figure 5-9. Fixed storage installation of liquid nitrogen. *Courtesy of Air Reduction Company*

Handling and Storage of Liquefied Natural Gas

Within the last few years, liquefied natural gas (LNG) has become an essential commodity important to the overall energy needs of the nation. A report prepared for the Department of Transportation states that:

It has become indispensable in maintaining adequate supplies of natural gas in many urban areas during peak demand, and it is about to become an important source of imported energy. Without it, a very severe adverse economic impact would be felt. The gas industry not only would be unable to meet the essential needs of other industries and institutions, but would also be unable to provide gas for home heating and other uses during the winter months. The use of natural gas cannot be readily reduced by the substitution of other forms of energy, and liquefaction is essential in preventing marked reductions in its availability to urban centers.

LNG first became important to northern U.S. cities, because an increased use of natural gas placed excessive demands on the pipelines bringing the fuel from the southern gas fields during the high use periods. By liquefying the gas from the pipeline during the summer, supplies could be stored for later regasification during high demand periods. Location of such liquefaction and storage facilities near the terminus of pipelines has allowed fuller utilization of our pipeline systems. At the present time there are approximately 50 peak-shaving plants in the United States capable of storing more than 12 million barrels (bbl) of the liquid (about 40 billion SCF equivalent). There are also about 50 satellite storage facilities.[6]

As an example of the type of problem that can be created by the storage of liquefied natural gas is the fire that occurred at the East Ohio Gas Company in Cleveland, Ohio.[7]

At approximately 2:40 P.M. on October 20, 1944, a disastrous fire occurred at the Liquefaction, Storage, and Regasification (LS&R) Plant of the East Ohio Gas Co., Cleveland, as the result of failure of an insulated cylindrical tank in which liquefied natural gas was stored at less than 5 psi pressure and at a temperature of –250°F (–156.7°C). Eye witnesses of the initial stages of the failure of the tank reported that streams of liquid or fog issued from its side, and that almost immediately the tank opened and discharged its entire contents of liquefied natural gas over the plant area and adjoining property at a lower elevation. The gas or vapor was ignited almost immediately after failure of the tank. Liquefied gas entered storm sewers, mixed with air, was ignited, and exploded. The explosions damaged the sewer system and the paving of some of the streets. The intense fire in the vicinity of the plant caused great loss of life and property. Final tabulation of fatalities by the coroner's office indicated that 128 had been killed. The East Ohio Gas Company paid 109 death claims. The difference between claims paid and fatalities estimated by the coroner's office may have been due to difficulties of identification of bodies or to the possibility that some of the

fatalities may have been transients or persons having no relatives. Estimates of the number injured ranged from 200 to 400. Property damage was estimated by the company at $6,800,000.

The primary hazards of LNG are fundamentally the same as those that have been dealt with for many years in the transmission and distribution of natural gas. LNG is a liquid fuel that readily vaporizes, and ignites and burns when mixed with air. The fire that can occur as a result of an accidental release can be injurious to people and damaging to property.

LNG safety considerations differ from those concerned with the natural gas that flows through pipelines, mains, and service lines, however. Because LNG is stored as a liquid at a very cold temperature (–260°F) or –162.2°C, special materials must be used to contain it, and thermal differences in piping and containment systems must be considered. Economics also require that LNG be stored in a few large insulated containers rather than in many small ones. Care must also be exercised to prevent leaks, since large quantities of flammable gas can thus be released accidentally. The fire hazard with LNG is much greater than with the more conventional natural gas operations, because of the large quantities involved and the need to employ more sophisticated engineering techniques in facility design, construction, and inspection.

LNG storage tanks consist of double shell construction (Figure 5-10). The annulus between the inner and outer tank (both walls and roofs) is insulated with expanded perlite. A resilient fiberglass blanket is installed on the outside of the inner shell to eliminate perlite compaction, which occurs during thermal cycling. Foamed glass blocks are used to support and insulate the bottom of the tank. The tank foundations are designed to be heated by electric elements to prevent freezing of the moisture in the surrounding soil, which could result in frost heave.

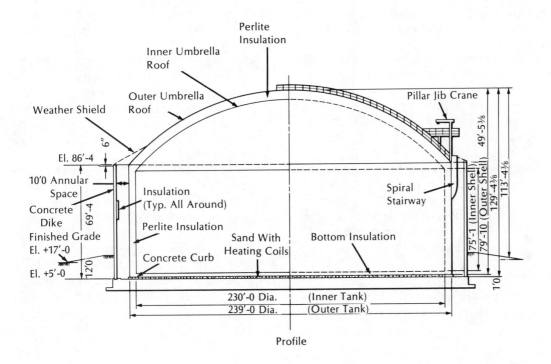

Figure 5-10. Inner Umbrella Roof, Perlite Insulation, Outer Umbrella Roof, Weather Shield, Pillar Jib Crane, El. 86'-4, 10'0 Annular Space, Concrete Dike, Finished Grade El. +17'-0, El. +5'-0, Insulation (Typ. All Around), Perlite Insulation, Concrete Curb, Sand With Heating Coils, Bottom Insulation, Spiral Stairway, 49'-5⅜, 75'-1 (Inner Shell), 79'-10 (Outer Shell), 129'-4⅜, 113'-4⅜, 69'-4, 12'0, 1'0, 230'-0 Dia. (Inner Tank), 239'-0 Dia. (Outer Tank), Profile

Figure 5-10. LNG storage tank. *Courtesy of U.S. Department of Interior*

The tanks are fitted with pressure relief valves discharging to the atmosphere to protect the tanks from overpressure. In addition, there are vacuum relief valves to prevent excessive buildup of negative pressures. Tank operating pressure is usually 15.5 psia, with pressure relief valves set at 1.5 psig.

All piping connections into or out of the inner tank are through the roof, with no penetrations of the inner shell below the liquid level.

LNG is changed to a gas by raising its temperature and then pumping it through the pipeline. Emergency response personnel must be familiar with LNG storage facilities as well as water transport vessels and docking areas.

COMPRESSED GASES

In the discussion in Chapter 2 of temperature-pressure-volume relationships, it is explained that for an ideal gas, the pressure times the volume at one point is equal to the pressure times the volume at the second point. Also appropriate to review here is

Charles' law, which states that if the volume of a gas is kept constant and the temperature is increased, the pressure increases in direct proportion to the increase in absolute temperature.

One final point to review is a summary of the discussion of vapor pressure. When a liquid evaporates in a confined area, the vapor that is developed exerts a pressure on its surroundings, just as any gas in a confined vessel causes a pressure based on its temperature and volume.

A compressed gas is defined as any material or mixture having in the container an absolute pressure exceeding 40 psi at 70°F (21.1°C) or, regardless of the pressure at 70°F (21.1°C), having an absolute pressure exceeding 104 psi at 130°F (54.4°C).

Shipping Containers

There are basically two techniques for shipping compressed gases. They can 'be shipped as a liquid kept in that state by maintaining a pressure, or as a gas under high pressure. Both methods of storing and shipping can present problems to emergency response personnel.

Storage cylinders and transporting vehicles vary a great deal in size. Cylinders can be as small as a home propane heating torch cylinder to a 33,000-gallon tank car of liquefied petroleum gas. In between, there are all sizes and shapes to accommodate a rapidly expanding market.

In the medical field, compressed gases come in cylinder sizes ranging from an A cylinder to an H cylinder. The relative sizes are illustrated in Figure 5-11. These cylinders contain from 5 cubic feet to 250 cubic feet of product, depending on the pressure created. The cylinders are transported on open, stakebody trucks. However, when a truck like this is involved in an accident, leakage can easily occur.

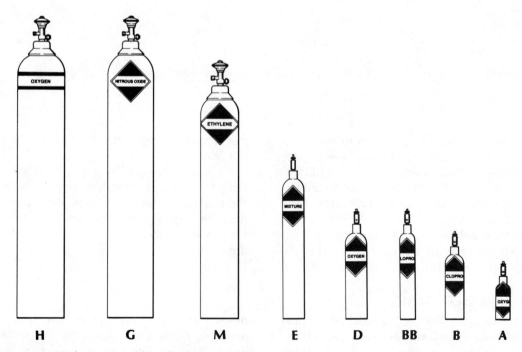

Figure 5-11.   Relative sizes of medical gas containers

Transportation by tank truck can range in size from several hundred gallons to several thousand gallons. Transportation by rail (Figure 5-12) increases the single car capability to over 30,000 gallons (1,140,000 liters).

Finally, water transportation has opened up a whole technique for moving large quantities of compressed gases. For example, in the case of liquefied natural gas, the new shipping vessels are capable of transporting extremely large quantities.

Color Coding of Containers

There is no required color coding of compressed gas cylinders by product. Some manufacturers use their own code or they may use only a single color for all products. This obviously makes the job of identification much more difficult. Emergency response personnel must visit local distributors to determine if they use any kind of coding.

There is at least one company, however, Air Products, which uses a standard code for their medical cylinders. This is the same code recommended by the Bureau of Standards and under consideration by the Compressed Gas Association. The product and its corresponding color are listed in Table 5-3.

Label Information

Compressed gas cylinders are labeled in accordance with either the Department of

Figure 5-12. Compressed gas tank car

Transportation requirements (covered in Chapter 6) or the standards of the Compressed Gas Association. Either label is acceptable. The major problem with the nonflammable compressed gas label is that some very important properties are not indicated. This means that emergency response personnel must determine the actual product and not rely solely on any visible label.

Table 5-3. Color Coding of Air Products Compressed Gas Medical Cylinders

Product	Chemical Symbol	Cylinder Color
Carbon Dioxide	CO_2	Gray
Carbon Dioxide–Oxygen	CO_2-O_2	Gray and Green
Cyclopropane	C_3H_6	Orange
Ethylene	C_2H_4	Red
Helium	He	Brown
Helium–Oxygen	He–O	Green and Brown
Nitrous Oxide	N_2O	Blue
Oxygen	O_2	Green
Air	—	Yellow

Anhydrous Ammonia

Anhydrous ammonia is essentially ammonia without water. Its chemical formula is NH_3 and it is composed of approximately 82 percent nitrogen and 18 percent hydrogen by weight. It is a colorless alkaline gas possessing a sharp pungent odor.

Ammonia gas irritates the skin and mucous membranes. At 50 parts per million (ppm) its odor is detectable and this is the maximum allowable exposure in an 8-hour day. At 5,000 ppm it is usually fatal. Table 5-4 summarizes the health hazards from exposure to ammonia.

Anhydrous ammonia freezes at –108°F (–77.8°C) and boils at –28°F (–33.3°C) if the container is open to the air. To prevent ammonia from boiling away under normal temperature conditions, the product is confined in a pressure vessel. The physical properties of anhydrous ammonia are listed in Table 5-5.

Even though anhydrous ammonia is shipped as a nonflammable compressed gas, emergency response personnel must be aware that it will autoignite at 1562°F. This means that a storage tank exposed to heat during an incident can ignite and create further problems. Emergency response personnel must not be lulled into a false sense of security by the nonflammable label.

Upon release from a pressure vessel into the atmosphere, liquid ammonia vaporizes and expands rapidly. As a vapor, it occupies about 850 times its liquid volume, and it absorbs heat as it changes state to a gas. Because of the cooling effect of expansion, water vapor in the air freezes and creates a white cloud for a short period of time. The cloud is frozen vapor because the ammonia is still colorless.

Table 5-5.   Physical Properties of Anhydrous Ammonia

Boiling Point	–28°F (–33.3°C)
Freezing Point	–108°F (–77.8°C)
Vapor Density	0.6
Specific Gravity of Liquid	0.7
Flammable Limits	16% to 25%
Ignition Temperature	1562°F (850°C)

Table 5-4.   Health Hazards from Exposure to Ammonia

Ammonia Vapor, (ppm)[a]	Percent	General Effect	Exposure Period
50	1/200 of 1	Odor detectable by most persons.	Prolonged, repeated exposure produces no injury.
100	1/100 of 1	No adverse effects for average worker.	Maximum allowable concentration for 8-hour working exposure.
400 to 700	4/100 of 1 to 7/100 of 1	Nose and throat irritation; eye irritation with tearing.	Infrequent short (1 hour) exposures ordinarily produce no serious effect.
2,000 to 3,000	2/10 of 1 to 3/10 of 1	Convulsive coughing; severe eye irritation.	No permissible exposure. May be fatal after a short exposure.
5,000 to 10,000	1/2 of 1 to 1	Respiratory spasm; rapid asphyxia.	No exposure permissible. Usually fatal.

[a]Parts of ammonia per million parts of air.

There are two types of storage vessels for anydrous ammonia. The first is a nonrefrigerated type, in which the ammonia is stored at atmospheric temperature and under moderate pressure of up to 250 psig. The second type uses refrigerated tanks, which allow storage under relatively low pressures. These tanks can vary in size from 60-gallon (228-liter) applicator tanks to 90,000-gallon (3,420,000-liter) storage tanks.

All pressure tanks containing anhydrous ammonia have safety valves designed to relieve excess pressure. This valve, shown in Figure 5-13, is installed directly into the vapor space of the tank so that it can discharge vapor when the pressure exceeds the setting.

Discharge lines are also equipped with back pressure check valves (Figure 5-14) to ensure that flow takes place in only one direction. This valve will ensure that the product will not flow in the wrong direction should a break in the fill piping occur.

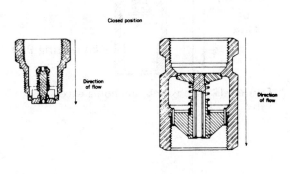

Figure 5-14. Anhydrous ammonia back pressure check valve. *Courtesy of The Fertilizer Institute*

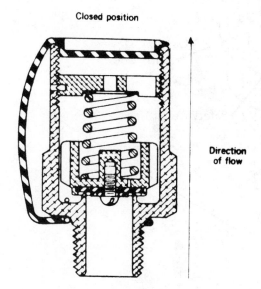

Figure 5-13. Anhydrous ammonia safety relief valve. *Courtesy of The Fertilizer Institute*

An excess flow valve (Figure 5-15) is also installed in the line. It is designed to close automatically at a predetermined flow rate to prevent the release of a large amount of product if a break should occur downstream from the valve. Once the system is made vapor tight by repairing the break, a small equalizing part in the excess flow valve automatically equalizes the pressure on both sides of the valve. A spring in the valve then returns it to the open position and normal flow resumes.

On large storage tanks, a liquid level gauge is installed, so that the amount of product remaining in the tank can be determined.

Railroad tank cars containing anhydrous ammonia are equipped with two liquid withdrawal valves, in line with the length of the car (Figure 5-16). Vapor valve(s) are located at a 90° angle to the axis of the car.

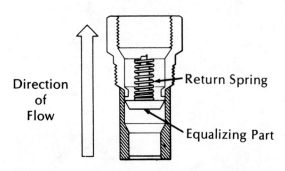

Direction
of
Flow

Return Spring

Equalizing Part

Figure 5-15. Anhydrous ammonia excess flow valve

Other storage facilities include a nurse tank (Figure 5-17) and an applicator tank (Figure 5-18). Note the locations of the various safety devices, as well as the withdrawal valves.

A typical plant operation is shown in Figure 5-19. Emergency response personnel should include information about their local anhydrous ammonia storage facility in preincident planning in accordance with the material provided in Chapter 7.

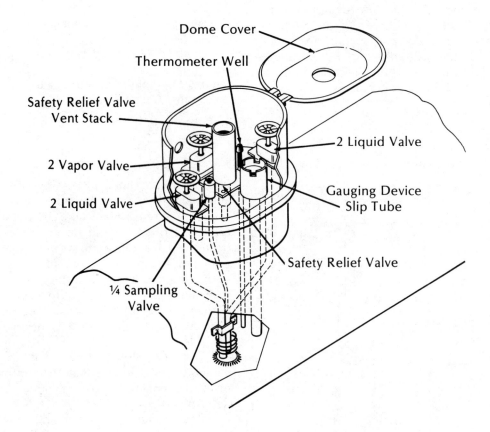

Dome Cover

Thermometer Well

Safety Relief Valve
Vent Stack

2 Liquid Valve

2 Vapor Valve

Gauging Device
Slip Tube

2 Liquid Valve

Safety Relief Valve

¼ Sampling
Valve

Figure 5-16. Fittings and connections on anhydrous ammonia tank cars. *Courtesy of The Fertilizer Institute*

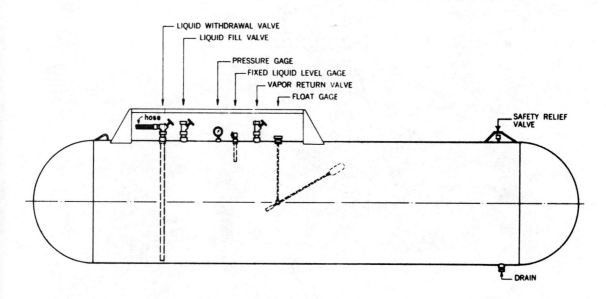

Figure 5-17. Typical ammonia nurse tank. *Courtesy of The Fertilizer Institute*

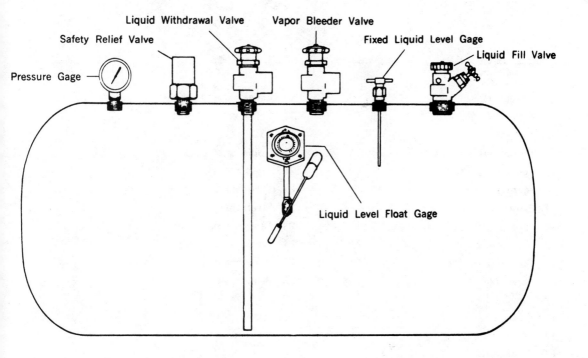

Figure 5-18. Anhydrous ammonia applicator tank. *Courtesy of The Fertilizer Institute*

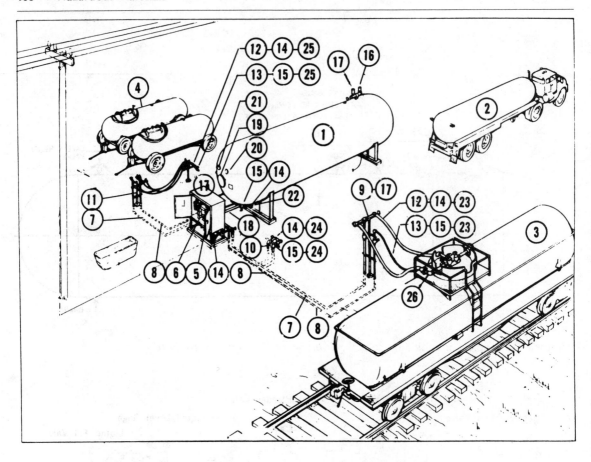

Figure 5-19. Typical anhydrous ammonia plant layout. *Courtesy of The Fertilizer Institute*

1. Storage tank
2. Transport truck
3. Tank car
4. Nurse tank Wagons
5. Compressor cabinet
6. Four-way valve
7. Extra heavy liquid piping
8. Extra heavy vapor piping
9. Tank car riser
10. Transport connections
11. Nurse tank riser
12. Liquid hoses
13. Vapor hose
14. Liquid valve
15. Vapor valve

16. Relief valve manifold
17. Relief valve
18. Back pressure check valve
19. Liquid level gage
20. Pressure gage
21. Thermometer
22. Steel plug
23. Tank car nipple
24. Transport coupling
25. Hose end valves and couplings
26. Bleeder valve

Anhydrous ammonia readily dilutes with the application of water. The best emergency handling technique therefore is to use large amounts of water. One volume of water will absorb 200 volumes of anhydrous ammonia at 68°F (20.0°C).

When combating a leak, emergency response personnel must wear full protective clothing and ensure that there are no openings at the ankles or wrists. Rubber or plastic gear is the best protection.

Chlorine

As described in Chapter 6, chlorine is considered a nonflammable compressed gas. However, under the new placarding system, a separate CHLORINE placard has been developed.

Chlorine is one of the chemical elements. While the product is not flammable, it reacts with many substances and in certain cases can be considered an oxidizer. The gas has a characteristic odor, with a greenish-yellow color. While only slightly soluble in water, chlorine gas is 2½ times heavier than air. Liquid chlorine is about 1½ times as heavy as water. Table 5-6 contains a summary of the physical

properties of chlorine. One volume of liquid chlorine, when vaporized, will yield about 460 volumes of gas.

With its characteristic odor, the presence of chlorine is easily detectable. However, from a health standpoint, a small amount of the gas can combine with moisture on the skin or in the lungs to form hydrochloric acid. This corrosive liquid can cause severe damage. Table 5-7 lists the exposure limits for chlorine.

Chlorine can be shipped in a variety of containers and by several different transportation modes. Small cylinders, up to 150 pounds (67.5 kilograms) capacity, are used for small quantities. One-ton containers (2,000 pounds or 900 kilograms) are used in industrial processing as well as in water treatment processes. Motor trucks with capacities of 15 to 20 tons (13.5 to 18 metric tons) as well as tank cars of 16, 30, 55, 85, and 90 tons (14.5, 27, 49.5, 76.5, and 81 metric tons) are required by high-volume users. Finally, barges with 600 to 1,200-ton (540 to 1,080 metric tons) capacity move the product on the waterways.

The flow of chlorine gas is dependent upon the internal vapor pressure within the tank. This pressure, in turn, is dependent upon the temperature of the liquid. Because

Table 5-6. Physical Properties of Chlorine

Boiling Point	−30°F	(−34.4°C)
Freezing Point	−150°F	(−101.1°C)
Vapor Density	2.48	
Specific Gravity	1.5	
Expansion	1 volume of liquid chlorine equals the weight of 458 volumes of gaseous chlorine	

Table 5-7. Health Hazards from Exposure to Chlorine

Chlorine Vapor, (ppm)	Percent	General Effect
4	$\frac{1}{2500}$ of 1	Odor detectable by most persons
15	$\frac{1}{667}$ of 1	Irritation to throat
50	$\frac{1}{200}$ of 1	Hazardous in a few breaths
1000	$\frac{1}{10}$ of 1	Rapidly lethal

chlorine has such a low boiling point (–30°F or –34.4°C) as the liquid vaporizes, there is a considerable cooling effect. If the flow rate is too great, the cooling effect of the vaporization process will lower the temperature of the remaining chlorine. This, in turn, will reduce the flow rate.

On the other hand, discharge of the liquid can take place at very high rates. Flows of up to 400 pounds (180 kilograms) per hour are possible from 1-ton (900 kilograms) cylinders.

In a single-unit tank car, there are five valves mounted inside the dome (Figure 5-20). Four of the valves are for product withdrawal, while the fifth, in the center, is the relief valve. The liquid withdrawal valves (Figure 5-21) are in line with the long axis of the tank car, while the gas withdrawal lines are perpendicular to the tank axis.

On each liquid flow line there is a rising ball, excess flow valve (Figure 5-22). This valve is designed to close when the chlorine flow exceeds 7,000 pounds (3,150 kilograms) per hour. This ensures that if there is a line break, automatic shutdown will occur.

The safety relief valve is a spring-loaded type. It is set to operate at between 225 to 375 psig. If the pressure is relieved below the relief valve setting, it will automatically close. The relationship between temperature and vapor pressure is shown in Table 5-8.

Chlorine tank barges do not have uniform valve arrangements. While the capacity of the barges is great, each will have two or three relief valves set at 300 psig. The excess flow valves are set to close if the flow exceeds 15,000 pounds (6,750 kilograms) per hour.

Figure 5-21. Valve for liquid chlorine withdrawal from a tank car

Table 5-8. Chlorine Vapor Pressure Versus Temperature Relationship

| Temperature | | Pressure, |
°F	°C	psig
–30	–34.4	0
–10	–23.3	8.29
0	–17.8	13.81
20	–6.7	27.84
40	4.4	46.58
60	15.6	70.91
80	26.7	101.76
100	37.8	140.20
120	48.9	186.95
140	60.0	243.33
160	71.1	310.35
180	82.2	389.17
200	93.3	480.97
220	104.4	587.13

Figure 5-20.   Dome arrangement in a tank car

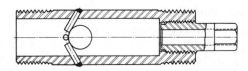

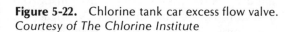

Figure 5-23.   Chlorine "C" kit in use

Figure 5-22.   Chlorine tank car excess flow valve. *Courtesy of The Chlorine Institute*

As soon as a chlorine leak is detected, it must be promptly handled. Because of the corrosive nature of chlorine when combined with moisture, a simple leak will grow progressively worse. Full protective clothing and self-contained breathing apparatus are absolutely necessary. All exposed skin must be fully protected to ensure that the chlorine does not react with skin moisture to cause severe burns.

When working to close a leak, always work in pairs. Work upwind and above the leak if possible. Since chlorine gas is heavier than air, the leak will flow close to the ground. Special control kits are available for sealing chlorine leaks in the various containers.

For the 100- and 150-pound (45 to 67.5 kilograms) cylinders a chlorine "A" repair kit is used. For 1-ton (900 kilograms) cylinders a chlorine "B" repair kit is used, while tank truck and tank car leaks can be stopped with the parts of the Chlorine "C" kit (Figure 5-23). All of the kits operate on the principle of capping off leaking valves or sealing off the leak in a side wall. In the "B" kit, there are capping devices for the fusible plug. Each kit contains the tools, wrenches, and specialized devices necessary for stopping the leak. However, there is no breathing apparatus or protective clothing contained in the kit. It should now be obvious to emergency response personnel that the time to learn where the three different kits are stored in your area and how to use the equipment is long before the emergency. Preplanning their use will prevent confusion during the incident. The location of kits for your area is available from the Chlorine Institute.

While chlorine is considered nonflammable, it reacts like oxygen under certain circumstances. Sometimes, this reaction can take place with explosive violence. At ordinary temperatures, dry chlorine will react with aluminum, arsenic, gold, mercury, selenium, tellurium, tin, and titanium. In addition, at temperatures above 250°F (121.1°C) chlorine will react with iron, lead, nickel, silver, and steel.

If a chlorine container is leaking, there are certain actions that emergency response personnel can take. First, if the liquid product is leaking, an attempt should be made to turn the container so that only the gas is escaping. Use one of the chlorine kits to close off the leak. Do not use water at the leak site as this will only cause corrosion and increase the leak.

If the container is not leaking, but exposed to fire, large amounts of water are necessary to keep the tank cool. As shown in Table 5-8, pressure rapidly increases with temperature. If not kept cool, the fusible disks will rupture as the pressure increases, causing a rapid release of liquid and/or gas.

If a leak cannot be stopped, evacuation downwind is essential. Since the gas is so much heavier than air, all low spots must be checked before allowing evacuated individuals to return. The Chlorine Institute has published a pamphlet entitled "Estimating Area Affected by a Chlorine Release,"[8] which is available to emergency response personnel.

Oxygen

Oxygen as a liquid is discussed under Cryogenics at the beginning of this chapter. As a liquid it is shipped as an oxidizer. In the gaseous state, oxygen is shipped as a nonflammable compressed gas.

Table 5-10. Physical Properties of Oxygen

Boiling Point	−297° F	(−182.8° C)
Freezing Point	−361° F	(−218.3° C)
Vapor Density	1.1	
Expansion	1 volume of liquid oxygen expands to 875 volumes of gas	

Oxygen is necessary to support life. Without oxygen human life could not exist. Since the combustion process also needs oxygen, it is consumed during burning. This, in turn, depletes the oxygen available for human use. Table 5-9 details the health hazard due to a lack of oxygen.

Oxygen is a colorless, odorless, and tasteless gas at ordinary temperatures. It is slightly denser than air and only slightly soluble in water. The physical properties of oxygen are shown in Table 5-10.

Oxygen as a gas is generally stored in small cylinders. When larger quantities of the gas are needed, they are usually transported and stored as a cryogenic liquid. Medical oxygen comes in standard sizes from 5.3 cubic feet through 250 cubic feet. In addition, oxygen gas for industrial uses such as cutting torches comes in the same size containers. The valves on these cylinders have a frangible disk and a fusible plug safety device. Upon overpressurization or under heat conditions, these will open to relieve the pressure.

Since oxygen supports combustion, a release of the product will intensify a fire. In addition, if gaseous oxygen is released, it will react with certain other chemicals to create oxides and peroxides. Some of these compounds are then unstable or highly reactive. A leak should be stopped. A water fog stream can be used to dissipate and disperse the vapor.

Table 5-9. Health Hazard Due to Lack of Oxygen

Percent Oxygen Available	Symptoms
20–100	Normal breathing can occur
12–15	Muscle coordination reduced
10–14	Rapid fatigue while working and faulty judgment
6–8	Collapse
0–6	Death occurs in 6 to 8 minutes

FLAMMABLE GASES

Flammable gases are ready to burn without any outside heating necessary. This means that they do not have a flash point. However, flammable gases do have a *flammable range*, the limits of the air-fuel ratio during which combustion will occur.

Several flammable gases are shipped as liquids by keeping the gas under pressure at normal temperatures. The most common of these gases are propane, butane, propylene, and butadiene.

Flammable gases can also be shipped in the gaseous state, under pressure. The most common shipped in this form include methane, ethane, and acetylene. The flammable ranges for these gases are listed in Table 5-11.

Handling Flammable Gas Incidents

Flammable gas emergency incidents will involve gas that is either leaking and not ignited or a leak that is burning. The generalized procedures for handling the incident are detailed below.

Unignited Leak

1. Evacuate personnel downwind from the leak (Figure 5-24). Remember, when approaching the scene, do not drive apparatus through the vapor cloud. Keep spectators, unnecessary emergency response personnel, and sightseers away from the scene.
2. Begin the attack from upwind and out of the vapor cloud.

Figure 5-24. Evacuation downwind from the gas leak

Table 5-11. Flammable Ranges of Gases

Gas	Shipping Form	Lower Flammable Limit, percent in air	Upper Flammable Limit, percent in air
Propane	Liquid	2.4	9.5
Butane	Liquid	1.9	8.4
Propylene	Liquid	2.0	11.1
Butadiene	Liquid	2.0	11.5
Methane	Gas	5.3	14.0
Ethane	Gas	3.0	12.5
Acetylene	Gas	2.5	81.0

3. Identify product leaking.

4. Using reference material and material in this text, determine best method of attack. For example, if the gas is water soluble, fog streams can be used. Other gases can be dispersed by the correct use of interlocking fog streams.

5. If possible, close valves to stop flow of gas. This is where preplanning comes in. Using the information contained in the various chapters to follow on modes of transportation as well as information gathered on visits to the storage facilities in your area will enable you to determine how to best shut off the leak.

Ignited Leak

1. As a general rule, a gas leak that has ignited should not be extinguished unless the leakage can be immediately stopped. This must be carefully followed because the vapors from an unignited leak can travel over a wide area, ignite from a remote source, and cause extensive injury and property damage.

2. Any surfaces that are exposed to the gas fire must be kept cool. If the exposure is a pressurized container, then a BLEVE is a possibility. Large quantities of water are necessary to cool the vessel. (See explanation of a BLEVE in the following paragraph.)

3. Under the cover of protective streams, attempt to shut off the fuel supply.

4. If the valve cannot be closed, then consideration should be given to controlled burning to allow the fuel to be consumed.

BLEVE

A BLEVE is an acronym for a Boiling Liquid Expanding Vapor Explosion. BLEVEs can occur in pressurized vessels containing the liquefied version of the gas under pressure. Following is the sequence of events that causes a BLEVE to occur:

1. A tank or container with a liquefied gas under pressure has a combination of the liquid and gas inside (Figure 5-25). The tank contains a relief valve to relieve *normal* excess pressure. The tank could also contain a flammable liquid in a closed container. There will also be gas in the vapor space of the tank, under somewhat less pressure.

2. The flame of a fire external to the tank or container impinges on the tank as an exposure (Figure 5-26).

3. The exposure fire causes a heat rise within the tank. The liquid begins to heat up, creating an increased amount of gas. This, in turn, increases the pressure within the tank, opening the relief valve when the set pressure is reached (Figure 5-27).

4. If the exposure fire continues to heat the liquid, more and more gas will be generated and escape from the relief valve.

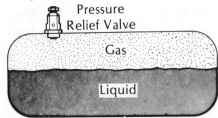

Figure 5-25. Liquefied gas under pressure

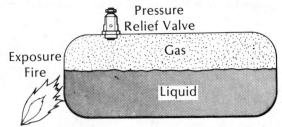

Figure 5-26. Flame impingement on liquefied gas tank

Eventually, the liquid level will fall below the point at which the flame is contacting the container. If the heat continues, then weakening of the metal will occur (Figure 5-28), because the liquid is no longer there to remove the heat.

5. Finally, when the metal has fatigued sufficiently so that the internal pressure exceeds the breaking strength, a BLEVE occurs (Figure 5-29). The breakup occurs violently, releasing the remaining liquid at a single instant. The parts of the shell can rocket for thousands of feet.

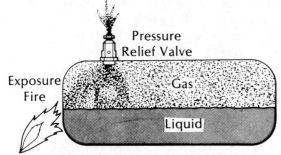

Figure 5-27. Relief valve operated by increased pressure

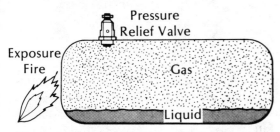

Figure 5-28. Exposure fire striking only gaseous area

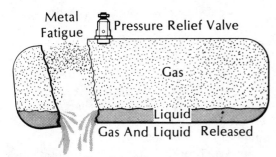

Figure 5-29. Conditions in which BLEVE occurs

Closed Containers

While the term BLEVE is used to describe a specific occurrence due to an exposure fire on a pressurized liquefied gas container, violent rupture can also occur in gas containers. This rupture occurs when the external heat causes a pressure buildup within the container faster than it can be reduced by the relief valve. When the pressure exceeds the strength of the weakest part of the container, rupture occurs, and there is a sudden release of the confined gas. As in the case of the BLEVE, the container can rocket many thousands of feet.

Liquefied Petroleum Gas

The term *liquefied petroleum gas* or LPG is used to describe any product composed of any of the following hydrocarbons or mixtures of them:

1. Propane
2. Propylene
3. Butane (isobutane, normal butane)
4. Butylenes

Although these are gases at normal atmospheric pressures, they are shipped and stored under pressure as a liquid. While LPG gases are considered nontoxic, they can cause asphyxiation in high concentrations due to exclusion of oxygen. In slightly lower concentrations, the elimination of oxygen can cause nausea and headache.

LPG, when produced, is virtually odorless. In order to assist in the discovery of leaks, an odorant is added. As an example of the problems that can be created, when converting propane from the liquid to gaseous state, it expands about 270 to 1.

The physical properties of LPG are listed in Table 5-12.

The liquefied petroleum gases can be stored or shipped in containers that range in size from small ones for household use as a

Table 5-12. Physical Properties of LPG

| Product | Ignition Temperature | | Flammable Limits, % | Specific Gravity | Boiling Point | | Vapor Density |
	°F	°C			°F	°C	
Propane	871	466.1	2.4 to 9.5	0.5	−44	−42.2	1.5
Propylene	927	497.2	2.0 to 11.1	—	−54	−47.8	1.5
Butane							
Iso	864	462.2	1.8 to 8.4	0.56	11	−11.7	2.0
Normal	761	405.0	1.9 to 8.4	0.58	31	− 0.6	2.0
Butylenes	725	385.0	1.6 to 10.0	—	19 to 39	−7.2 to +3.9	1.9

heating torch to giant 48,000 gallon (190,000 liters) water capacity tank cars. In addition, they can be shipped by pipeline between producers, terminals, and refineries. The storage and distribution system is shown in Figure 5-30.

In cylinder-filling plants a chart is used to determine the maximum weight for filling cylinders. This chart shows the permissible number of pounds of gas of a given specific gravity that may be placed in a given type cylinder. This becomes extremely important because LPG tanks are never completely filled. Approximately 15 percent of the gross capacity of the tanks is used as space for vapor to form.

For large containers, a rotary fixed-length dip tube or magnetic gauge is usually used to indicate the quantity that may be placed in the container for various temperatures and specific gravities. The temperature of the LP gas is measured by means of a thermometer, and the specific gravity is given on the bill of lading, waybill, or shipping ticket. On the basis of these two figures, the gauge is set at the maximum volume and the container is then filled to that point. The maximum volume is indicated when a frosty liquid spray emerges from an opening at the end of the gauge.

Smaller containers are often fitted with a fixed-length dip tube, the length of which is determined by necessary calculations. The maximum volume is indicated when a frosty liquid spray emerges from an opening located

at the top of the dip tube. At that point the liquid going into the container is shut off and the dip tube outlet is closed. The sliding dip tube (slip tube) and the rotary gauge are other gauges commonly used.

All cylinders for LPG are provided with overpressurization devices. Spring-loaded relief valves (Figure 5-31) and/or fusible plugs are used. Some large storage tanks contain two, three, or even four relief valves in one assembly. In this way one of the relief valves can be removed for testing or repair without emptying the product from the storage tanks.

Vinyl Chloride

Vinyl chloride is a colorless, invisible gas under normal conditions. At low concentrations, it is odorless. Vinyl chloride is shipped as a liquid under pressure or under refrigeration (below 7°F or −13.9°C), or in some cases under pressure and refrigeration.

Vinyl chloride is used in the manufacture of products made of polyvinyl chloride, such as wrapping film, electrical insulation, pipe, conduit, and records. Since it is used only in manufacturing, emergency response personnel must be concerned with it during transportation or storage at a plant. It is normally shipped by railroad tank car or barge. As with the other gas containers, they have the full range of safety relief valves.

Vinyl chloride can be an extreme health hazard. Current evidence indicates that exposure to vinyl chloride fumes can cause cancer many years after exposure. In its refrigerated form, the extreme cold can result in thermal "burns". The concentration and exposure hazard from vinyl chloride is shown in Table 5-13.

The physical properties of vinyl chloride are contained in Table 5-14.

As with the leaks or fires in the other gases, the best method is to attempt to control the leak. However, because of the health hazard, full breathing equipment is essential.

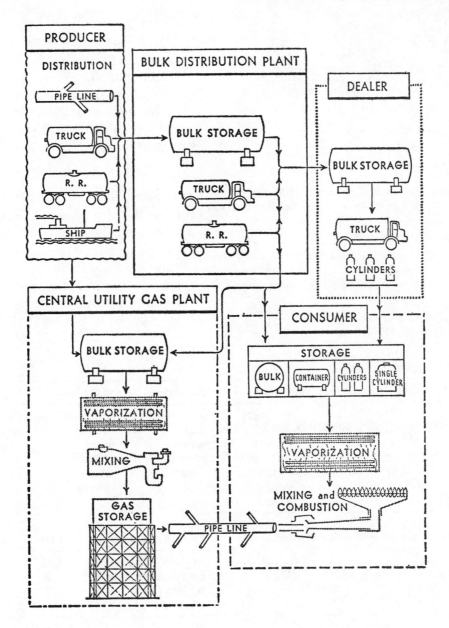

Figure 5-30. LPG distribution and storage system. *Courtesy of the American Insurance Association*

Figure 5-31. LPG relief valve

Natural Gas

The major ingredient of natural gas is methane. This gas receives wide distribution by pipeline and distribution mains for both consumer and industrial use.

Methane is considered nontoxic, but it can kill by the exclusion of oxygen. In its natural state, methane is odorless and therefore an odorant is added.

The physical properties of methane are listed in Table 5-15.

The Michigan Gas Association has published a booklet that gives suggested actions for emergency response personnel during leaks or fires from natural gas.[9] The following procedures are quoted from their text.

Gas Escaping, Burning Indoors

1. Clear building of all occupants.
2. Let the gas flame burn until the gas supply can be shut off.
3. Shut off the gas supply at the valve ahead of the meter or at the underground service line valve, if possible.
4. Protect nearby combustibles.

Gas Escaping, Burning Outdoors

1. Clear the area of people, barricade the area, and stand by at a safe distance.
2. Let the gas burn. Make no effort to extinguish the flame itself.
3. Quench fire in, or prevent ignition of, nearby combustible material by using water. Fog spray is preferable to straight streams.

Table 5-13. Health Hazards of Vinyl Chloride

0-1 ppm	1-5 ppm	260 ppm	500 ppm	10,000 ppm	20,000 ppm
No odor. Considered safe.	No odor. Unsafe without respirator for more than 15 minutes.	Can detect sweet etherlike odor above 260 ppm, which will dull the sense of smell. Unsafe.	Dizziness may occur in several minutes. Unsafe.	Unconsciousness may occur in several minutes. Unsafe.	Death may occur in several minutes. Acutely toxic.

Table 5-14. Physical Properties of Vinyl Chloride

Ignition Temperature	882°F	(472.2° C)
Flammable Limits	3.6% to 33%	
Specific Gravity	0.9	
Boiling Point	7°F	(-13.9° C)
Vapor Density	2.2	

Table 5-15. Physical Properties of Methane

Ignition Temperature	999°F	(537.2° C)
Flammable Limits	5.0% to 15.0%	
Boiling Point	-259°F	(161.7° C)
Vapor Density	0.55	

4. Keep metallic structures in the flame area cooled with water.

Gas Escaping, Not Burning Indoors

1. Clear the building of all occupants.

2. Shut off the gas supply at the valve ahead of the meter or at the underground service line shutoff valve, if possible.

3. Eliminate sources of ignition.

4. Ventilate the building only by opening doors and windows.

5. Stay out of the building at a safe distance and keep others out and away from the area.

6. Stand by until the gas company has controlled the leak and the building is free of gas.

Gas Escaping, Not Burning Outdoors

1. Clear the area of people, barricade the area, and stand by at a safe distance.

2. Close windows, doors, or other openings in nearby buildings if gas is escaping at or near building walls.

3. Remove or extinguish all open flames, prohibit smoking in the area, and take any other precautions necessary to prevent ignition.

4. Restrict or reroute traffic until gas company personnel bring the gas flow under control.

5. Keep fire hoses and equipment ready in

the event escaping gas ignites.

Gas Escaping, Not Burning, Location Unknown

1. DO NOT allow persons to reenter buildings or areas suspected of containing escaping gas until the gas leak is under control and the building has been thoroughly ventilated.

2. DO NOT allow open flames, or smoking, or spark producing devices in either open or closed areas if the presence of combustible gas is known or suspected.

3. DO NOT ring doorbells, operate electric light switches, or use telephones in areas in which the presence of escaping gas is suspected or known.

4. DO NOT use power fans or exhausters to ventilate buildings.

5. DO NOT open doors to stoves, furnaces, or other closed appliances containing a live flame or live coals.

6. DO NOT open a CLOSED gas valve at any time or at any place. Gas valves should not be opened by anyone except GAS COMPANY EMPLOYEES.

7. DO NOT enter a pit, vault, manhole, or sewer unless absolutely necessary, and only after proper safety precautions have been taken.

8. DO NOT operate flashlights unless approved for hazardous atmospheres.

FLAMMABLE LIQUIDS

Flammable liquids are among the most common hazardous products stored, transported, and used. Everything from lawnmowers in a home garage to military jet tankers carrying thousands of gallons of gasoline contain flammable liquids. To this must be added solvents and alcohols, which are also flammable liquids.

In accordance with the Department of Transportation definition, a flammable liquid is one that has a flash point below 100°F (37.8°C).

Since there are hundreds of different flammable liquids, they will be covered only in a general manner. Information on specific hazards can be obtained from the various reference sources. A few of the common liquids are explained here in detail.

Handling a Flammable Liquid Incident

Flammable liquid incidents can involve a leak with or without ignition. The following generalized procedures are given for handling the incident:

Unignited Leak

1. Evacuate personnel downwind and downhill from the leak. Use caution in locating and positioning apparatus and personnel. Keep sightseers and nonrequired emergency response personnel away from the scene.
2. Identify the leaking product.
3. If possible, attempt to contain the leak within natural or artificial barriers (Figure 5-32). Try to divert flow from exposures. Try to prevent liquid from entering sewer system.
4. Eliminate possible ignition sources.
5. Attempt to close valves or stop the leak. Personnel making the attempt should be protected from a possible ignition of the vapors. If possible, the spill should be

covered with a foam to reduce vapor production. Special devices can be used to attempt to close the hole or leak, as demonstrated in Figure 5-33.

Ignited Leak

1. Keep personnel and apparatus upwind and on higher ground than the liquid.
2. Identify the leaking product.
3. If possible, attempt to contain the leak as in procedure (3) given for unignited leaks.
4. Attempt to stop the leak as in procedure (5) given for unignited leaks.

Figure 5-33a.   Leak stopper

Figure 5-32.   Artificial flammable liquid polyurethane barrier

Figure 5-33b.   Leak stopper expanded against opening

5. Use water streams to keep storage tanks cool to prevent a BLEVE. Flush burning liquids out from under storage tanks. Remember, unmanned master streams can be used to cool tanks.

6. Stay away from the ends of the storage tanks, as indicated in Figure 5-34. Since the tank can swivel in event of a BLEVE, an area 30° from the horizontal should also be kept clear. It is important to note that this does not imply that an attack from the sides is safe. Tanks have been known to swivel 90° and overrun side positions. Use unmanned streams wherever possible.

7. Coordinate use of fog streams. One crew should not try to flush product out from under the tank while the other crew pushes it back.

8. The change in pitch of the escaping gas from the relief valve can indicate a build-up of pressure. Move personnel back should this occur.

9. Protect the steel supports of storage tanks to prevent weakening and collapse due to heat.

10. Apply correct extinguishing agent for product involved.

11. Always keep personnel safety in mind. Make sure an escape path is always available. Keep backup lines ready. Keep

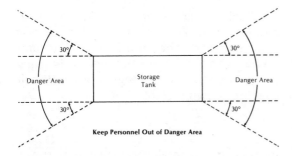

Figure 5-34. Fire attack on flammable liquid storage tank

apparatus headed in the direction of escape.

Gasoline

Gasoline is composed of a blend of several hydrocarbons with other additives to make the product usable with current engines. The hydrocarbon portions of gasoline are colorless. While the fumes are not harmful, continued breathing of the vapor can cause lung irritation.

The physical properties of gasoline are listed in Table 5-16.

Ethyl Alcohol

Ethyl alcohol is one of the most common flammable liquids found in the home. It is the component of whiskey, when it is mixed with water and flavorings. It is also used in many industrial and manufacturing processes.

This particular type alcohol does not have an adverse impact on health, except when consumed to excess. However, from the point of view of emergency response personnel, it will not present a great health hazard. In large quantities, the liquid can cause collapse of the respiratory system. The vapor in high concentrations can cause some irritation.

The physical properties of ethyl alcohol are listed in Table 5-16. Since it is very water soluble, large amounts of water can be used to dilute the liquid.

Benzene

Benzene is a clear, colorless liquid which is widely used as a chemical intermediate and as a constituent of such products as gasoline, paint removers, fumigants, insecticides, rubber cement, and various solvents. It has a strong, pleasant odor. Other names for benzene are benzol and coal naphtha.

Breathing high concentrations of benzene vapors will have an immediate, harmful effect on life. A 2 percent by volume of the vapor in air can kill in 5 to 10 minutes. Exposures of 50 ppm to 5,000 ppm for periods of time can result in feelings of euphoria followed by giddiness, headache, nausea, and if exposure continues, death. Breathlessness, nervous irritability, and unsteadiness in walking may persist for 2 to 3 weeks after the acute exposure. In addition to the problems of the vapor, the liquid will cause a drying and reddening of the skin.

The physical properties of benzene are listed in Table 5-16.

Table 5-16. Physical Properties of Selected Flammable Liquids

Product	Flash Point, °F	Flash Point, °C	Ignition Temperature, °F	Ignition Temperature, °C	Flammable Limits, %	Specific Gravity	Boiling Point, °F	Boiling Point, °C	Vapor Density
Gasoline	-36 to -45	-37.78 to -42.78	536 to 853	280.00 to 456.11	1.4 to 7.6	0.8	100 to 400	37.78 to 204.44	3 to 4
Ethyl Alcohol	55[a]	12.78	793	422.78	4.3 to 19	0.8	173	78.34	1.6
Benzene	12	-11.11	1,040	560.00	1.3 to 7.1	0.9	176	80.00	2.8

[a]Depends on proof of liquid (percent of alcohol in water); 55° F (12.8° C) is for 100% alcohol. Flash point increases as amount of alcohol decreases.

COMBUSTIBLE LIQUIDS

The Department of Transportation defines a combustible liquid as one with a flash point between 100°F (37.8°C) and 200°F (93.3°C). Normally, liquids in this classification will not produce vapors at ordinary temperatures, but emergency response personnel must still exercise caution, because flammable vapors are produced under fire conditions. Do not get lulled into a false sense of security, thinking that no hazard is present when the material spills or leaks.

The procedures for handling a spill or fire of a combustible liquid are the same as for a flammable liquid, given in the preceding section.

Kerosene

Kerosene is a mixture of several hydrocarbons. Known by several names, range oil, number 1 fuel oil, and coal oil, it is pale yellow to white in color, with a strong odor.

The only health problem presented by kerosene is skin irritation from the liquid. Otherwise, the vapors are not particularly irritating. Obviously, prolonged breathing of high concentrations should be avoided.

Table 5-17 lists the physical properties of kerosene.

Fuel Oil

Fuel oils are formulated in various amounts of thickness to be used in specific type appliances. The various grades are numbered, with the higher number designating the heavier fuel. The designations and their uses are shown in Table 5-18.

FLAMMABLE SOLIDS

With the increasing technological advances of modern times, a never-ending number of uses are being found for combustible solids. These materials, which may be metallic or nonmetallic, are used in more and more industries.

The alkali metals are being used as heat transfer media in thermoelectric devices, as

Table 5-17. Physical Properties of Selected Combustible Liquids

Product	Flash Point, °F	Ignition Temperature °F	Ignition Temperature °C	Flammable Limits, %	Specific Gravity	Boiling Point °F	Boiling Point °C
Kerosene (No. 1 Fuel Oil)	100 minimum	410	210.0	0.7 to 5	< 1	300 to 575	148.9 to 301.7
Fuel Oil							
No. 2	100 minimum	494	256.7	—	< 1	—	
No. 4	130 minimum	505	266.8	—	< 1	—	
No. 5	130 minimum	—		—	< 1	—	
No. 6	150 minimum	765	407.2	—	1	—	

Table 5-18. Fuel Oil Designations

Number	Designations	Uses
1	Kerosene	Small appliance fuel, cleaner, degreaser, solvent. When mixed with gasoline, used as a jet fuel.
1D	Light diesel fuel	Vehicle fuel
2	—	Domestic fuel oil
2D	Medium diesel fuel	Vehicle fuel
4	Light industrial	Industrial furnaces
5	Medium industrial	Industrial burners
6	Heavy industrial	Industrial burners

lubricants, and hydraulic fluids. Flammable metals are numerous in foundries, metal-working plants, electroplating and electro-polishing operations. Other flammable solids are used in agriculture as pesticides and fertilizers, and the military and construction industry use thousands of tons of solid explosives.

Flammable Metals

Metals are defined as chemical elements or alloys of such elements that have luster, can conduct heat and electricity, have tensile strength, and will form oxides. Many metals have fire hazardous properties and nearly all metals will burn under certain conditions.

Some metals, such as magnesium, are highly combustible, because of the ease of ignition of thin sections, fine particles, or molten metal. The same materials in large solid forms may be comparatively difficult to ignite. Other metals which are not normally thought of as combustible, such as aluminum and steel, will ignite and burn when in a finely divided form. Therefore, the size, shape, quantity, alloy, and degree and intensity of the ignition source are important factors when evaluating metal combustibility.

Dust clouds of metals may burn with explosive force. Since the dusts are easily ignited, some merely by friction, and are more reactive when in powdered form, some are shipped and stored under inert gas or liquid regulations to reduce fire risks. In occupancies where metal dust has accumulated, careful direction of hose streams is necessary.

Nonmetallic Flammable Solids

Nonmetallic flammable solids are those solids that have an unusually high burning rate, need special extinguishing techniques, or produce extremely toxic fumes while burning.

Examples of nonmetallic flammable solids include camphor, red phosphorus, and sulfur.

Handling a Flammable Solid Incident

The successful control and perhaps extinguishment of a metal fire depends on the state and quantity of the metal, the extinguishing agent available, and the skill and experience of the fire fighter in applying it. No one agent will effectively control or extinguish all metal fires. Small fires can be handled by using special dry powders. An adequate supply of dry powders, however, may not be available for larger fires. Combustible metals burning in large quantities react violently with water; thus, water generally should be used for exposures only. Because the combustion products and smoke are toxic and/or corrosive, full protective clothing and self-contained breathing apparatus must be worn.

Aluminum

Aluminum will burn when in dust, powder, or flake form from such operations as grinding, buffing, or polishing. As the dust is suspended in air, it may explode and cause serious damage. Carbon dioxide, dry chemical, and water are ineffective and will stir up the dust. Carbon tetrachloride will cause the dust to explode. Extinguish with fine dry sand or dry powder.

Antimony

Antimony, a silvery white metal in fine particles, has a low ignition temperature and burns with a bright bluish flame. In molten state it oxidizes and reacts with water to evolve hydrogen; the hydrogen then reacts with the antimony to form highly toxic stibine gas. Heated antimony emits toxic fumes.

Beryllium

Beryllium, a light steely metal, is less flammable than other metals, but the powder can form explosive mixtures with air. It is extremely toxic, and inhalation of dust or fumes or contact with the skin can be fatal. Approximately 25 percent of those exposed die. Symptoms such as skin irritation, burning eyes, unhealing ulcers, coughing, short breath, or loss of weight can be delayed. Early fluorescent tubes contained beryllium and should be handled as a poison. Extinguish with dry powder, *not water*, and wear full protective clothing, including self-contained breathing apparatus. After a fire, shower, wash hair, and clean fingernails. Contaminated clothing should be washed separately and equipment scrubbed thoroughly.

Cadmium

Cadmium is a grayish-white powder or bluish-white soft metal used for metal coating. It reacts with oxidizing agents and releases hydrogen, and will ignite in dust form. When heated, it oxidizes and emits dense, brown poisonous fumes. High concentrations can cause death by swelling of the lungs.

Camphor

Camphor is a white or colorless crystalline substance and is used in medicines, toothpastes, embalming fluid, and the manufacture of cellulose-nitrate plastics and insecticides. The vapors, which may be irritating, are similar to naphthalene. For extinguishment use carbon dioxide, dry chemical, or standard foam.

Cesium

The most reactive of all metals, cesium is pyrophoric. It reacts with cold water explosively and evolves hydrogen. Cesium burns without a visible flame and with little smoke. It is the strongest base known and will deteriorate flesh on contact.

Lithium

Lithium, a silvery metal, is the lightest solid element. It reacts with water or air exothermically and with hydrogen at ordinary temperatures. Near its melting point, lithium will ignite and burn with an intense white flame. A white, opaque smoke that can fill an area is emitted. Although water is decomposed there is not enough heat generated to ignite the hydrogen. However, molten or burning lithium reacts explosively with water or moisture laden materials to produce lithium hydroxide, which is caustic to eyes, skin, and mucous membranes. Lithium can burn in air, oxygen, carbon dioxide, or nitrogen which means that extinguishers propelled by nitrogen or carbon dioxide cannot be used. Lithium also reacts violently with foam, vaporizing liquids, and dry chemical extinguishing agents. Met-L-X can be used on spill fires and Lith-X or G-1 for spill fires or fires in depth.

Magnesium Chips or Ribbons

In the ribbon form, magnesium ignites quickly; loose chips from sawing, drilling, or turning ignite readily, but a compact pile of chips is more difficult to ignite. Fine chips in the presence of water may ignite spontaneously. These fires can be controlled by covering the magnesium with a layer of Class D extinguishing powder. Since the usual means of application is by shovel, this procedure is generally limited to smaller fires. Larger fires can be attacked with special appliances developed to expel the powder from a safe distance. Extinguishing agents include graphite, G-1, Metal Fyr, X-8, M-1, and Met-L-X powders. Since they do not absorb moisture from the air, storage is not a problem. These products extinguish by smothering, forming a hard cake over the chips, or by cooling them. Water would be used to protect exposed areas. The impact of a stream of water on the burning magnesium would cause considerable scattering of molten and burning particles.

Magnesium Dust

Because magnesium is subject to violent dust explosions, this is the most serious hazard. Any small fire can serve as an ignition source where there is any quantity of magnesium suspended in the air. The result would be an immediate explosion with little chance of initiating effective fire fighting. In occupancies where magnesium dust is present, care should be used in directing hose streams so that dust on shelves, fixtures, ledges, or beams is not disturbed.

Magnesium Solid

There is often difficulty in obtaining or applying extinguishing agents to solid magnesium. For this reason, water can be used for control if there are no immediate exposures and ample water supplies. Fires involving ingots, castings, sheets, or extruded shapes can be fought with water, using straight streams at low nozzle pressures. The water will cool the magnesium mass below its ignition temperature if applied quickly. Control will be brought about by directing the stream on the exposed magnesium, but not on metal involved in the fire. Since streams striking the burning magnesium will cause a violent reaction, they must be directed from a safe distance on the unburned metal. Cooling water that flows into the burning mass will also cause reactions. Water can also be used for extinguishment if the intention is to remove the fuel by accelerated burning. With large quantities of burning magnesium heavily involved in fire inside a building, an exterior attack should be mounted with heavy streams used from safe positions. Particular attention must be given to exposures and safe operating distances while burning magnesium is within reach of the streams because of possible violent reactions.

Naphthalene

Naphthalene smells like mothballs and comes in volatile white crystals or flakes. This product is used in the manufacture of dyes, resins, fungicides, explosives, and, of course, mothballs. Since it is toxic, avoid contact with the liquid or vapors, which are irritating to skin, eyes, and respiratory tract. Self-contained breathing apparatus must be worn. Water, carbon dioxide, dry chemical, or ordinary foam can be used in extinguishment. When using water or foam on molten naphthalene, expect extensive frothing.

Phosphorus, Red

Red phosphorus is a reddish-brown powder used as a source of pure phosphorus for the manufacture of phosphoric acid, fertilizers, insecticides, and matches. It will explode when mixed with oxidizing materials. Fumes of burning phosphorus are very irritating, but only slightly toxic unless high concentrations are encountered. Extinguish by flooding with water, being careful not to scatter the burning material. Then cover with wet sand or dirt. Extreme care must be used in overhaul because reignition can occur and at high temperatures red phosphorus can revert to more hazardous white phosphorus.

Phosphorus, White or Yellow

White or yellow phosphorus is a colorless or white, soft waxy solid that turns yellow on exposure to light. White phosphorus is used in poisons and incendiary bombs. Generally found in stick form, it ignites spontaneously on contact with air at or above 86°F or at room temperature when finely divided (Figure 5-35). It is explosive when mixed with oxidizing materials.

Solid phosphorus causes severe burns on contact with the skin. Small amounts inhaled or ingested can be fatal. The highly irritating fumes of burning phosphorus are slightly toxic, but high concentrations are dangerous. Avoid skin or eye contact with phosphorus as it is corrosive, and wear fire-resistant full protective clothing and self-contained breathing apparatus. Extinguish with copious amounts of water until the phosphorus has solidified. Phosphorus has a high specific gravity, is not soluble in water, and will not burn under water. Use care not to scatter the burning material. Then cover it with wet sand or dirt. Care must be used in overhaul to avoid reignition.

Figure 5-35. Yellow phosphorus in stick form ignites spontaneously into running spill fire when it dries out

Potassium

Potassium is a silvery white metal that is somewhat more reactive than sodium. It ignites in warm air and gives off an irritating greenish smoke. Potassium reacts violently on contact with water; this reaction evolves hydrogen, which will ignite. Water, carbon dioxide, vaporizing liquid, or foam should not be used. Met-L-X or G-1 dry powders are recommended for spill fires or fires in depth.

Plutonium

Plutonium is a pyrophoric, toxic, radioactive metal which may ignite spontaneously or by friction. Do not use water, but extinguish with inert gas.

Rubidium

Rubidium is a silvery, slightly radioactive metal which is very reactive. It ignites spontaneously in air and burns with a blue flame, which leaves a dark brown to black residue. Rubidium reacts similarly to sodium and decomposes water, forming a strong alkali that will destroy human flesh.

Sodium

Sodium is a very light, waxy-looking silver metal that loses its luster and turns grayish-white in air. It may be found where chemicals, drugs, soaps, cosmetics, hollow valve stems, and nonglare street lights are manufactured or in refineries and school chemistry laboratories. Finely divided and molten sodium will ignite in air.

Sodium reacts with vaporizing liquid, dry chemical, foam, and carbon dioxide extinguishing agents. At normal temperatures sodium decomposes water explosively, forming sodium hydroxide (lye), a caustic, and releasing hydrogen, which ignites. Sodium in contact with the skin causes alkali and thermal burns. Burning sodium releases highly irritating sodium oxide fumes. The fumes form caustic soda, which is injurious to skin, eyes, and mucous membranes. Flush the skin with copious amounts of water for one-half hour if contacted. Full protective clothing and self-contained breathing apparatus are imperative at sodium emergencies. Extinguish by using dry powder, sodium chloride, graphite, sand, or other *dry* inert substances.

Sulfur

Sulfur is a solid in the form of a powder or yellow crystals and used to make various compounds, in rubber production, as a fungicide, and in the manufacture of black powder. It is readily ignitable; the dust or vapors form explosive mixtures with air; and sulfur in contact with oxidizing materials forms explosive mixtures. The product of sulfur combustion (sulfur dioxide) is a toxic gas. Extinguish with fog lines, avoiding straight streams, which will scatter the molten sulfur or dust. Sulfur is not water soluble and has a specific gravity of 2.0.

Thorium

Thorium metal is used in lamp mantles and alloys of magnesium. In dry powder form it has a low ignition temperature and can be exploded. It is subject to ignition from an electrostatic spark. Do not use water for extinguishment, but an inert gas such as helium or argon.

Titanium

Titanium is a lightweight, high-strength metal for equipment requiring high resistance to chemical or atmospheric corrosion. It is widely used in aircraft construction. The powder form is pyrophoric and can burn in atmospheres other than air; chips and turnings are subject to spontaneous heating, especially when covered with oil. Small amounts can be extinguished by submerging in water. Water should not be applied because of the violent reaction, but dry sand, graphite, or certain dry powders can be used effectively.

Uranium

Uranium is used as a fuel for nuclear reactors. It gives off radioactivity and products of combustion that are a serious health hazard. In dust form it is pyrophoric; slight moistening accelerates spontaneous heating. Extinguish by submersion or flooding.

Zirconium

Zirconium is a dark-gray powdered metal used in vacuum tubes, flashbulbs, flares, fireworks, detonators, and metal alloys. Fine particles may be ignited by heat, static electricity, or friction, and dust clouds may ignite spontaneously with explosive force. In the powder form it is commonly handled wet because it is more difficult to ignite; however, once ignited, it will burn more violently than dry. Dry sand, G-1, Met-L-X, or large volumes of foam will control zirconium fires.

London, Ontario, Molten Zinc Fire, December 2, 1970

A fire in a London, Ontario, die casting plant could have developed into considerable disaster, had fire fighters used water indiscriminately.[10] There was a fire in the ductwork, which spread through the blower pipe. Directly below the blower pipe was 60,000 pounds (27,000 kilograms) of molten zinc.

Fearing that water cascading down onto the zinc would break the surface, react violently, and cause a serious explosion, fire fighters worked carefully from ladders. In less than an hour the fire was extinguished, without involvement of the zinc. An awareness of the properties of the flammable solid helped to avoid injury and property damage.

Anchorage, Alaska, Tank Truck Fumes,[11] 1976

An aircraft fuel delivery truck had a faulty internal valve which was in need of repair. The truck had a 10,000-gallon (40 cubic meters) capacity and had 6 to 12 inches (15 to 20 centimeters) of fuel in the bottom.

A repair worker, wearing SCUBA type breathing apparatus without a face piece, entered the tank through a fill hole. Another individual remained on the top of the tanker. Shortly after the repair worker entered, the worker on the top noted that there was no noise from the breathing apparatus. The fire department was called. The repair worker on the inside was face down in the remaining pool of fuel.

The first individual to respond, an airport fire fighter guard, entered the tank without breathing apparatus, managed to reach the victim, and then collapsed unconscious. Other fire fighters now arriving on the scene began the rescue of all personnel on the inside.

The rescuers working on the inside with masks, located the fire fighter guard in the fourth compartment. A mask was passed in for the unconscious victim and the rescue was made. Then, using a mask with a long hose and keeping the bottle outside the tank, rescuers were able to recover the body of the repairman.

The toll was very high. The repair worker was dead. The initial rescuer had severe inhalation burns to the mouth and throat as well as chemical burns of the body. The next two rescuers were treated for inhalation problems. Finally, the last four rescuers were treated for chemical burns and inhalation of fumes.

Hazards presented by flammable liquids and gases result not only from their combustibility, but also from the toxic nature of their fumes. Personnel must have specialized equipment available or at least have noted its location well in advance of the incident.

QUESTIONS FOR REVIEW AND DISCUSSION

1. Define the term cryogenics and explain why emergency response personnel must be familiar with it.

2. Explain the relationship between critical temperature and critical pressure.

3. How are cryogenic containers insulated?

4. What are the hazards to emergency response personnel from liquid hydrogen? liquid helium? liquid oxygen? liquid nitrogen? liquid argon? liquefied natural gas?

5. What methods are used to ship compressed gases?

6. Explain the two different types of labels that can be used on compressed gas cylinders.

7. What are the hazards to emergency response personnel from anhydrous ammonia? chlorine? oxygen?

8. From an ignition point of view, how do flammable liquids differ from flammable gases?

9. Prepare a list of the flammable gases shipped or stored in your area, including their flammable limits.

10. Outline the basic steps which must be taken when handling an unignited leak of a flammable gas.

11. Outline the basic steps which must be taken when handling an ignited leak of a flammable gas.

12. Define a BLEVE and explain how it happens.

13. Describe the different gases which make up liquefied petroleum gas and list their physical properties.

14. Describe the various techniques which are used to protect compressed gas cylinders from overpressurization.

15. What are the hazards to emergency response personnel from vinyl chloride?

16. Outline the basic steps which must be followed when handling a flammable liquid incident.

17. Explain the significance of the attack position when handling an incident involving a closed container exposed to fire.

18. What is the difference between a flammable and a combustible liquid as defined by the Department of Transportation?

19. What factors affect the ability of a flammable solid to burn?

20. Outline the steps for handling a flammable solid incident.

21. Select one of the flammable solids listed in the chapter and prepare its (a) physical properties, (b) health hazards, (c) storage methods, (d) uses, (e) locations where it can be found.

NOTES

1. R. E. Cole, R. S. Magee, and J. W. Hollenberg, *Hydrogen Storage and Transfer*, National Technical Information Service AD-A016 256, August 25, 1975 (Washington, D.C.: Stevens Institute of Technology), p. 16.

2. Ibid., pp. 38–44.

3. Ibid., p. 37.

4. P. M. Ordin, "Review of Hydrogen Accidents and Incidents in NASA Operations," *9th Intersociety Energy Conversion Engineering Conference*, San Francisco, August 1974, p. 442.

5. L. H. Cassutt, F. E. Maddocks, and W. A. Sawyer, "A Study of the Hazards in the Storage and Handling of Liquid Hydrogen," in *Advances in Cryogenic Engineering*, vol. 5 (New York: Plenum Press, 1960), p. 55.

6. D. Allan, et al., *Technology and Current Practices for Processing, Transferring and Storing Liquefied Natural Gas* (Cambridge, Mass.: Arthur D. Little, Inc. December 1974), p. 1.

7. M. A. Elliott, et al., *Report on the Investigation of the Fire at the Liquefaction, Storage, and Regasification Plant of the East Ohio Gas Co.*, Cleveland, Ohio, October 20, 1944.

8. A. E. Howerton, *Estimating Area Affected by a Chlorine Release*, no. R71 (New York: Chlorine Institute, 1969).

9. Michigan Gas Association, *The Fireman and Gaseous Fuels* (Ann Arbor, Mich., 1966).

10. Frank Bell, "Water Can Blow It!" *Fire Command*, May 1971, p. 24.

11. James R. Evans, "Inside-tank Rescue Claims Life, Injures Fire-Fighters," *Western Fire Journal*, February 1977, pp. 18–19.

6
Identification of hazardous materials

When emergency response personnel arrive on the scene of a hazardous materials incident, one of the first things to be determined is the product involved. Hazardous products being transported within the United States are subject to the requirements of the U.S. Code of Federal Regulations as administered by the Department of Transportation (DOT).

If the product is a drug, markings are also regulated by the Food and Drug Administration. Pesticides and agricultural chemicals are marked in accordance with the requirements of the Environmental Protection Agency. Emergency response personnel must be familiar with these varied federal regulations so that they can interpret product information.

In addition to these requirements, several local governments have adopted individual marking requirements for hazardous materials stored under their jurisdiction. The most common system used under these conditions is the National Fire Protection Association's 704 marking labels. Even if this system is not mandatory, some industrial firms adopt it

voluntarily for use within their facility. Emergency response personnel must, therefore, also be familiar with its meaning.

Shipping papers for those materials in transit are another source of information for the emergency response team. Types of documents and the technique for interpreting them are described in this chapter.

Many reference sources are available to provide the incident commander with specific information on the hazards involved. These sources range from special telephone numbers which emergency response personnel can call to receive information, to text and reference books. Each of these sources has both strong and weak points. They can provide certain information quickly, while other particular problems might involve long delays. For this reason, the incident commander must be familiar with all available reference sources to be able to decide quickly which ones to use in a particular incident. A discussion of the various information sources is contained in this chapter.

FEDERAL DEPARTMENT OF TRANSPORTATION REGULATIONS

For all modes of transport, the U.S. Code of Federal Regulations requires markings and/or labeling of containers carrying generic and certain specified hazardous materials. These requirements vary by mode of shipment, as explained in Chapter 1. Under the law there are two basic requirements for identifying hazardous materials shipping containers: placards and labels. Placards are affixed to the transporting vehicle on all four sides. Labels, on the other hand, are attached to the container holding the material on only one side. The major reasons for placing labels and placards on packages and vehicles are:

1. To provide an immediate warning of potential danger.
2. To inform the emergency response personnel of the nature of the hazard.
3. To indicate any required preventive action.
4. To minimize possible injurious effects if exposure to the product does occur.

Placarding

The placards to be used on transporting vehicles, trains, and trucks are specified by the Materials Transportation Bureau in the Code of Federal Regulations Title 49, Part 172, Subpart F, Paragraph 172.500.[1] A summary of the placarding requirements is included in the following paragraphs. Remember, the information following is just a summary, because there are many exceptions to these generalized requirements. Consult the actual code for a full legal interpretation.

1. A placard shall only be used when it accurately represents the material being transported as well as its hazard.
2. No other signs with the same shape, color, design, or content of any placard can be used on the vehicle or container. This en-

sures that there is no confusion for emergency response personnel.

3. Placards must be placed on each end and each side of the transporting vehicle or container, with the placards specified in the regulations.
4. A freight container, motor vehicle, or rail car containing two or more classes of materials requiring different placards may be placarded DANGEROUS in place of the separate placarding required. However, if 5,000 pounds (2,250 kilograms) or more of one class is loaded at one facility, the placard specified for that class must be applied.
5. Motor vehicles or freight containers having less than 1,000 pounds (450 kilograms) aggregate gross weight of some classes of hazardous materials do not have to be placarded. This exemption includes rail cars loaded with freight containers or motor vehicles. However, this exemption does not apply to portable tanks, cargo tanks, or tank cars or to transportation by air or water.
6. No motor carrier may transport a hazardous material in a motor vehicle unless the placards required for the hazardous material are properly installed and located.
7. No rail carrier may accept a rail car containing a hazardous material for transportation unless the placards for the hazardous material are properly installed and located.
8. Any freight container having a capacity of over 640 cubic feet (19.2 cubic meters) must be placarded except for the exemptions noted above.
9. Each placard must be visible from the direction it faces, except if coupled to another vehicle or car.
10. Each placard must be located at least 3 inches (7.6 centimeters) from any advertising or writing.

Labeling

The label required on the containers is specified by the Materials Transportation Bureau in the Code of Federal Regulations Title 49, Part 172, Subpart E, Paragraph 172.400.[2] A summary of the labeling requirements is included in the following paragraph. As in the information on placarding, the following is just a generalized summary, which has not taken into account the many exceptions.

1. You cannot offer for transportation nor can a carrier accept any package bearing a hazardous label unless:
   a. The package contains a hazardous material.
   b. The label represents a hazard of the hazardous material in the package.
2. You cannot transport a package with markings similar in color, design, or shape to a label to prevent confusion of package contents.
3. Multiple labeling requirements are quite complex. They are spelled out in detail in the next section.
4. Packages containing samples of chemicals for laboratory analysis which have not been classified may be labeled in accordance with the shipper's tentative class assignment.
5. Each label must be printed or affixed to the package near the shipping name of the material. However, the shipper can affix the label to a tag or use another suitable means if:
   a. The package has no radioactive material and has dimensions smaller than those of the required label.
   b. The package is a compressed gas cylinder.
   c. The package has an irregular surface to which a label cannot be satisfactorily affixed.
6. When two or more labels are required, they must be located next to each other.
7. Each label must be placed against a contrasting background or have a solid or dotted line outer border.
8. Each label must be displayed on at least two sides or two ends, excluding the bottom, of each package or freight container having a volume of between 64 and 640 cubic feet (1.9 and 19.2 cubic meters).
9. A label must not be obscured by markings or attachments.

Placards and Labels in Use

The classes of hazardous materials that can be transported are defined by the Department of Transportation, Materials Transportation Bureau regulations. These classes are explosives, flammable, oxidizing, corrosive, poisons, compressed gas, radioactive, and etiologic. The particular labels and placards for each of these classes of materials and their use for a combination of materials are discussed in the following paragraphs. In addition, special hazard labels as well as other regulated materials (ORM) are included in the discussion.

The Code of Federal Regulations, Title 49, Part 172, paragraph 172.101[3] provides a list of hazardous materials and their required labels and placards. A sample page is shown in Figure 6-1. However, if a hazardous material has more than one dangerous property and is not listed in paragraph 172.101, it is classed in accordance with only the most dangerous of its multiple properties. The most dangerous of its properties is determined in the order of the following list:

1. Radioactive
2. Poison A (gas)
3. Flammable gas
4. Nonflammable gas
5. Flammable liquid
6. Oxidizer
7. Flammable solid
8. Corrosive liquid

9. Poison B (liquid)
10. Corrosive solid
11. Irritating materials
12. Combustible materials

Note that both explosive and etiologic agents must be so identified at all times and use required labels and placards.

While the placarding and labeling requirements, which are discussed in detail in the following paragraphs, are designed to provide information to emergency response personnel, one of the problems is that some shippers do not meet the legal requirements. This creates difficulties in product identification for emergency response personnel in the event of a transportation accident.

The Office of Hazardous Materials Operations of the Department of Transportation reported on several known violations in their July 1976 Newsletter.[4] These include:

1. Failure to notify pilot of hazardous materials on board.

2. Accepting shipment of Class B poison improperly packaged and marked.

3. Failure to report leakage of a corrosive liquid on an airplane.

4. A hospital that offered a shipment of corrosive liquid, N.O.S., which was improperly packaged, labeled, and marked.

5. A shipper who improperly marked and labeled a package of hazardous materials.

6. A motor freight carrier who transported poison in the same vehicle with foodstuffs and transported a prohibited combination of hazardous materials.

7. A shipper who offered shipment of a flammable liquid with no shipper certification, improper description on shipping papers, improper marking of containers, and no labels.

8. A scientific firm that shipped a package containing titanium tetrachloride (corrosive) by air not properly packaged and marked, and not properly packed to prevent the escape of liquid.

Just because the package or shipping documents do not indicate that a hazardous material is on board is no reason for the incident commander to throw caution aside. Each shipment must be carefully checked to ensure that no hazardous materials are being transported.

§172.101   38

§172.101   Hazardous Materials Table—Continued

(1) */W/A	(2) Hazardous materials descriptions and proper shipping names	(3) Hazard class	(4) Label(s) required (if not excepted)	(5) Packaging		(6) Maximum net quantity in one package		(7) Water shipments		
				(a) Exceptions	(b) Specific requirements	(a) Passenger-carrying aircraft or railcar	(b) Cargo only aircraft	(a) Cargo vessel	(b) Passenger vessel	(c) Other requirements
•	Compound, tree or weed killing, liquid	Combustible liquid	None	173.118a	None	No limit	No limit	1,2	1,2	
•	Compound, tree or weed killing, liquid	Corrosive material	Corrosive	173.244	173.245	1 quart	1 quart	1,2	1,2	
•	Compound, tree or weed killing, liquid	Flammable liquid	Flammable liquid	173.118	173.119	1 quart	10 gallons	1,2	1	
•	Compound, tree or weed killing, liquid	Poison B	Poison	173.345	173.346	1 quart	55 gallons	1,2	1,2	
•	Compound, tree or weed killing, solid	Oxidizer	Oxidizer	173.153	173.154 173.229	25 pounds	100 pounds	1,2	1,2	
•	Compound, vulcanizing, liquid	Corrosive material	Corrosive	173.244	173.245	1 quart	1 quart	1,2	1,2	

Figure 6-1. Sample page from paragraph 172.101 of Title 49, Part 172, Code of Federal Regulations

Explosives

There are three classes of explosives: A, B, and C. Class A is the most hazardous and Class C the least hazardous. In Class A explosives there are nine types, ranging from a Type 1 (solid explosive, such as black powder) to a Type 9 (propellant explosive, such as solid rocket propellant). The definitions for the various classes of explosives as well as some examples of the categories they fit into are given in Table 6-1.

Department of Transportation regulations are very explicit about placarding and labeling of shipments of hazardous materials. Table 6-2 describes and illustrates the DOT vehicle placards and package labels required for explosives. Any special requirements that must be taken into account are also listed.

Flammable Materials

Flammable materials are broken down by the Department of Transportation into flammable liquids, flammable solids, flammable gases, and combustible liquids. The definitions for these various classes are given in Table 6-3.

Table 6-4 describes and illustrates the placarding and labeling required by DOT regulations for flammable materials in transport. Any special requirements are also listed. Figures 6-2 and 6-3 show examples of proper placarding and labeling. Figure 6-4 is an example of improper placarding.

Oxidizers

In addition to identifying the regular oxidizing materials, there are special requirements for identifying organic peroxides and oxygen. The definitions for these various classes are given in Table 6-5. Table 6-6 describes and illustrates the placarding and labeling required by DOT regulations for these materials.

Table 6-1. Classes of Explosives

Hazard Class	Definition	Examples
Explosive	Any chemical compound, mixture, or device, the primary or common purposes of which is to function by explosion, i.e., with substantially instantaneous release of gas and heat, unless such compound, mixture, or device is otherwise specifically classified in the regulations.	
Class A	A detonating or otherwise maximum hazard.	Black powder, dynamite, priming devices, blasting caps, cannon ammunition, boosters.
Class B	Those explosives which, in general, function by rapid combustion rather than detonation.	Special fireworks, flash powders, liquid propellant explosives for rocket motors, smokeless powder for small arms.
Class C	Certain types of manufactured articles which contain class A or class B explosives or both as components but in restricted quantities, and certain types of fireworks.	Small arms ammunition, explosive cable cutters, safety fuse, paper caps, common fireworks.

Table 6-2. Placard and Label Requirements for Explosives

Hazard Class	Vehicle Placard	Package Label	Quantity	Special Requirements
Explosives A	EXPLOSIVES A (orange with black lettering). For rail transport, the EXPLOSIVES A placard must be placed against a white square background with a black border. 	EXPLOSIVES A (orange with black lettering) 	Any amount	Explosives A and B loaded together are placarded and labeled EXPLOSIVES A. Should not be loaded or stored with any other placarded materials.
Explosives B	EXPLOSIVES B (orange with black lettering)	EXPLOSIVES B (orange with black lettering)	Any amount	Should not be loaded or stored with poisons.
Explosives C	FLAMMABLE (red with white lettering) 	EXPLOSIVES C (orange with black lettering) 	Any quantity by rail; over 1,000 pounds (450 kilograms) by truck.	Should not be loaded or stored with poisons. If combined load and 5,000 pounds (2,250 kilograms) picked up at one facility, then a placard is needed.

Table 6-3. Classes of Flammable Materials

	Definition	Examples
Flammable Liquid	Any liquid having a flash point below 37.8°C (100°F) with the exception of a mixture having one component or more with a flash point above 37.8°C and making up at least 99 percent of the total volume.	Pentane, gasoline, methyl ethyl ketone, isobutylamine
Flammable Solid	Any solid material, other than one classed as an explosive, which, under conditions normally incident to transportation is liable to cause fires through friction, retained heat from manufacturing or processing, or which can be ignited readily and when ignited burns so vigorously and persistently as to create a serious transportation hazard.	Hafnium, fish meal, burnt fibers, safety matches, phosphorus, trisulfide
Flammable Solid —Dangerous When Wet	Same definition as flammable solid above with the additional fact that water will accelerate the reaction.	Phosphorus sesquisulfide, magnesium scrap, lithium silicon, sodium amide
Flammable Gas	Any material or mixture having in the container an absolute pressure exceeding 40 psi at 70°F (21.11°C); or regardless of the pressure at 70°F (21.11°C); having an absolute pressure exceeding 104 psi at 130°F (54.44°C); or any liquid flammable material having a vapor pressure exceeding 40 psi at 100°F (37.78°C); and provided that either a mixture of 13% or less (by volume) with air forms a flammable mixture or the flammable range with air is wider than 12% regardless of the lower limit.	Anhydrous monomethylamine, methane, methyl chloride, liquefied petroleum gas, ethylene
Combustible Liquid	Any liquid that has a flash point at or above 37.8°C (100°F) and below 93.3°C (200°F) except any mixture having one component or more with a flash point at 93.3°C or higher that makes up at least 99% of the total volume of the mixture (Note that the Code of Federal Regulations, Title 49, paragraph 173.115, states the following: Markings such as "NONFLAMMABLE" or "NONCOMBUSTIBLE" should *not* be used on a vehicle containing a material that has a flash point of 200°F (93.3°C) or higher.[5])	Pine oil, plastic solvent, petroleum naphtha, petroleum distillate, ink, fuel oil

Table 6-4.   Placard and Label Requirements for Flammable Materials

	Vehicle Placard	Package Label	Quantity	Special Requirements
Flammable Liquid	FLAMMABLE (red with white lettering). The word GASOLINE in white lettering on red background may be used when transporting that commodity. 	FLAMMABLE LIQUID (red with black lettering), plus POISON (black lettering on white background) if the material has that property. 	Any quantity by rail; over 1,000 pounds (450 kilograms) by truck.	Should not be loaded with poisons or explosives. If combined load with 5,000 pounds (2,250 kilograms) picked up at one facility, then a FLAMMABLE placard is needed in addition to other placards.
Flammable Solid	FLAMMABLE SOLID (black lettering on white background with seven vertical red stripes). FLAMMABLE (white letters on a red background) is also permissible to be used in place of the FLAMMABLE SOLID placard. 	FLAMMABLE SOLID (black lettering on white background with seven vertical red stripes) plus POISON (black lettering on white background) if the material has that property. 	Any quantity by rail; over 1,000 pounds (450 kilograms) by truck	Should not be loaded with poisons or explosives. If combined load and 5,000 pounds (2,250 kilograms) picked up at one facility, then a FLAMMABLE SOLID placard is needed in addition to other placards.

Table 6-4. *(Continued)*

Hazard Class	Vehicle Placard	Package Label	Quantity	Special Requirements
Flammable Solid, Dangerous When Wet	FLAMMABLE SOLID (black lettering on white background, with seven vertical red stripes and blue triangle with white W)	FLAMMABLE SOLID (black lettering on white background, with seven vertical red stripes) plus DANGEROUS WHEN WET (blue background with black lettering)	Any quantity	Should not be loaded with poisons or explosives.

Flammable Gas	FLAMMABLE GAS (white lettering on a red background)	FLAMMABLE GAS (black lettering on red background)	Any quantity by rail; over 1,000 pounds (450 kilograms) by truck	Should not be loaded with poisons or explosives. If combined load and 5,000 pounds (2,250 kilograms) picked up at one facility then a FLAMMABLE GAS placard is needed in addition to other placards.

Table 6-4. *(continued)*

Hazard Class	Vehicle Placard	Package Label	Quantity	Special Requirements
Combustible Liquid	COMBUSTIBLE (white letters on red background). The words FUEL OIL (white letters on red background) may be used in place of the word COMBUSTIBLE when that commodity is carried.  FLAMMABLE (white letters on red background) is also permissible to use in place of the COMBUSTIBLE placard. 	None required.	Any quantity by rail; over 110 gallons (416 liters) by truck.	If combined load and 5,000 pounds (2,250 kilograms) picked up at one facility then a COMBUSTIBLE placard is needed in addition to other placards.

Figure 6-2. Flammable placards must be able to survive an accident and also be visible from any angle

Figure 6-3. Pure sodium container with labels affixed

Table 6-5. Classes of Oxidizing Materials

Hazard Class	Definition	Examples
Oxidizer	A substance that yields oxygen readily to stimulate the combustion of organic matter.	Sodium bromate, sodium chlorate, silver nitrate, potassium peroxide
Organic Peroxide	An organic compound containing the bivalent –0–0 structure and which may be considered a derivative of hydrogen peroxide where one or more of the hydrogen atoms have been replaced by organic radicals.	Lauroyl peroxide, isopropylpercarborate, dicumyl peroxide, dry caprylyl peroxide solution
Oxygen	An odorless, colorless gaseous chemical element, essential to life, which supports combustion. At extremely low temperatures the gas liquefies.	Oxygen

Corrosives and Poisons

Corrosives and poisons are divided by the Department of Transportation into four major classes as well as one special class. In addition to the corrosive category, there are poisons A and B and irritant. The special category is for chlorine. The definitions for these various classes are given in Table 6-7. Table 6-8 describes and illustrates placarding and labeling requirements. Difficulties arise when identification markings do not conform to regulations. Figures 6-5 through 6-7 show some actual practices that emergency personnel may meet.

Figure 6-4. Improper placards on rail car, using a combination of an old placard and label

Table 6-6. Placards and Labels for Oxidizing Materials

Hazard Class	Vehicle Placard	Package Label	Quantity	Special Requirements
Oxidizer	OXIDIZER (black letters on yellow background)	OXIDIZER (black letters on yellow background)	Any quantity by rail; over 1,000 pounds (450 kilograms) by truck	If combined load and 5,000 pounds (2,250 kilograms) picked up at one facility, then an OXIDIZER placard is necessary in addition to other placards.

Table 6-6. *(continued)*

Hazard Class	Vehicle Placard	Package Label	Quantity	Special Requirements
Organic Peroxide	ORGANIC PEROXIDE (black lettering on yellow background) 	ORGANIC PEROXIDE (black lettering on yellow background) 	Any quantity by rail; over 1,000 pounds (450 kilograms) by truck	If combined load and 5,000 pounds (2,250 kilograms) picked up at one facility, then an ORGANIC PEROXIDE placard is necessary in addition to other placards.
Oxygen (liquid)	OXYGEN (black lettering on yellow background) 	OXIDIZER (black lettering on yellow background) Label is not required if cylinder is marked in accordance with CGA Pamphlet C-7 or if the cylinder is permanently mounted. The word OXIDIZER may be replaced by the word OXYGEN as long as all other elements of the label remain the same.	Any quantity by rail; over 1,000 pounds (450 kilograms) by truck	If combined load and 5,000 pounds (2,250 kilograms) picked up at one facility, then an OXYGEN placard is necessary in addition to other placards.
Oxygen (gas)	NON-FLAMMABLE GAS (white lettering on a green background) 	NON-FLAMMABLE GAS (black lettering on a green background) 	Any quantity by rail; over 1,000 pounds (450 kilograms) by truck	If combined load and 5,000 pounds (2,250 kilograms) picked up at one facility, then a NON-FLAMMABLE GAS placard is necessary in addition to other placards.

Table 6-7. Classes of Corrosives and Poisons

Hazard Class	Definition	Examples
Corrosive	A liquid or solid that causes visible destruction or irreversible alterations in human skin tissue at the site of contact, or in the case of leakage from its packaging, a liquid that has a severe corrosion rate on steel.	Acetyl bromide, ammonium hydroxide, sulfuric acid, nitric acid (40 percent or less)
Poison A	Poisonous gases or liquids of such a nature that a very small amount of the gas or vapor of the liquid mixed with air is dangerous to life.	Bromacetone, cyanogen, diphosgene, hydrocyanic acid
Poison B	Poisonous liquid or solid (including pastes and semisolids), which are known to be so toxic to humans as to afford a hazard to health during transportation.	Silver cyanide, potassium arsenate (solid), parathion (liquid)
Irritating	A liquid or solid substance, which upon contact with fire or when exposed to air gives off dangerous or intensely irritating fumes.	Brombenzyleyanide, chloracetophenone, tear gas candle, xylyl bromide
Chlorine	A greenish-yellow gas which is an element.	Chlorine

Table 6-8. Placards and Labels for Corrosives and Poisons

Hazard Class	Vehicle Placard	Package Label	Quantity	Special Requirements
Corrosive	CORROSIVE (white lettering on a black background and black drawing on a white background) 	CORROSIVE (white lettering on a black background and black drawing on a white background) 	Any quantity by rail; over 1,000 pounds (450 kilograms) by truck	If combined load and 5,000 pounds (2,250 kilograms) picked up at one facility, then a CORROSIVE placard is necessary in addition to other placards.
Poison A	POISON GAS (black lettering on white background). For rail transport, the POISON GAS placard must be placed against a white square background with a black border. 	POISON GAS (black lettering on white background) 	All quantities	Do not load or store with any flammable, oxidizers, corrosives, or explosives.

Table 6-8. *(continued)*

Hazard Class	Vehicle Placard	Package Label	Quantity	Special Requirements
Poison B	POISON (black lettering on white background)	POISON (black lettering on white background)	Any quantity by rail; over 1,000 pounds (450 kilograms) by truck	If combined load and 5,000 pounds (2,250 kilograms) picked up at one facility, then a POISON placard is necessary in addition to other placards. Do not load or store with any flammable, oxidizer, corrosives, or explosives.
Irritating	DANGEROUS (black lettering on white background with red triangles in upper and lower corners)	IRRITANT (red letters on white background)	Any quantity by rail; over 1,000 pounds (450 kilograms) by truck	If combined load and 5,000 pounds (2,250 kilograms) picked up at one facility, then a DANGEROUS placard is necessary in addition to other placards.

Figure 6-5. Corrosive label on the outside container. (Consider the problems that would be created when the empty box is used to store other goods.)

Table 6-8. *(continued)*

Hazard Class	Vehicle Placard	Package Label	Quantity	Special Requirements
Chlorine	CHLORINE (black letters on white background if over 110 gallons [416 liters]) NON-FLAMMABLE GAS (white letters on a green background if 110 gallons or less)	NON-FLAMMABLE GAS (black letters on green background) plus POISON (black letters on white background). The word POISON may be replaced by the word CHLORINE as long as all other elements of the label remain the same. The CHLORINE label then replaces both the NON-FLAMMABLE and POISON labels.	All quantities must be placarded either CHLORINE (over 110 gallons [416 liters]) or NON-FLAMMABLE GAS (110 gallons or less [416 liters])	If combined load and 5,000 pounds (2,250 kilograms) picked up at one facility, then a CHLORINE placard is necessary in addition to other placards. Do not load or store with any flammable, oxidizers, corrosives, or explosives.

Figure 6-6. Hydrocyanic acid (Poison A) car is placed back on the track after derailment. (Note how difficult the placards are to read.)

Compressed Gas

The two kinds of compressed gases are flammable and nonflammable. The requirements for using the flammable gas placard are covered in a preceding section on Flammable Materials. This section deals with non-flammable compressed gases. The definition for this class as well as placard and label requirements are shown in Table 6-9.

Figure 6-7. Poison B (methyl bromide) without proper placards on truck

Table 6-9. Definition, Placards, and Labels for Compressed Gas

Hazard Class	Definition			Examples
Compressed gas	Any material or mixture having in the container an absolute pressure exceeding 40 psi at 70°F (21.1°C) or, regardless of the pressure at 70°F (21.1°C), having an absolute pressure exceeding 104 psi at 130°F (54.4°C)			Nitrous oxide, nitrogen fertilizer solution, neon, hydrogen chloride, carbon dioxide

Hazard Class	Vehicle Placard	Package Label	Quantity	Special Requirements
Compressed Gas	NON-FLAMMABLE GAS (White letters on green background)	NON-FLAMMABLE GAS (black letters on green background)	Any quantity by rail; over 1,000 pounds (450 kilograms) by truck	If combined load and 5,000 pounds (2,250 kilograms) picked up at one facility, then a NON-FLAMMABLE GAS placard is necessary in addition to other placards.

Radioactive Materials

The Department of Transportation breaks down radioactive materials into three classes: I, II, and III. Under this system class I presents the least hazard, while class III is the most dangerous. The definitions for these classes are given in Table 6-10.

Table 6-11 describes and illustrates the DOT required placards and labels for radioactive materials. Radioactive labels must have the following information included:

1. *Contents*. The name of the radionuclides as taken from the listing established by DOT.

2. *Number of Curies*. Units shall be expressed in curies (ci), millicuries (mci), or microcuries (μci). For a fissile material, the weight in grams or kilograms of the fissile radioisotope may be inserted.

3. *Transportation Index*. The number placed on a package to indicate the degree of control to be exercised by the carrier during transportation. The index is determined by the highest of either (1) or (2) below. The index should not exceed 10 on any one vehicle, except for exceptions spelled out in the regulations.

Table 6-10. Classes of Radioactive Materials

Hazard Class	Definition	Examples
Radioactive I	Packages which may be transported in unlimited numbers and in any arrangement, and which require no nuclear criticality safety controls during transportation.	Package containing any radioactive material having a measurement of 0.5 millirem or less per hour at each point on the external surface of the package.
Radioactive II	Packages which may be transported together in any arrangement but in numbers which do not exceed an aggregate transport index of 50. No nuclear criticality safety control by the shipper during transportation.	Package containing any radioactive material measuring more than 0.5 but not more than 50 millirem per hour at each point and not exceeding 1.0 millirem per hour at 3 feet (0.9 meter) from each point on the external surface of the package.
Radioactive III	Shipments of packages which do not meet the requirements of class I and II and which are controlled to provide nuclear criticality safety in transportation by special arrangements between the shipper and the carrier.	Package containing any radioactive material which (1) measures more than 50 millirems per hour at each point or exceeds 1.0 millirem per hour at 3 feet (0.9 meter) on the external surface, (2) contains large quantity as defined by DOT.

(1) The highest radiation dose rate in millirem per hour at 3 feet (0.9 meter) from any accessible external surface of the package or,

(2) For class II packages only by dividing the number 50 by the number of similar packages which may be transported together.

Table 6-11. Placards and Labels Required for Radioactive Materials

Hazard Class	Vehicle Placard	Package Label	Quantity	Special Requirements
Radioactive I	None Required	RADIOACTIVE I (black lettering on white background with red I)	Any package meeting the requirements in Table 6-10	None
Radioactive II	None Required	RADIOACTIVE II (black lettering with white and yellow background and red II)	Any package meeting the requirements in Table 6-10	None
Radioactive III	RADIOACTIVE (black lettering on white and yellow background)	RADIOACTIVE III (black lettering on white and yellow background with red III). If the radioactive material also meets the definition of additional hazardous materials, then a label for each hazard must be applied.	Any package meeting the requirements in Table 6-10 must be labeled. Any transportation vehicle with a label III material must be placarded.	Do not load or store with class A explosives.

Dangerous

A freight container, motor vehicle, or rail car containing 1,000 pounds (450 kilograms) or more of two or more classes of materials in the following list *may* be placarded DANGEROUS in place of the separate placarding specified for each of those classes. However, when 5,000 pounds (2,250 kilograms) or more of one of the materials in the list is loaded at one facility, the required individual placard must be applied. Figure 6-8 shows an example of this.

FLAMMABLE
CHLORINE (over 110 gallons) (416 liters)
FLAMMABLE SOLID
(not dangerous when wet)
POISON
NON-FLAMMABLE GAS
COMBUSTIBLE
FLAMMABLE GAS
ORGANIC PEROXIDE

OXYGEN
OXIDIZER
CORROSIVE

The materials in the following list must contain their specific placards, no matter what the quantity and independent of any mixed loads:

EXPLOSIVES A
EXPLOSIVES B
FLAMMABLE SOLID (dangerous when wet)
RADIOACTIVE
POISON GAS

Empty Vehicles and Packaging

Each empty tank car must be placarded with an EMPTY placard that corresponds to the material the tank car last contained. The EMPTY placard is the same as the standard placard except that the symbol at the top is replaced by the word EMPTY. Tank cars placarded COMBUSTIBLE do not have to be placarded EMPTY when unloaded. In addition, if the tank car is sufficiently cleaned of residue and purged of vapor to remove any hazard, there is no requirement for an EMPTY placard.

When a Poison A material has been unloaded from a rail car, the POISON GAS—EMPTY placard must be applied against a white square background with a black border.

When any packaging that originally required a hazardous material label is offered for transportation when empty, it must have the label either removed, obliterated, or completely covered when the package is empty. One way to do this is to cover the existing label with an EMPTY label.

Any empty package which had contained radioactive materials must not have any radiation exceeding 0.5 millirem per hour. The EMPTY label (see Figure 6-9) must be affixed to all packages which contained any radioactive materials.

Figure 6-8. Box car containing over 5,000 pounds (2,250 kilograms) of a corrosive and a non-flammable gas. (Note, however, that instead of using the required dual placards, labels were used in their place.)

Figure 6-9. Miscellaneous special labeling

Special Labeling Requirements

Each package containing a viable micro-organism or its toxin, which causes or may cause human disease, must contain an ETIO-LOGIC AGENTS label (see Figure 6-9). This label, in addition to warning emergency response personnel of a problem, contains the telephone number of the Communicable Disease Center in Atlanta, Georgia (404-633-5313), which can be called for assistance. The label has predominantly red printing on a white background. Examples of etiologic agents are:

Anthrox   Polio
Botulism   Rabies
Cholera   Tetanus
Encephalitis

Multiple labeling of a package is necessary if the material it contains has more than one hazard class and one of its classifications is Explosive A, Poison A, or Radioactive. In addition, a Poison B liquid that also meets the definition of a flammable liquid must be labeled POISON and FLAMMABLE LIQUID. Some examples of the multiple labeling requirements are:

Chemical	Label
Acrolein inhibited	FLAMMABLE LIQUID and POISON
Ammunition, chemical, explosive, with Poison A material	EXPLOSIVE A and POISON GAS
Arsine	POISON GAS and FLAMMABLE GAS
Boron trifluoride	POISON and NON-FLAMMABLE GAS
Bromine trifluoride	POISON and OXIDIZER
Nitrogen peroxide, liquid	OXIDIZER and POISON GAS

Certain hazardous materials when transported by air can, by law, be carried only on cargo aircraft. However, it should be noted that the Airline Pilots Association has refused to fly passenger planes carrying hazardous materials even though the law says they can. At this time, therefore, hazardous materials are not carried on passenger airlines. The shipper must still, however, put a CARGO AIRCRAFT ONLY label on certain commodities to be shipped by air. The label is orange with black lettering. Examples of commodities which are restricted to cargo planes are:

Trifluorochlorethylene
Vanadium oxytrichloride
Sulfur trioxide, stabilized
Sulfuric acid, spent
Phosphorus, amorphous, red
Oleum (sulfuric acid fuming)

A substance is considered a magnetized material if, when packaged for transportation by air, it has a magnetic field strength of 0.002 gauss or more at a distance of 7 feet (2 meters) from any point of the package. In addition, if the material can affect the airplane's instruments, it is considered a magnetic material. All magnetized packages offered for shipment by air must contain a MAGNETIZED MATERIAL label (see Figure 6-9), which has blue lettering on a white background.

Each metal barrel or drum containing a flammable liquid having a vapor pressure between 16 and 40 psia at 100°F (37.78°C) must have affixed a BUNG label in addition to the FLAMMABLE LIQUID label. The BUNG label (Figure 6-9) is white with black lettering.

For export shipments, other countries require that certain additional labels be affixed to the package. These labels are used in addition to those required by the United States Department of Transportation regulations. The labels used for export are:

Label	Label Description
SPONTANEOUSLY COMBUSTIBLE	Red and white background with black lettering
IRRITANT	White background with black lettering and black skull and crossbones. (This can be used to replace the regular IRRITANT label on export shipments.)
DANGEROUS WHEN WET	Black lettering on blue background

Other Regulated Material (ORM)

Other regulated material (ORM) includes any material that does not meet the definition of a hazardous material, other than a combustible liquid having a capacity of 110 gallons (416 liters) or less. There are four classes of ORM, designated ORM-A, -B, -C, and -D. No labels or placards are required for these materials. The definitions and some examples of each class are given in the following paragraphs.

An ORM-A material is one that has an anesthetic, irritating, noxious, or toxic property, which can cause extreme annoyance or discomfort to passengers and crew in the event of leakage during transportation. Some examples are:

> Bromochloromethane
> Trichloroethylene
> Camphene
> Solid carbon dioxide
> Carbon tetrachloride
> Ethylene dibromide
> Chloroform
> Napthalene
> DDT

An ORM-B material is a material (including a solid when wet with water) capable of causing significant damage to a transport vehicle or vessel from leakage during transportation, including materials that can corrode aluminum. Some examples are:

> Ferric chloride
> Copper chloride
> Calcium oxide (unslaked lime, quicklime)
> Metallic mercury

An ORM-C material is a material (other than an ORM-A and ORM-B) with inherent characteristics that make it unsuitable for shipment unless properly identified and prepared for transportation. Consult the Code of Federal Regulations, paragraphs 173.910 through 173.1085 for identification and shipping requirements of ORM-C material. Some examples are:

> Battery parts
> Calcium cyanide (not hydrated)
> Castor beans
> Copra
> Cotton batting
> Excelsior
> Fish meal
> Fish scrap
> Hay or straw
> Metal borings
> Oakum
> Oiled materials
> Pesticides

> Petroleum code
> Rosin
> Rubber curing compound
> Sulfur

An ORM-D material is a material, such as a consumer commodity, that presents a limited hazard during transportation because of its form, quantity, and packaging (all have strict requirements for type and weight of outside packaging). Remember, there are no labels or placards required for this class of materials:

1. Flammable Liquids
   a. In inside metal containers, each having a rated capacity of 1 quart (0.95 liter) or less
   b. In inside containers, each having a rated capacity of 1 pint (0.47 liter) or less
   c. In inside containers, each having a rated capacity of 1 gallon (3.785 liters) or less, when the flammable liquid has a flash point of 73°F (22.79°C) or higher (almost all alcoholic beverages)

2. Corrosive Liquids
   a. In bottles, each having a rated capacity of 1 pint (0.47 liter) or less
   b. In metal or plastic containers, each having a rated capacity of 1 pint (0.47 liter) or less
   c. In metal or plastic inside containers, each having a rated capacity of not over 1 quart (0.95 liter), provided the liquid mixture contains 10 percent or less corrosive material

3. Corrosive Solids
   a. In earthenware, glass, plastic, or paper containers each having a net weight of 5 pounds (2.25 kilograms) or less

b. In metal, rigid fiber or composition cans, or cartons having a net weight of 10 pounds (4.5 kilograms) or less

4. Flammable Solids

  a. In inside containers having a net weight of 1 pound (0.45 kilogram) or less

  b. Charcoal briquettes in packages having a net weight of 65 pounds (29.25 kilograms) or less

5. Oxidizers

  Inside containers having a rated capacity of 1 pint (0.47 liter) or less for liquids or 1 pound (0.45 kilogram) or less for solids

6. Organic Peroxides

  a. Inside containers which are securely packed and cushioned and containing not over 1 pound (0.45 kilogram) or 1 pint (0.47 liter) of the material

  b. In closed tubes having a maximum fluid capacity of 1/6 ounce (4.66 grams) each

7. Poison B Liquids or Solids

  a. Inside containers each having a rated capacity of 8 ounces (224 grams) or less by volume for liquids

  b. Inside containers containing 8 ounces (224 grams) or less solid material net weight

8. Compressed Gases

  a. Inside containers having a water capacity of 4 fluid ounces or less

  b. Solutions of materials and compressed gases with limited capacities and internal pressures, the contents of which are not flammable, poisonous, or corrosive

Exemptions to Label and Placard Requirements

After reading the regulations for placard-ing and labeling, emergency response personnel are likely to think that they have at least a usable system for gaining initial information. Unfortunately, this is not true. As already explained, shipments of 1,000 pounds (450 kilograms) or less of many commodities by truck do not require any external placard-ing. Small quantities of ORM-D materials do not even require labeling, yet in greater quantities they must be both labeled and placarded.

Finally, there is another way for the regulations to be bypassed. The Materials Transportation Bureau of the Department of Transportation can grant exemptions to the Code of Federal Regulations requirements. These exemptions can be of a temporary nature or may be made on a permanent basis. In order to have the exemptions granted, however, notice must be filed in the Federal Register and comments invited. Since very few emergency response personnel regularly read the Federal Register, it is safe to assume that their views are very seldom heard.

In the Federal Register, Vol. 42, No. 152 of August 8, 1977 there was a notice of "Individual Exemptions, Conversion to Regulation of General Applicability." The introduction to the proposed changes states:

> SUMMARY: The Materials Transportation Bureau is considering amending the regulations governing the transportation of hazardous materials to incorporate a number of changes based on existing exemptions which have been granted to individual applicants allowing them to perform particular functions in a manner that varies from that specified by the regulations. Adoption of these exemptions as rules of general applicability would provide wider access to the benefits of transportation innovations recognized as effective and safe.[6]

Examples of some of the proposed changes are given in Table 6-12.

Table 6-12. Examples of Proposed Changes to Code of Federal Regulations

Regulation Affected	Nature of Exemption or Application	Nature of Proposed Amendment
173.306 (d)	Requests authority to ship a non-flammable, compressed gas in authorized DOT specification cylinders in the trunks of passenger automobiles when the container is part of a tire inflator system.	To add paragraph (d) (4) to read: (4) A cylinder which is part of a tire inflator system in a motor vehicle, charged with a nonliquefied, non-flammable compressed gas is excepted from the requirements of pts. 170-189 of this subch. except: (i) Unless otherwise authorized by the Department, each cylinder must be in compliance with one of the cylinder specifications in pt. 178 and authorized for use in Sec. 173.302 for the gas it contains. (ii) Each cylinder must be in compliance with the filling requirements of Sec. 173.301. (iii) Each cylinder must be securely installed in the trunk of the motor vehicle and the valve must be protected against accidental discharge
173.65 (d)	Authorizes shipment of triaminotrinitrobenzene (TATB), trichlortrinitrobenzene (TCTNB), and hexanitrostilbene (HNS) as reagents in accordance with 49 CFR 173.65 (d). (Modes 1, 2, and 4.)	To add to paragraph (d): The following materials may be shipped dry, in quantities not exceeding 4 ounces (112 grams) in 1 outside package, by rail freight, or highway, as drugs, n.o.s., or medicines, n.o.s., without any other requirements when in securely closed bottles or jars cushioned to prevent breakage: (17) Triaminotrinitrobenzene (18) Trichlortrinitrobenzene (19) Hexanitrostilbene Not authorized for transportation by air.

Placard Interpretation

So far in this section, the legal and technical requirements for the various placards have been explained. However, as emergency response personnel, you will normally have to work in the other direction. You are given the placard as a piece of information as you begin your size-up. Table 6-13, therefore, is designed to summarize the possible hazard indicated by the placard. Note that the quantity qualifications apply only to truck transportation. All quantities carried by rail must be placarded. The combination placarding requirements apply to both truck and rail transportation.

Table 6-13. Summary of Hazards

Placard	Cargo Could Possibly Contain
EXPLOSIVES A	Any quantity of explosives A or explosives A and B loaded together
EXPLOSIVES B	Any quantity of explosives B
FLAMMABLE SOLID W	Any quantity of a flammable solid which is dangerous when wet
RADIOACTIVE	Any quantity of radioactive material which contains a radioactive class III label
POISON GAS	Any quantity of a poison A material
DANGEROUS	1,000 pounds (454 kilograms) or more of a combination of materials
	1,000 pounds (454 kilograms) or more of an irritating material carried by itself
	5,000 pounds (2,250 kilograms) or more of an irritating material picked up at one location and carried in combination with other hazardous materials
FLAMMABLE	1,000 pounds (450 kilograms) or more of an explosives C material
	1,000 pounds (450 kilograms) or more of a flammable liquid
	1,000 pounds (450 kilograms) or more of a flammable solid which is not dangerous when wet
	Over 110 gallons (416 liters) of a combustible liquid
	5,000 pounds (2,250 kilograms) or more of any of the above materials picked up at one location, and carried in combination with other hazardous materials
CHLORINE	More than 110 gallons (416 liters) of chlorine
	5,000 pounds (2,250 kilograms) or more of chlorine picked up at one location and carried in combination with other hazardous materials
FLAMMABLE SOLID	1,000 pounds (450 kilograms) or more of a flammable solid which is not dangerous when wet
	5,000 pounds (2,250 kilograms) or more of a flammable solid which is not dangerous when wet and is picked up at one location and carried in combination with other hazardous materials
POISON	1,000 pounds (450 kilograms) or more of a class B poison or fluorine
	5,000 pounds (2,250 kilograms) or more of a class B poison or fluorine which is picked up at one location and carried in combination with other hazardous materials

Table 6-13. (continued)

Placard	Cargo Could Possibly Contain
NON-FLAMMABLE GAS	1,000 pounds (450 kilograms) or more of a non-flammable gas
	110 gallons (416 liters) or less of chlorine
	1,000 pounds (450 kilograms) or more of gaseous oxygen
	5,000 pounds (2,250 kilograms) or more of non-flammable gas or gaseous oxygen which is picked up at one location and carried in combination with other hazardous materials
COMBUSTIBLE	Over 110 gallons (416 liters) of a combustible liquid
	5,000 pounds (2,250 kilograms) or more of a combustible liquid which is picked up at one location and carried in combination with other hazardous materials
FLAMMABLE GAS	1,000 pounds (450 kilograms) or more of a flammable gas
	5,000 pounds (2,250 kilograms) or more of a flammable gas which is picked up at one location and carried in combination with other hazardous materials
ORGANIC PEROXIDE	1,000 pounds (450 kilograms) or more of an organic peroxide
	5,000 pounds (2,250 kilograms) or more of an organic peroxide which is picked up at one location and carried in combination with other hazardous materials
OXYGEN	1,000 pounds (450 kilograms) or more of liquefied oxygen
	5,000 pounds (2,250 kilograms) or more of liquefied oxygen which is picked up at one location and carried in combination with other hazardous materials
OXIDIZER	1,000 pounds (450 kilograms) or more of an oxidizer
	5,000 pounds (2,250 kilograms) or more of an oxidizer which is picked up at one location and carried in combination with other hazardous materials
CORROSIVE	1,000 pounds (450 kilograms) or more of a corrosive material
	5,000 pounds (2,250 kilograms) or more of a corrosive material which is picked up at one location and carried in combination with other hazardous materials

PESTICIDE LABELING

A pesticide is a chemical or mixture of chemicals used to control any living thing considered to be a "pest." Pesticides are covered in detail in Chapter 3. Pesticide labeling requirements are established by the Environmental Protection Agency (EPA).

Emergency response personnel must become familiar with the basic information on a pesticide label so that the potential hazard can be recognized during an emergency situation.

Each pesticide label, regulated by the EPA, must contain one of three signal words: DANGER, WARNING, or CAUTION (Figures 6-10, 6-11, and 6-12). The signal word provides basic information on how dangerous the product is if the emergency response team should come in contact with the pesticide.

The most toxic materials use the signal word DANGER, the next most toxic use WARNING, and finally, the signal word CAUTION is used for the least toxic.

The words EXTREMELY FLAMMABLE on a pesticide indicate the additional hazard which the material will exhibit.

The pesticide label will also contain information on the ways the chemical can enter the body, as well as requirements for its storage and disposal. In addition, first aid information is provided for some of the more toxic chemicals. Where an antidote for the particular pesticide poisoning is known, the name of the antidote is included on the label.

NATIONAL FIRE PROTECTION ASSOCIATION 704 SYSTEM

The National Fire Protection Association, through its committee system has developed a standard system for identifying the hazards of all chemicals. The system is basically a voluntary one for use by industry. However, some local governments have adopted the system as a requirement for identifying all hazardous chemicals stored in their area of responsibility. Emergency response personnel could, therefore, come in contact with the label. The identification system, as developed by the NFPA Committee, is listed as standard 704. Hence, the labeling technique is known as the 704 System (Figure 6-13).

The system identifies the hazard in terms of three categories:

Health

Flammability

Reactivity (instability)

The severity of the hazard is indicated by a numerical designation from 0 to 4. The designations graduate from 0 for a product that has no special hazard, to a 4, which indicates a severe hazard.

The label system uses a diamond shape, which is divided into four sections. The box to the left of the diamond is for the health hazard; the top box for the flammability hazard; and the right box for the reactivity hazard. The bottom box is used to indicate any special hazard such as reactivity to water (W), radioactivity,

an oxidizing agent (OXY), or polymerization (P).

In addition to the physical locations of the boxes within the diamond, each box is color coded for easier identification:

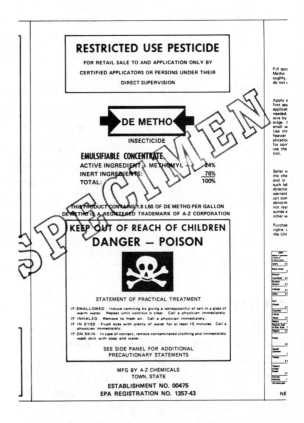

Figure 6-10. Pesticide label with DANGER signal word. *Courtesy of Federal Environmental Protection Agency*

Health—blue background

Flammability—red background

Reactivity—yellow background

Special hazards—white background

The degree of hazard indicated by each numerical rating is summarized below:

HEALTH

4 Too dangerous to enter vapor or liquid even with air packs

3 Extremely dangerous: Use full protective clothing

2 Hazardous: Use breathing apparatus

1 Slightly hazardous

0 Like ordinary material

FLAMMABLE

4 Extremely flammable

3 Ignites at normal temperature

2 Ignites when moderately heated

1 Must be preheated to burn

0 Will not burn

REACTIVE

4 May explode; vacate area if materials are exposed to fire.

3 Strong shock or heat may cause an explosion; use monitors from behind explosion resistant barriers.

2 Violent chemical change possible; use

hose streams from a distance.

1 Unstable if heated.

0 Normally stable.

MY-OWN

**TRIPLE
ACTION**

ROSE AND
FLOWER CARE

A. Systemic Insecticide—6 week action
B. Weed and Grass Preventer
C. Fertilizer 8-12-4

Ingredient Statement as an
Insecticide-Herbicide

Active ingredients:
O, O-diethyl S-[2-(ethylthio)
ethyl] phosphorodithioate 1.000%
Trifuralin (a, a, a-trifluoro-2, 6-
dinitro-N, N-dipropyl-p-toluidine) 0.174%
Inert Ingredients: 98.826%
    Equivalent 20 lbs. SECTOFF per ton.
    per ton.
    Equivalent 3.48 lbs. trifuralin per ton.

WARNING: KEEP OUT OF REACH OF CHILDREN

NET WEIGHT 84 OUNCES

Figure 6-11. Pesticide label with WARNING signal word. *Courtesy of Federal Environmental Protection Agency*

UNITED NATIONS CLASSIFICATION SYSTEM

The United Nations organization has established a standardized class number system for hazardous materials. Some countries have adopted the system and require that all imported goods be properly labeled. The system divides hazardous materials into nine groups, each identified by a number. The number, if it is placed on a label, appears in the lower point

of the diamond. The numbers are:

1. Explosives: class A, B, and C

2. Non-flammable and flammable gases

3. Flammable liquids

4. Flammable solids, spontaneously combustible substances, and water-reactive substances

**HERBICIDE
WETTABLE POWDER**

ACTIVE INGREDIENT: weed out *(triazoic acid)*   80.0%
INERT INGREDIENTS: 20.0%
TOTAL: 100.0%

KEEP OUT OF REACH OF CHILDREN

CAUTION

STATEMENT OF PRACTICAL TREATMENT

In case of contact, wash skin with plenty of soap and water. Get medical attention if irritation persists.

PRECAUTIONARY
STATEMENTS

Hazard to Humans
and Domestic Animals

Harmful if swallowed, inhaled, or absorbed through the skin. Avoid breathing dust or spray mist. Avoid contact with skin, eyes, or clothing. Wash thoroughly after handling. Remove and wash contaminated clothing before reuse.

Environmental Hazards

Keep out of lakes, streams, or ponds. Do not apply when weather conditions favor drifts from target area.

A-Z
A-Z Chemicals, Inc.
Chemcity, Minnesota 55888

EPA Reg. No. 102357-41
EPA Est. 102357-MN-1

Net Weight 5 Pounds

Figure 6-12. Pesticide label with CAUTION signal word. *Courtesy of Federal Environmental Protection Agency*

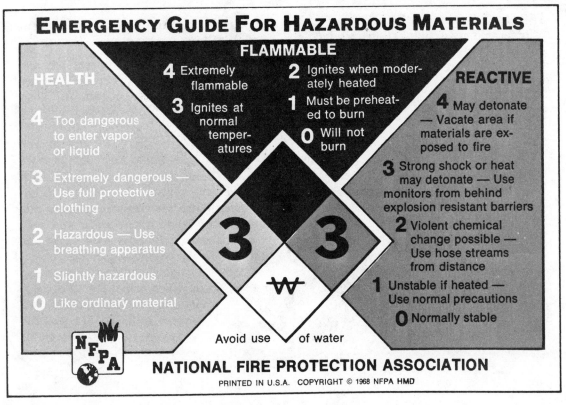

Figure 6-13. NFPA 704 labeling system

5. Oxidizing materials and organic peroxides

6. Poisons: class A, B, and C

7. Radioactive I, II, and III

8. Corrosives

9. Miscellaneous materials which can present a hazard during transport, but are not covered by other classes

MILITARY PLACARDING

The military has established its own placarding system. Emergency response personnel must become familiar with the system if there is a military base within their response area. The explosive hazards are divided into four classes (Figure 6-14):

Figure 6-14. Military class markings. *Courtesy of U.S. Army*

Class 1: Mass Detonation Hazard
   Subject to mass detonation, producing blast pressure and fragments. Do not fight fire unless a rescue attempt is being made.

Class 2: Explosion with Fragment Hazard
   When subjected to fire, boxed items explode progressively.

Class 3: Mass Fire Hazard
   Burns with intense heat.

Class 4: Moderate Fire Hazard
   Principally a fire hazard. Minor explosions may project fragments up to 500 feet (150 meters).

In addition to the class markings, the military uses three hazard symbols (Figure 6-15):

Chemical Hazard
   Highly toxic chemical agents (blue background, red border, and red fire fighter)
   Harassing agents (blue background, yellow border, and yellow fire fighter)
   White phosphorous munitions (blue background, white border, white fire fighter)

Apply No Water

Wear Protective Mask or Breathing Apparatus

Figure 6-15.   Military hazard symbols. *Courtesy of U.S. Army*

MANUFACTURING CHEMISTS ASSOCIATION LABELING SYSTEM

   The Manufacturing Chemists Association has a labeling system called "Labels and Precautionary Information (LAPI)." This system has now been adopted as an American National Standards Institute (ANSI) Standard Z129.1. LAPI labels provide the following information to emergency response personnel:

1. Signal words designating the degree of hazard.

2. Affirmative statement of hazards.

3. Precautionary measures covering actions to be followed or avoided.

4. Instructions in case of contact or exposure.

5. Instructions for container handling and storage.

6. Instructions in case of fire, spill, or leak.

SHIPPING PAPERS

   While the placard or label system gives the emergency response team a broad category for the material, much more detailed information is needed. The name of the specific product being transported must be determined. Here, the incident commander must locate and interpret the shipping papers to gather the required information. The papers can include a shipping order (Figure 6-16), bill of lading (Figure 6-17), manifest, or waybill (Figure 6-18).

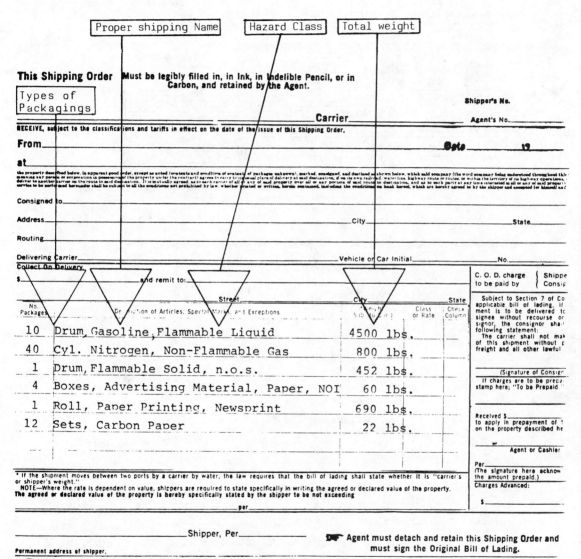

Figure 6-16. Sample copy of a shipping order

GUIDELINE EXHIBIT A

E-05675   4(2/77)   Page 1 of 2

G-77-C   REV. 7/76
STRAIGHT BILL OF LADING – SHORT FORM – ORIGINAL – NOT NEGOTIABLE

SID NUMBER MUST BE SHOWN ON ALL FREIGHT BILLS AND CORRESPONDENCE

| Carrier | KLINE | SCAC–SCAC–SCAC–SCAC–S / –SCAC–SCAC–SCAC–SCAC– | Date | Shipment Identification Number | SID–SID–SID–SID–SID–SID–SID–SID–S / D–SID–SID–SID–SID–SID–SID–SID |

RECEIVED, subject to classifications and tariffs in effect on date of issue of this Bill of Lading.

At: SPLC–SPLC–SPLC–SPLC–S

Du Pont Reference(s)

Carrier's No.

FROM E. I. DU PONT DE NEMOURS & CO. (INC.)
SPLC–SPLC–SPLC–SPLC–S   DU PONT S/P CODE
ELASTOMERS   DEPARTMENT

Customer's Order Number(s)

If charges are to be prepaid, write or stamp here, "To be Prepaid."

TO BE PREPAID

IF PREPAID
MAIL FREIGHT BILLS TO:

The property described below, in apparent good order, except as noted (contents and condition of contents of packages unknown), marked, consigned, and destined as indicated below, which said carrier (the word carrier being understood throughout this contract as meaning any person or corporation in possession of the property under the contract) agrees to carry to its usual place of delivery at said destination, if on its route, otherwise to deliver to another carrier on the route to said destination. It is mutually agreed, as to each carrier of all or any of said property over all or any portion of said route to destination, and as to each party at any time interested in all or any of said property, that every service to be performed hereunder shall be subject to all the terms and conditions of the Uniform Domestic Straight Bill of Lading set forth (1) in Uniform Freight Classification in effect on the date hereof, if this is a rail or a rail-water shipment, or (2) in the applicable motor carrier classification or tariff if this is a motor carrier shipment.
    Shipper hereby certifies that he is familiar with all the terms and conditions of the said bill of lading, including those on the back thereof, set forth in the classification or tariff which governs the transportation of this shipment, and the said terms and conditions are hereby agreed to by the shipper and accepted for himself and his assigns.

CONSIGNED TO

ABC FOAM INC.

SPLC–SPLC–SPLC–SPLC–S
City **CHESTER**   DU PONT DESTINATION CODE

State **PA**   County   Zip

Route **KLINE**

SCAC–SCAC–SCAC–SCAC–SCAC–SCAC–SCAC
Delivering Carrier (If Rail)

For help in CHEMICAL EMERGENCIES
Involving Spill, Leak, Fire or Exposure
CALL DAY OR NIGHT
Continental United States Excluding Washington, D.C.
(800) 424-9300 (TOLL-FREE)
Washington, D.C. 483-7616
Outside continental U.S. (202) 483-7616

Repetitive Pattern Code
R   R
P   P
C–RPC–RPC–RPC–RPC–RPC–RPC–RPC–RPC

Vehicle Initials & Number

DUNS–DUNS–DUNS–DUNS–DUNS–DUNS–D / U–DUNS–DUNS–DUNS–DUNS–DUNS–

Subject to Section 7 of conditions of applicable bill of lading, if this shipment is to be delivered to the consignee without recourse on the consignor, the consignor shall sign the following statement.

The carrier shall not make delivery of this shipment without payment of freight and all other lawful charges.

E. I. du Pont de Nemours & Co.

Per_____
        (Signature of Consignor)

If lower charges result, the agreed or declared value of the within described commodities is hereby specifically stated by the shipper to be not exceeding 50 cents per pound per article

DU PONT CODE COMM.	CONT.	NUMBER OF PACKAGES	TYPE OF PACKAGES	DESCRIPTION OF MATERIALS, SPECIAL MARKS, AND EXCEPTIONS	STCC CODE	WEIGHT SUBJECT TO CORRECTION
		10	DRS	TOLUENE DISOCYANATE		5100 lbs
				POISON B		
				PLACARDS REQUIRED		
				HYLENE TM		

SPECIAL CONDITIONS

____ SHL   Shipper Load
____ COU   Consignee Unload
____ EXC   Exclusive Use
____ SSF   Single Shipment
DTL ____   Minutes Detention
           Other (specify)

GEN. LEDGER

SUB. ACCOUNT

		GROSS PRODUCT WEIGHT	5100	MODE	EQUIP.	TERMS
		GROSS PALLET WEIGHT (____PALLETS @____LBS/PALLET)		SPEC.CIR. SPEC.TRAF.DATA		
		GROSS SHIPPING WEIGHT				

THE DESCRIPTION AND WEIGHT INDICATED ON THIS BILL OF LADING ARE CORRECT SUBJECT TO VERIFICATION BY THE RAIL ROAD INSPECTION BUREAUS ACCORDING TO AGREEMENT

This is to certify that the above-named materials are properly classified, described, packaged, marked and labeled, and are in proper condition for transportation, according to the applicable regulations of the Department of Transportation.

E. I. du Pont de Nemours & Company, Shipper   _____ Agent

Per___**SAM SPADE**_____   Per_____   **1**

Permanent post-office address of shipper,

        DEEPWATER, N.J.

Figure 6-17. Sample copy of a bill of lading

PLACE SPECIAL SERVICE
PASTERS HERE   X30   **50 - THE BALTIMORE AND OHIO RAILROAD COMPANY -50**

B&O-C&O FORM
MADE IN U.S.A.

DANGEROUS

FREIGHT WAYBILL
To be used for Single Consignments, Carload and Less Carload

CAR INITIALS AND NUMBER	KIND	WEIGHT IN TONS			LENGTH OF CAR		MARKED CAPACITY OF CAR	
		GROSS	TARE	NET	ORDERED	FURNISHED	ORDERED	FURNISHED
NATX 34642	T			75				

STOP THIS CAR	C. L. Transferred to	DATE 3/14/73	WAYBILL No. 192859

Consignee and Address at STOP

At _____
At _____

TO No.	STATION	STATE OR PROV.
RDG	EPHRATA.	PA.

FROM No. 780   CALMYSN. W. VA, STATION STATE OR PROV.
B/A   781   PINE GROVE. W. VA.

Route (Show each Junction and Carrier in Route order to destination of waybill)

B&O   C-RUN   WM   LURG   RDG

FULL NAME OF SHIPPER, AND, FOR C. O. D. SHIPMENTS, STREET AND POST OFFICE ADDRESS, AND INVOICE NUMBER.

CONSOLIDATED GAS SUPPLY CORP
CLARKSBURG. W. VA.

← When shipper in the United States executes the no recourse clause of Section 7 of the B/L insert "Yes"

Show "A" if Agent's Routing or "S" if Shipper's Routing   S

RECONSIGNED TO	STATION	STATE OR PROV.

ORIGIN AND DATE, ORIGINAL CAR, TRANSFER FREIGHT BILL AND PREVIOUS WAYBILL REFERENCE AND ROUTING WHEN REBILLED.

DANGEROUS

AUTHORITY _____
CONSIGNEE AND ADDRESS

PENGITE COMPANY.
C/O UGITE GAS INC

FINAL DESTINATION AND ADDITIONAL ROUTING

WEIGHED

C.O.D.

AT _____
GROSS _____
TARE _____
ALLOWANCE _____
NET _____

AMOUNT $
FEE $
TOTAL $

ON C. L. TRAFFIC—INSTRUCTIONS (Regarding Icing, Ventilation, Milling, Weighing Etc. If Iced, Specify to Whom Icing Should Be Charged.)

EVENT THIS CAR IS SHOPPED NOTIFY
CONSOLIDATED GAS SUPPLY CORP
CLARKSBURG. W. VA.

PREPAID

⚓ IF CHARGES ARE TO BE PREPAID, WRITE OR STAMP HERE "TO BE PREPAID"

COMMODITY CODE	TRAILER NO.	L/E	TRAILER NO.	L/E	Indicate by symbol in block labeled * how weights were obtained. R - Railroad Scale, S - Shipper's Tested Weights, E - Estimated Weight and Correct, T - Tariff Classification or Minimum				
29 121 11					*	WEIGHT	RATE	FREIGHT	ADVANCES	PREPAID

T/C	LIQUEFIED PETROLEUM GAS						
	FLAMMABLE COMPRESSED GAS PROPANE	150372	.49	736.82			736.82
	SHELL 34055						
	TEMP 56						
	GALS LOADED AT 60 F 31994						
	OUTAGE 12½ 2263						
	SP GR 0.508						
	ETHYL MERCAPTAN 4.5 LB						
	DANGEROUS PLACARDS APPLIED		≪ DOT S. P. 5900≫				

DESTINATION AGENT'S FREIGHT BILL NO.

CAR TRIP LEASED TO CONSIGNEE

Outbound Junction Agent Will Show Junction Stamps in Space and Order Provided. Additional Junction Stamps & all Yard Stamps to be placed on back hereof.

DESTINATION AGENT WILL STAMP HEREIN STATION NAME AND DATE REPORTED

FIRST JUNCTION	SECOND JUNCTION	THIRD JUNCTION	FOURTH JUNCTION

Figure 6-18.   Train waybill

As a general rule, all of the shipping papers will contain:

1. Shipper's name and address
2. Consignee's name and address
3. Proper shipping name
4. Proper classification of the shipment
5. Total quantity by weight or volume
6. A certification by the shipper that the shipment has been properly prepared.

The truck bill of lading shown in Figure 6-17 is read as follows:

The E. I. DuPont De Nemours and Company, Elastomers Department is sending a shipment to ABC Foam, Inc., of Chester, Pa. The shipment consists of 10 drums of a poisonous liquid "toluene Diisocyanate." The weight of the shipment, picked up in one spot, is 5,100 pounds (2,295 kilograms).

Note that the number for CHEMTREC (the Chemical Transportation Emergency Center) appears right on the bill of lading.

The shipping papers are kept in the cab of the truck by the driver. They are sometimes marked in a special manner (Figure 6-19). On trains, the conductor keeps all of the papers.

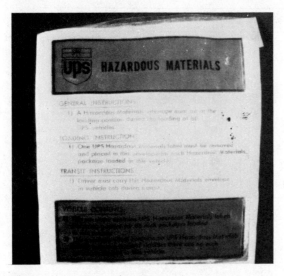

Figure 6-19. Special container for carrying hazardous materials shipping papers

They may therefore be found in the caboose or the engine, depending on the location of the conductor. On barges, the papers are maintained in the pilot house of the tugboat. For air shipments, the papers are kept in the cockpit of the cargo plane.

For shipment by rail, there is a freight waybill for each car in the train. The conductor has these waybills generally arranged in sequence, starting with the first car behind the engine, which would be the first waybill. The fifth car from the front would be the fifth waybill in the stack. At a derailment, you can determine the last car at both ends which remained upright. Once the waybills for the last upright cars are located, those papers in between represent the derailed cars. A quick search of these papers will indicate if there are any hazardous commodities on board.

The waybill shown in Figure 6-18 is read as follows:

The "Car Initials and Number" (NATX 34642) indicate the owner (National Tank Car) and the car serial number (34642). If the car is carrying a hazardous chemical, the indication is stamped on the waybill as close to the car number as possible (DANGEROUS, EXPLOSIVE, POISON GAS, RADIOACTIVE). The left side of the form provides the information on where the material is going (Ephrata, Pa.), while the right side shows the shipper (Consolidated Gas Supply Corp., Clarksburg, W. Va.). At the bottom of the waybill is the description of the material being carried. In this case, it is T/C (tank car) of Liquefied Petroleum Gas, Flammable Compressed Gas Propane. There are 31,994 gallons (128 cubic meters) of propane at 60°F (15.66°C).

Approximately 40 products, when carried by rail, must have the product name stenciled on the outside of the tank vehicle in letters at least 4 inches (10 centimeters) high. This requirement is in addition to the required placard. In Figure 6-20, the stencil ethylene dibromide is clearly visible. Emergency response personnel must check all stenciled items on the rail car for any clues to the material which it contains.

One of the problems with using shipping papers for information was pointed out in a study prepared for the Coast Guard by Arthur D. Little, Incorporated.

> Unfortunately, the "proper shipping name" (or similar designations) specified in the regulations is not sufficient to identify many hazardous chemicals, particularly those whose "names" refer to the type of commodity (e.g., alcohol, corrosive, oxidizing agent) rather than to the specific chemical. Designations for proprietary materials also inhibit ready identification. Thus the "proper shipping name" may sometimes be inadequate for identifying the spilled material.[7]

One other document that could be of assistance during a rail incident is called a "consist" or "train list." (The name for this list varies from rail line to rail line. Personnel must check their local area for the correct name.) This is a computer-generated list of all of the freight cars in the train, in sequence starting at the head or end with the engine. The commodity as well as hazardous marking are noted on the consist. A copy of this list is located at the freight yard where the train originated, as well as at the yard where it finishes its trip. In addition, the conductor may keep a copy in the caboose.

By consulting the shipping papers, emergency response personnel can find out additional information above and beyond that given by labels and placards.

Figure 6-20.   Rail car stenciled with commodity name, ethylene dibromide

REFERENCE SOURCES

The large numbers of hazardous materials that could be involved in an incident make it impossible for an individual to memorize all possible procedures. For this reason, emergency response personnel must be familiar with a wide variety of reference sources. Each source can provide the incident commander with another piece of the puzzle of identification of the product and the hazards. The time to assemble the information and understand its use is before the incident and not while it is on-going.

Chemical Transportation Emergency Center (CHEMTREC)

The Chemical Transportation Emergency Center (CHEMTREC) is funded and staffed by the Manufacturing Chemists Association. Around the clock there is an individual on duty to provide assistance to emergency response personnel in handling a hazardous incident. CHEMTREC will provide information not only on transportation accidents, but any time there is an incident involving hazardous materials.

The CHEMTREC communicator was chosen specifically for the ability to remain calm during an emergency and not because of any special chemical knowledge. The communicator will thus not interpret any information, but will read the material prepared by other experts.

CHEMTREC is set up to provide two very important functions. First, if the product has been identified, information from the files will be read by the communicator. Second, if the product is unknown, but other facts such as shipper, manufacturer, or trade name are known, the communicator can tap many other sources to obtain information.

Once the manufacturer of the product is known, they will be contacted to provide direct, expert information to the scene. In addition, if the incident is severe enough, the manufacturer will send expert help directly to the scene. The shipper is also notified so that they, too, can provide on-the-scene assistance.

Mutual aid programs exist for some products, whereby one producer will service field emergencies involving another producer's product. In such cases, initial referral may be in accord with the applicable mutual aid plan rather than direct to the shipper. Arrangements of this sort are established on chlorine through the Chlorine Institute and on pesticides through the National Agricul-tural Chemicals Association (NACA).

The former has CHLOREP, the Chlorine Emergency Plan, in which the nearest producer responds to a problem. NACA has a Pesticide Safety Team Network (PSTN) of some 40 emergency teams distributed throughout the country. CHEMTREC serves as the communication link for both programs.

To reach CHEMTREC in an emergency situation you can call toll free from anywhere in the continental United States. From all other locations dial direct to the number shown. You must use a long distance access code if one is required in your area. Remember, however, that these are emergency numbers and should not be used to obtain general information.

TOLL FREE (Continental United States)
(800) 424-9300
Washington, D.C. Area   483-7616
Outside Continental United States
(202) 483-7616 (collect)

The CHEMTREC communicator will need certain basic information from the incident commander in order to be able to provide information. It should be written down so that it can be supplied quickly. Have a form already preprinted with the information that will be needed. A sample form is shown in Figure 6-21. Provide space for the following information:

Name of Caller
Call Back Number
Shipper
Manufacturer
Container Type
Rail Car or Truck Number
Carrier Name
Consignee
Product Involved
Problem
Location of Incident

MONTGOMERY COUNTY FIRE AND RESCUE SERVICES

CHEMTREC FIELD INCIDENT REPORT FORM

the dispatcher

Obtain the following information if possible and relay to ~~E.O.C.~~

1. Products involved: *Mercuric Cyanide*

2. Quantity *6 25-pound boxes*

3. Type of Container: *Fiberboard box with metal liner*

4. Mixed or single load *Mixed load- other commodities include, several 55-gallon drums of a Flammable liquid*

5. Type of accident (Auto, train, storage etc.) *Truck and car - one box has left the truck and split open on ground*

6. Time accident occurred *0230 hours*

7. Number & types of injuries: *Truck driver has walked in spilled material - Car driver trapped in vehicle*

8. Weather and temperature conditions: *Cloudy with rain threatening - temperature 62 °F*

9. In populated or open area *Fairly populated area*

10. Carrier name and type of carrier(s) *Murray's Trucking Co No address as yet available*

11. Rail Car Number: *N/A*

12. Truck Trailer Number: *MAI 64246*

13. Being sent to: *Bowling Green Kentucky*

14. Being sent from: *Collegeville, Pa*

15 Bill of lading or Waybill No. *88127BG*

Figure 6-21. Sample form for gathering CHEMTREC information

If the incident commander cannot locate the shipping papers and the identity of the material is unknown, CHEMTREC can utilize the name of the shipper or manufacturer, and/or the rail car or truck number to trace the cargo back to its point of origin (Figure 6-22). Here the tank car number is DUPX 9940. On trucks the tractor number, trailer number, and/or license plate can provide CHEMTREC with information with which to begin the search.

Figure 6-22. Tank car number (DUPX 9940, which can be used to determine the commodity)

One other important function for CHEMTREC is to convert trade names of products to their common chemical names. They have access to the components of over 50,000 trade-named products. Thus, if the incident commander only knows the trade name, the hazards of the product can still be determined. Figure 6-23 shows a tank car with the trade name "Genetron" stenciled right on the car.

One word of caution when talking to the CHEMTREC communicator. Remember that giving the names of products over the radio tends to be very confusing. Words and letters can sound virtually the same. A mistake in a single letter can spell the difference between a successful conclusion and a disaster. Spell

chemical names phonetically. Have the dispatcher repeat it back to you before calling CHEMTREC. Ensure that the dispatcher spells the name to the communicator. While time is of the essence, take the time to get the correct information back to CHEMTREC.

You probably remember the children's game called "telephone" in which you line up five or six people in a row. The first one reads a message to the second who passes it on verbally to the third and so forth until it reaches the last person. The last person then repeats the message out loud. Very rarely is the message even close to what was written at the beginning. Getting information to and from CHEMTREC involves many complex items. They must be written down and double-checked to ensure that they do not change as they pass from individual to individual. If a public service telephone is available, the incident commander should call CHEMTREC direct, to reduce the possibility of error.

Pesticide Safety Team Network (PSTN)

The Pesticide Safety Team Network (PSTN), sponsored by the National Agricultural Chemicals Association, provides information, personnel, and equipment to handle clean-up and decontamination of pesticide incidents. There are currently over 40 teams spread throughout the United States, which have been broken down into 10 units (Figure 6-24).

The individual reporting the incident is contacted and advised what immediate steps to take. Then, if the situation warrants, either the manufacturer or the on-duty safety team will respond.

The team is notified by the CHEMTREC communicator when an incident has occurred in the field. Therefore, to request this important service, all that is necessary is to call the CHEMTREC emergency telephone number.

Figure 6-23. Rail car stenciled with trade name "Genetron," which can be used to determine properties of the commodity

If the incident involves a Chevron chemical, they can be called 24 hours a day collect at (415) 233-3737.

Transportation Emergency Assistance Program (TEAP)

In Canada, the Canadian Chemical Producers' Association operates a Transportation Emergency Assistance Program (TEAP). Using regional teams, assistance is provided either by phone or by direct field response. TEAP can be reached by using the following regional numbers:

Atlantic Provinces and Eastern Quebec	(819) 537-1123
Southwestern Quebec	(514) 373-8330
Eastern Ontario	(613) 348-3616
Central Ontario	(416) 356-8310
Southwestern Ontario	(519) 339-3711
Northern/ Western Ontario	(705) 682-2881
Manitoba, Saskatchewan, Alberta	(403) 477-8339
British Columbia	(604) 929-3441

Oil and Hazardous Materials–Technical Assistance Data Systems (OHM–TADS)

The Environmental Protection Agency has developed the Oil and Hazardous Materials–Technical Assistance Data Systems (OHM–TADS) to provide immediate information on hazardous substances to emergency response personnel. The OHM–TAD system records up to 126 different facts for over 1,000 different substances.

Among the facts recorded are common name, synonyms, trade names, containers, flammable limits, flash point, various action levels, as well as personnel safety precautions. In cases where only color or other physical characteristic is known, OHM–TADS can be used to search for a substance that matches the known description.

To reach the EPA, 24 hours a day, call the Emergency Response Center, (800) 424-8802.

Chemical Hazards Response Information System (CHRIS) of the Coast Guard

The United States Coast Guard has developed the Chemical Hazards Response Information System (CHRIS) for use during emergencies involving the waterborne transport of hazardous chemicals. This system was created on the basis of an evaluation of the decision processes and procedures applicable to spills of hazardous chemicals into navigable waterways.

Under the CHRIS system, four handbooks were produced. These are:
Manual 1. A Condensed Guide to Chemical Hazards

Provides chemical-related information to be used during the early stages of an incident involving the accidental discharge of a hazardous chemical. Provides descriptive infor-

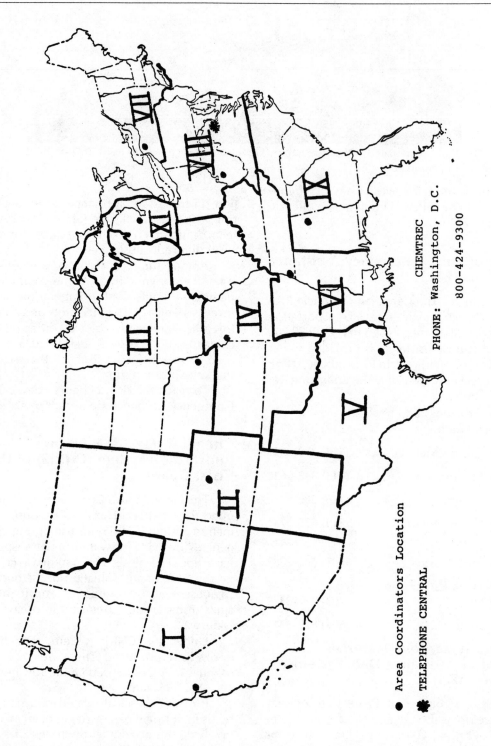

Figure 6-24. Pesticide safety team network locations

mation on the hazardous nature of the chemical; recommends immediate actions to take to safeguard life and property; and provides precautionary advice on the chemical, physical, and biological hazards posed by the material.

Manual 2. Hazardous Chemical Data

Provides specific information on the chemical properties of the hazardous material. It is very technical in nature, requiring a background in chemistry to understand the material. It does give detailed, quantitative information on health hazards, flammability, and physical properties.

Manual 3. Hazard Assessment Handbook
(and Computer System)

Provides trained field personnel and other hazardous material specialists with methods and procedures for estimating the magnitude and location of the threat presented by the potential or actual release of a hazardous chemical. Provides an approach to the selection and use of quantitative methods for predicting the many physical processes that may occur upon the release of a wide variety of hazardous chemicals. It includes procedures for predicting the rate of release of the chemical from its container, the movement and dispersal of the chemical in water and/or air, the thermal radiation from fires, and the area over which the resulting toxic, thermal, and explosive effects may threaten surrounding areas.

Manual 4. Response Methods Handbook

Presents methods for selecting specific response procedures based on the chemical discharged and the existing conditions during the incident. Provides information on commodity transfer; containment and motion control; removal; and chemical, physical, and biological treatment, dispersal, and flushing.

Chemical Data Guide for Bulk Shipment by Water

The *Chemical Data Guide for Bulk Ship-*

ment by Water, CG-388, was prepared by the Coast Guard and revised in 1976. This fifth edition of the guide provides information on 279 different hazardous chemicals. Data provided include synonyms, physical properties, fire and explosion hazard, health hazard, reactivity data, and spill and leak procedure.

National Fire Protection Association Texts

The National Fire Protection Association (NFPA) publishes two texts which can provide the incident commander with immediate information at the scene of a chemical incident. These are the *Fire Officer's Guide to Dangerous Chemicals* and the *Fire Protection Guide on Hazardous Materials*, which are available from National Fire Protection Association, 470 Atlantic Avenue, Boston, Massachusetts 02210.

Since both of these texts contain very technical material, emergency response personnel should become familiar with their contents before the incident. In this way, the incident commander will know exactly where to look instead of thumbing through the text, while confusion reigns at the incident.

This is particularly necessary for the *Fire Protection Guide on Hazardous Materials* because it is made up of five different NFPA texts. The information provided on the products includes:

1. Flash point and manufacturer by trade name
2. Properties of flammable liquids, gases, and solids
3. Flash point
4. Ignition temperature
5. Flammable limits
6. Specific gravity
7. Vapor density
8. Boiling point
9. Water solubility

10. Extinguishing methods
11. Hazardous chemical data
12. NFPA 704 symbol for the chemical
13. Fire and explosive hazard
14. Life hazard
15. Personal protection requirement
16. Fire fighting technique
17. Usual shipping containers
18. Storage facilities
19. Hazardous chemical reactions
20. Explanation of the 704 system

Other On-Site Guides

In addition to the previous guides listed, several others are published. These might provide further information for personnel supervising the control of the incident. Because costs change rapidly, none are given. Write directly to the addresses shown for further information:

Emergency Handling of Hazardous Materials in Surface Transportation
Bureau of Explosives
Association of American Railroads
1920 L Street
Washington, D.C. 20036

Emergency Services Guide for Selected Hazardous Materials
Department of Transportation
Office of Hazardous Materials
Washington, D.C. 20590

The Fire Fighters' Handbook of Hazardous Materials
Fire Fighters' Handbook
P.O. Box 31009
Indianapolis, Indiana 46231

Dangerous Materials Emergency Procedures Slide Rule
Able Fire and Safety Equipment Company
2737 W. Fulton Street
Chicago, Illinois 60612

Chemical Transportation Safety Index
Graphic Calculation Co.
234 James Street
Barrington, Illinois 60010

Dangerous Properties of Industrial Materials
Van Nostrand and Reinhold Co., Inc.
300 Pike Street
Cincinnati, Ohio 54202

Fire Officers Guide to Dangerous Chemicals
Fire Officers Guide to Emergency Action
National Fire Protection Association
470 Atlantic Avenue
Boston, Massachusetts 02210

Local Sources of Information

In your local area there are also many sources of help to which you can turn. Develop a list of them and their telephone numbers so that they can be readily contacted during an incident. These include:

1. Poison control centers
2. Medical personnel at hospitals
3. Chemical dealers and manufacturers
4. State and local environmental agencies
5. State and local civil preparedness agencies
6. State university agricultural agencies

Summary of Emergency Numbers

In addition to the 24-hour telephone numbers which have been discussed, emergency response personnel should prepare a list of other needed numbers:

CHEMTREC	(800) 424-9300
PSTN	(800) 424-9300
TEAP	
Atlantic Provinces and eastern Quebec	(819) 537-1123
Southeastern Quebec	(514) 373-8330
Eastern Ontario	(613) 348-3616
Central Ontario	(416) 356-8310
Southwestern Ontario	(519) 339-3711

Northern/
   Western Ontario   (705) 682-2881
Manitoba, Saskatchewan,
   Alberta   (403) 477-8339
British Columbia   (604) 929-3441
EPA and Emergency
   Response Center   (800) 424-8802
Public Health Service   (404) 633-5313
National Foam   (215) 363-1400
Department of
   Transportation   (202) 426-1830
Dow Chemical Company   (517) 636-4400
Chevron Chemical Company   (415) 233-3737
DuPont Company   (302) 774-7500
Poison Control Center   _____
State EPA   _____
Local EPA   _____
Emergency Preparedness
   Agency   _____
State DOT   _____

Local DOT   _____
Coast Guard Base   _____
Federal Aviation
   Administration Office   _____
Ordinance Demolition Team   _____
Pipeline Companies Which
   Go Through Area   _____

Gaining information on the actual material being carried is the first key job of the emergency response personnel at the scene of an incident. By knowing how to interpret the placards, labels, and shipping papers, the incident commander receives the first clue to identifying the hazard. Next, the many reference sources available must be used to determine facts about the chemical as well as suggestions for handling the incident. All of this must be planned in advance to reduce the confusion at the incident.

QUESTIONS FOR REVIEW AND DISCUSSION

1. Describe the difference between a placard and a label.

2. Describe the regulations for applying placards and labels.

3. Describe the physical characteristics of each label and placard, including color and symbol.

4. Explain how the required placard and label for a given chemical are determined.

5. Describe the circumstances in which dual labels or placards are required.

6. What circumstances require that special labels be applied?

7. Explain the use of the category "other regulated material (ORM)."

8. When are exemptions to the placarding and labeling system granted?

9. Explain the NFPA 704 system and detail how information can be determined from it.

10. Describe the military placarding system.

11. Obtain a set of rail and truck shipping papers. Prepare a description of all the information they contain.

12. Prepare a list of the information that emergency response personnel should supply to CHEMTREC when requesting assistance.

13. What information will CHEMTREC supply to emergency response personnel?

14. What supplementary sources can CHEMTREC call upon to contact emergency response personnel directly in the field?

15. Prepare a list of responsible agencies in your area who can be called upon for assistance at a hazardous materials incident.

16. Prepare a list of reference texts which emergency response personnel should have available in the event of a hazardous materials incident.

NOTES

1. *Code of Federal Regulations, Part 49, Parts 100 to 199,* Office of the Federal Register,

Superintendent of Documents, Washington, D.C., December 31, 1976, pp. 148–151.

2. Ibid., pp. 138–141.
3. Ibid., pp. 42–132.
4. *OHM Newsletter*, Materials Transportation Bureau, Department of Transportation, Washington, D.C., July 1976, pp. 3–5.
5. *Code of Federal Regulations, Part 49, Parts 100 to 199*, Office of the Federal Register, Superintendent of Documents, Washington, D.C., December 31, 1976, p. 236.
6. *Federal Register*, Superintendent of Documents, Washington, D.C., vol. 42, no. 152, August 8, 1977, pp. 40003–40006.
7. Donald S. Allen, *Chemical Hazards Response Information System for Multimodal Accidents (CHRISMA). (A Reevaluation of CHRIS for All Modes of Transportation)* (Cambridge, Mass., Arthur D. Little, Inc., NTIS, April 1975), p. 25.

7
Preplanning for incidents

Preplanning is one of the means by which emergency response personnel prepare to control an incident instead of waiting for something to happen and then rushing in blindly. Preplanning is the preparation made by the emergency response personnel for handling a specific incident which has not happened so that all personnel can do an intelligent and efficient job in an organized manner. The loss of lives and property can be reduced by planning ahead.

The preceding chapters have made it clear that hazardous materials are almost unlimited in number and characteristics; that a single hazardous material can react differently when subjected to various conditions; and that hazardous materials can be found in innumerable places in our communities where they are manufactured, stored, used, transported, and disposed of. All these facts make preplanning a very important, although a very diverse and difficult task.

FACTORS AFFECTING PREPLANNING

Preplanning is most crucial because at an actual incident, sizing up the situation may be difficult. Accessibility may be poor, vapor clouds may delay or obstruct the approach, darkness or smoke may obscure vision, radiated and/or convected heat can make a close approach difficult if not impossible, toxic fumes can be dangerous if inhaled or absorbed, and still decisions must be made to handle the situation. Decisions, must be made within a limited amount of time and often under considerable duress because of the

possible impending severity of the incident. Generally, with hazardous materials, these decisions will be better made not on what can be seen at the time, but on what was learned during preplanning.

The more information that the emergency response personnel have learned during preplanning, and the more information they have gained about the "unexpected" that may occur, the greater safety and protection they will be providing for themselves and other persons at the incident. Preplanning involves not only the gathering of information, but also the development of tentative action plans including tactical operations, apparatus placement, initial attack procedures, rescue, and command post initiation. These should be subsequently practiced in drills and reviewed, so that changes can be made if necessary to further minimize losses at the time of an incident.

Although some areas normally surveyed in preplanning are covered in this chapter, the emphasis will be on the special considerations necessary for hazardous materials. Contingency planning for a major disaster will be covered in Chapter 9.

PREPLANNING RESPONSIBILITY

Although planning initiated by a single company or shift will greatly improve their ability to cope with the hazardous materials emergency, the preferred method of planning is one that is organized through and endorsed by the administrators of the entire emergency response delivery system. This assures allotting time for the surveys, forms, and procedures to produce uniformity throughout the system, and review and dissemination as needed. Preplanning can often be enhanced by soliciting the cooperation of industry. Considerable help and technical assistance can be gained from the chemists, engineers, and safety personnel who work with the products daily. With the proper backing, special sections, such as the photo unit, can be utilized to assist in the planning. Also the manner of storage and retrieval of the finalized plans will become departmental policy.

COMPANY-LEVEL PREPLANNING

Company-level preplanning consists of a study to identify the location of hazardous materials, a pre-incident survey with inspection of the premises and/or transportation routes to determine special problems and resources, a tentative plan for dealing with the potential incident, practice in carrying out the plan, the evaluation and revision of the plan as necessary, and the printing and distribution of the final plan.

HAZARDOUS MATERIALS STUDY

The first step in preplanning is the identification of locations of hazardous materials in the local area. A study should be made by the first-due company to determine all fixed installations having hazardous materials as well as the transportation routes of hazardous materials through the response area. Installations with hazardous commodities and the hazardous material shipping routes should be indicated on a map to be distributed to all responding companies, appropriate chiefs, and fire alarm dispatchers. An example of a hazardous materials study map is shown in Figure 7-1.

Review of the places to check for hazardous materials discussed in Chapter 1 will be helpful in making the study to identify fixed locations. Information on fixed installations having hazardous materials can be found through company members' prior knowledge of such occupancies, a systematic drive through the response area, scanning the yel-

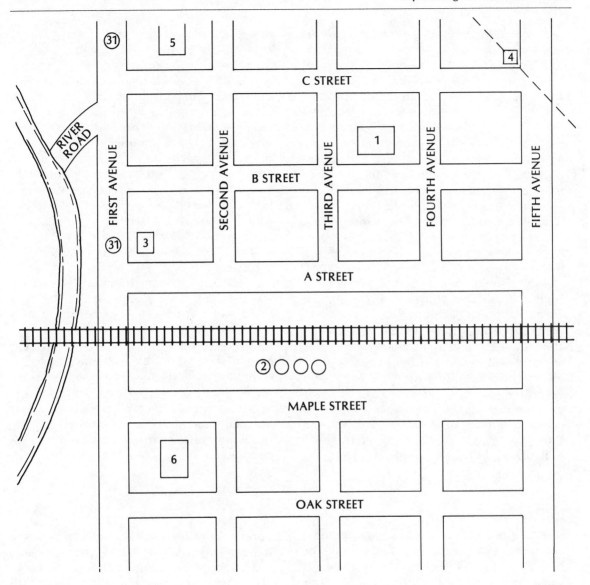

Figure 7-1. Map illustrating fixed installations and transportation routes where hazardous materials are found

Legend

1 High School (chemistry lab)
2 Bulk Plant (flammable gas and anhydrous ammonia)
3 Lawn and Garden Store
4 Pipeline Pump Station
5 Hospital

6 Shopping Center: Paint Store, Beauty Shop, Drug Store, and Grocery Store

|||||| Railroad

— 4 — Pipeline

31 Highway

low pages of the telephone book, and reviewing fire prevention inspection reports.

Most emergency response personnel are familiar with preplanning buildings and other stationary installations in the local area, but what about the commodities that pass through the jurisdiction? Major hazardous materials transportation routes (rail, highway, air, pipeline, or waterway) should also be preplanned. This will require a general study of the major roadways, railroads, airport approaches, waterways, and pipeline right-of-ways. This can be accomplished by studying the transportation modes. It may be necessary for a company to make a traffic count at a highway intersection of the placarded vehicles using the route over a set period of time. Highways can also be checked by visiting a weigh station or talking to local freight terminal dispatchers. Railroads can be surveyed along the right-of-way, by checking placards at classification yards, or by discussing shipments with a rail dispatcher.

Waterway transport of hazardous products is more difficult to determine. Local docks or terminals can be checked and the Coast Guard contacted concerning frequent bulk hazardous material shipments. The commodities transported by air must be checked at air freight terminals. The cargo containers for air freight are placarded; the shipping papers identifying the product would be with the agent. Information regarding pipeline routes can be gained from observing the pipeline right-of-way markers, asking pipeline companies, requesting information from the state department of transportation, and asking natural gas utilities about distribution line routes. The route of pipelines would be indicated as well as whether they are liquid or gas. The commodities change frequently, and it will only be possible to ascertain the general products transported from the pipeline company.

Companies on in-service fire hazard inspections and fire prevention inspectors must be ever alert to detect the presence of hazardous materials and report them to the first-response companies. Such reports assure a continual update of the hazardous materials study. In addition, an annual comprehensive review of the study should be made and any necessary changes incorporated. In conjunction with the study, a pre-incident survey should be made for each location and transportation route.

PRE-INCIDENT SURVEY

All pre-incident surveys should begin with a letter of introduction to the owner or occupant requesting permission for a visit. An example is shown in Figure 7-2. The letter should explain the purpose of the survey and stress that it will enable the department to render better service in the event of an incident and that it is not intended to find fire safety violations.

The survey should be scheduled at the convenience of the occupant or owner and every attempt should be made to gain full cooperation because the firm has valuable experience in dealing with the materials and the processes in which they use them. Information on health hazards and reactions of the materials when mixed, heated, or exposed to water should be sought, along with any special properties that would be significant during a fire or leak. The emergency response personnel should learn what special protective clothing will be needed at the location. In many cases chemicals will penetrate or destroy normal protective clothing and natural rubber or polyethylene may be needed. Also, in a manufacturing or processing area, perhaps some kind of ear protection against excessive noise levels may be required. Management should be consulted about special storage provisions, as well as where the inven-

ANNBURG
FIRE DEPARTMENT

Tom Hawk Industries
4545 Main Street
Annburg

Dear Sir:

Your fire department has a responsibility to protect you,
your employees, and your property from fire. In our efforts
to minimize losses within the community we have developed a
program of Pre-Incident Planning.

The intent of the program is to make the fire department
familiar with your property in the event of an emergency.
This will enable us to better cope with any situation which
may arise.

This program is designed to have members of the department
visit your facility and gather vital information for planning
purposes before an incident occurs. The material will be
formalized and made available to command personnel on the
emergency scene. This will allow the problem to be handled
in a more effective manner, in a shorter period of time, with
increased safety to the personnel at your facility, and the
members of the fire department.

We request that when you are contacted and an appointment is
set up for a Pre-Incident Planning survey that you assign a
responsible individual with thorough knowledge of your
facility to accompany and assist the fire company making the
survey. During the survey we will obtain the data necessary
to make sketches of your facility, take photographs of
important features, and complete the information forms
required. Tentative action plans will then be developed
for apparent or anticipated problems that could occur during
an emergency. The fire department may subsequently schedule
a drill on your premises to realistically test the plan. This
will be done at your convenience in order to cause no undue
concern or interruptions.

While it is the fire departments responsibility to prevent
fires through inspections and code enforcement, this is not
the purpose of this survey. Your assistance with our Pre-
Incident Planning Program will enable us to better serve you.

Truly yours,

John Campbell
John Campbell
Chief

Figure 7-2. Sample letter of introduction to explain purpose of hazardous materials pre-incident
survey

tory records are kept. If the firm has not already established the practice, it can be suggested that a duplicate copy of the inventory be maintained at a remote location where it could be easily obtained should an incident occur. Information should be requested from the occupant on whom to notify in case of emergencies and how to contact them. A number of records useful in planning can often be obtained from management. Among those to seek are plant maps, and diagrams of the sewers and drains, product pipelines, electrical system, and fire protection system.

Plant fire protection personnel and emergency response personnel should agree on the functions of both groups during an incident. The limitations and capabilities of each should be discussed so that the efforts of each can be complementary. Important decisions such as when critical processes should be shut down, power disconnected, or private protection valves closed can be planned with the occupant's assistance. The manager should also be questioned as to possible resources that could be provided to the community in the event of an incident or disaster outside his own facility.

The pre-incident survey is an inspection to gain familiarization and collect information while inspecting the premises or routes, making sketches, drawings and/or photographs of pertinent features that can be used in the formation of the tentative action plan and as references in the event of an actual incident.

The survey can be done by a single company or the entire first-response team. Recently some cities have developed programs that employ several full-time investigators who do all the surveying for the community. Company surveys have the advantage of familiarizing personnel directly with information of the location.

The pre-incident survey must consider the response route, accessibility, building-occupancy, exposures, water supply, storage, fire protection, and utilities.

Response Route

The survey should begin by inspecting the response routes and making notes of such obstructions as grade-level railroad crossings, bridges, traffic congestion, busy intersections, freeways and interstates, cloverleafs, hills, curves, narrow streets, and terrain (Figure 7-3). The location of water supplies and the possibility of responding from two different directions are important considerations. The approach to the scene must consider the position of fixed storage tanks, the probable position of tank trucks or rail cars at the time of a wreck, and relationship of those tanks to other tanks. How to proceed in the event of vapor clouds must be considered, because they could be an obstruction to response and the initiation of immediate actions.

Figure 7-3. Access to this site would be by the nearest bridge or grade crossing. Response from both directions would probably be necessary.

Accessibility

Accessibility to and into the building or complex is vital. Obstructions at the scene, such as the fences illustrated in Figure 7-4, could seriously hamper emergency operations. Considerations for accessibility to the incident scene are as follows:

Entrance ways should be checked to see if they are passable at all times. The possibility of approach from all sides must be considered, because many complexes are near rail-

Figure 7-4. Obstructions such as parked equipment, fences, and barrels of hazardous materials should be noted in preplans.

Figure 7-5. A loading rack berm is designed to limit the spread of a spill in case of a tank overfill.

roads and are therefore cut off on at least one side. The location of fences and locked gates should be noted. Since forcible entry may be required, the spot where forcible entry can be made quickly, easily, and with the least damage should be known.

Hazardous materials stored in the area in tanks or drums will be of concern and their locations must be indicated. If hoses or other equipment must be carried some distance, this should be noted since additional personnel may be necessary.

The ground slope can hamper approach if there are flowing liquids, and the prevailing wind direction could bring vapor problems if there is a leak. Both of these conditions should be observed and noted during the survey.

Inside the facility it is good to know where obstacles exist, such as fenced security areas, special hazards, doors to basements or other dropoffs, and any mantraps like chutes or open shafts. If a guard dog is on the premises, this should be noted along with the security system used.

Building-Occupancy

If the hazardous material location includes a building, its address, direction of frontage, and size should be recorded as well as major construction features, including roof details for ventilation. Special note should be taken of the type of heating plant and its location, the number of floors, and uses of attic, mezzanine, or basement areas. Any dewatering devices such as scuppers, drains, sump pumps, or usable doors can be important, as well as toilet locations. Loading rack berms (Figure 7-5) and drains should also be designated.

The occupancy of the building must be noted and whether the business is engaged in manufacturing, processing, or warehousing. The hazardous materials should be identified by classification, name (chemical and trade), and volume or amount so that additional information about the products can be obtained. The number of employees, their hours on the premises, and the location of any invalids or handicapped must be learned.

Exposures

In surveying the exposures, the locations of other buildings, tanks, or equipment must be recorded in addition to the life hazards. Special note should be made of nearby target hazards. The survey should consider the type of construction, susceptibility to fire, distance, and occupancy. The possible effects of weather conditions, radiated and convected heat, and prevailing wind direction should be taken into account. Exposures downwind are of major concern when fumes become a problem and intakes to ventilation systems may have to be shut down. Notes should be made of exposures downgrade in the event of possible spills, leaks, or liquid flows; details should also be given of exposures in a circular pattern in all directions which could be endangered in case of a rupture or explosion.

Where flammable liquid or gas tanks are exposed within a complex, as shown in Figure 7-6, in an external exposed group, or as vehicles parked nearby at night, the distance, direction, product, tank height and construction, and possible spread by grass or weeds will all be of interest. The gradient and prevalent wind direction should be noted for any exposed wires, streams, or sewers. The need for spill or runoff control equipment should be indicated.

Water Supply

The available water supply should be ascertained by fire flow tests if unknown. In-

formation on these processes can be obtained from the American Mutual Insurance Alliance, American Insurance Association, or International Fire Service Training Association. For determining hose needs, all hydrant locations and main sizes should be included on the water supply drawings, including the distance to the hydrants. Extensive hose requirements which may be necessary for water supply should be noted. Distances can be quickly and accurately obtained by using a low-cost range finder. Any available static water sources, including swimming pools (Figure 7-7), irrigation canals, ponds, creeks, rivers, cisterns, wells, and lakes that could be used for drafting, should also be recorded. Notes should be made of the need for special pumping operations such as relay pumping or tanker shuttles, which may be necessary to reach large truck terminal areas, a railroad classification yard, or remote areas along transportation routes and pipelines. Where the water system will allow dual pumping or require supplementary pumping, provisions should be made to accomplish this. Water supply preplanning should also consider how long the supply will be needed and whether the sources are of adequate duration.

Figure 7-6. Large, internally exposed complexes such as this flammable gas storage area pose special fire fighting problems.

Figure 7-7. Static water supply sources should be included in preplanning information. *Courtesy of William Eckman*

Storage

When gathering information on storage, in addition to the name, DOT classification, and possibly the chemical makeup, the amount and manner of packaging, storage location, reactivity of the material, and toxicity should be considered. Where in the building or on the premises is the material located? What is the type of storage? Is the material protected in any way or the area cut off? Are the hazardous materials segregated or any particular ones separated? What is the reactivity of the material? Will it react with oxidizers and ordinary materials found at the location? What are the toxicity characteristics of the material? What are the toxic symptoms to watch for, antidotes, and special precautions that must be taken during fire suppression and overhaul? If necessary, during operational size-up, the need for gas detecting or analyzing equipment or radiological monitoring devices must be noted. Any requirement for nonsparking tools should be indicated. Are special extinguishing agents required? Facility managers can be encouraged to mark their storage areas with the NFPA 704 system as shown in Figure 7-8 to alert emergency response personnel to the dangers.

Where large quantities or particularly exotic materials are found, the manufacturer should be identified in case a request for technical assistance should become necessary. If possible, chemical data sheets on the materials in frequent usage or storage should be obtained.

When horizontal tanks are in use, the type of support for the tanks should be noted. The support can be unprotected steel, a concrete saddle, or masonry saddle; some tanks are placed directly on the ground. In surveying vertical tanks, the roof construction should be recorded. The roof may be conical with a weak welded seam to relieve overpressure, a floating roof type, or a lifter type with a liquid seal. The pressure the tank is designed for should be determined, and recorded. Liquid tanks are categorized according to pressure as atmospheric, low, or pressure tanks. Construction features of cryogenic, compressed, and liquefied gas tanks should be noted. The location and arrangement of the tanks should be shown by diagrams or photographs.

Information on the piping layout, as shown in Figure 7-9, should be obtained, including where the pipes run, how they are protected or whether they are exposed to

Figure 7-8. NFPA 704 symbols alert responding emergency personnel to nature and degree of hazard.

Figure 7-9. All piping and valving arrangements, including color codes, should be noted in preplans

physical or fire damage, and what the color coding designates. Valve locations are important for adequate, safe control of flammable liquid and gas fires. Their location should be pinpointed so that they can be quickly reached under the duress of an emergency situation. Notes should be made of what they control, how they operate, and the type, manual lock, spring loaded, internal self-closing, external gate, locked, remotely controlled, and whether they operate properly and have fusible links. The type of vents on the tanks, whether they are of proper size to prevent overpressure, have a flame arrestor in good condition, and operate properly must be noted.

The storage of hazardous materials other than in bulk tanks is extremely important. How they should be stored to meet recommended practices or codes will be discussed in detail in Chapter 16. The materials that come under this category are:

1. Gases that could be flammable, toxic, corrosive, liquefied, compressed, or cryogenic in cylinders;

2. Liquids that could be flammable, combustible, toxic, corrosive, radioactive, oxidizing, or water-reactive and are stored in bottles, cans, drums, carboloys, plastic jugs, dip tanks, or unsuspected containers:

3. Solids that might be explosive, flammable, combustible, toxic, corrosive, radioactive, oxidizing, or water-reactive, stored in bags, barrels, drums, cartons, carts, or bulk;

4. Combustible fibers or plastics that give off toxic combustion products;

5. Dusts that are explosive, combustible, toxic, corrosive, oxidizing, or water-reactive, in containers similar to those listed above;

6. Chemicals that could be flammable, explosive, toxic, corrosive, oxidizing, water-reactive, are unstable, sensitive to heat, shock, friction, pressure, air, or mixing;

7. Explosives that could detonate or deflagrate, either Class A, B, or C.

If possible, information should be obtained on the expected results when materials are stored together and accidentally mixed during a spill or fire. Be especially alert for unmarked materials that could be hazardous at the time of a fire. Examples of such materials would include aerosol cans, pesticides, fertilizers, resins, plastics, or materials stored in small quantities such as ORM-D. Vehicles parked on the premises may also be loaded with hazardous materials. This may create an additional exposure problem, but in many cases they can simply be moved.

Fire Protection Features

Full information about the fire protection facilities on the premises and their locations is one of the most important factors in preplanning. That portion of the installation protected by sprinklers, standpipes, foam systems, or specialized extinguishing systems should be clearly indicated on plan sketches. Locations of the following should be marked: Automatic sprinklers with test, drain, and sectional valve controls and siamese connections, fire extinguishers, hose cabinets, fire doors, automatic venting and control switches, total flooding systems, and at bulk tank locations, dikes, dike drains, bonding arrangements, and protection against runoff. The conditions and maintenance of these devices are important to note, and when the devices were tested last. How the facilities can be utilized and what action is required for their usage will be vital to fire control. The installation and location of private hydrants, their capability, and the effect on other systems if they are used will all have important bearing on incident control.

Utilities

Locations of utility cutoffs for controlling

operations should be recorded and marked on diagrams. These include telephone lines, master switches for electrical power in buildings, to the site, and in overhead lines, ventilation and blower systems, natural gas shutoffs, water main controls, and hazardous material control valves from pumps, tank loading racks, conveyors, or pipelines. Be sure to include remotely controlled valves. In many instances, the electrical power will have to be shut down outside the fire area to eliminate the possibility of igniting fumes.

Transportation Routes

In surveying transportation routes of hazardous materials there are some considerations that are similar to surveying fixed installations and some that differ. Information learned during the hazardous materials study regarding shipment frequencies, containers, and amounts should be included in the preincident survey. All information should be recorded in a map format. The entire route is surveyed and mapped. Then when an incident occurs, the specific location can be placed on the map.

Maps of the transportation routes, highways, railroads, navigable waterways, pipelines, or airport runway approaches should be prepared. In some instances, this will be one map including all routes, while in others, because of the congestion, a series of maps will be required. In lieu of maps, aerial photographs of the routes could be used and would be beneficial, since all structures would also be indicated. Clear plastic overlays can be placed over the photos on which to mark information for planning and for use at incidents.

These route maps will indicate where the various modes of transportation traverse an area and whether it is rural, commercial, industrial, sparsely populated, or heavily populated. Thus, serious exposures and target haz-

ards can be indicated. Also worthy of noting on the maps would be hazardous material installations (truck terminals, rail yards, air freight terminals, docks, and pumping stations) and the frequencies of deliveries.

Evacuation corridors should be indicated on these maps, photographs, or overlays. This information will give those responding to an incident an indication of the problem at hand if evacuation is necessary. The corridor should indicate the mile evacuation distance.

Any accessibility or obstruction problems as would be found in reaching portions of a railroad, pipeline, or waterway should be indicated. Plans could then be made on how to deal with a situation in these remote sections. Areas where the routes are above grade or below grade should be indicated. Electrical lines should be shown, especially those of high voltage, which could be hazardous. Another concern along transportation routes has to do with runoff and spill control. Changes in topography such as hills, drainage ditches, and surface waterways should be marked on maps. Exposed sewers and manholes should be indicated. This information will indicate where it will be necessary to build a dam or dike if a hazardous material incident develops nearby.

All water sources that could be utilized by the fire department in a hazardous material incident must be noted. If they are static sources, access locations must be indicated and possible periods of nonuse. Since relays are necessary to move the large quantities of water that may be needed and since due to time and hose constraints, 2,500 feet (750 meters) is normally the maximum relay distance, this distance should be indicated. Any incident within that distance would automatically initiate an adequate response of fire fighters and equipment. Areas beyond this distance will necessitate tanker operations and plans can be developed to have a tanker task force respond.

DATA RECORDING

The information gathered during the pre-incident survey must be recorded on a form or series of forms that will display the data in a manner suitable for study, analysis of the problems, and the determination of the tentative action plan. This is after all the reason for the preplanning survey. In order to do this, two types of records will be necessary.

The first type is a series of maps, diagrams, sectional drawings, or photographs illustrating pertinent features and placing them in their proper perspective.

The second type is an orderly form detailing information on the premises. This form should be suitable for use in collecting the data during the pre-incident survey and must be standardized within the department or preferably throughout the mutual aid area. This form enables all responding officers to draw the information they need in a minimum of time since they are familiar with the form and its layout. A cover page with an index might be used, but the most practical is a synopsis on the first page covering the general topic areas of the survey and including a severity rating for each section (Figure 7-10). These forms should be simple yet complete in order to assist in the thoroughness of the survey and to help avoid overlooking any vital details. They should also be of sizable print for ease in reading. Numerous forms and procedures are available for gathering and recording this information. Hillandale, Silver Springs, and Baltimore, Maryland, all have excellent examples.

The detail that goes into the preplanning visuals can range from very basic line drawings to extremely detailed and sophisticated combinations of sketches and photos. In Long Beach, California, a plan or line drawing is combined with photographs of key fire protection or construction features of concern to initial responding emergency personnel. Also included is a map denoting the location of the preplanned complex. Long Beach has also acquired diagrams of storm sewers so that spills or runoff will not appear unknowingly at a distant remote location causing further unsuspected problems. Elk Grove Village, Illinois, uses a series of slides, including one of the completed diagram.

The sketching during the pre-incident survey and the subsequent finalized diagram will provide a considerable amount of information. They can be carried in a book by the responding companies and checked by the officer en route to the scene or quickly referred to for a review of the premises on arrival. Photos taken during the survey will be invaluable. They can be taken inside, outside, of the roof, or aerial views, as in Figure 7-11 to give the overall area layout.

Several sets of symbols are available for use in diagramming. Diagrams should show ample information in a clear, concise manner without becoming cluttered. Hazardous material locations, types and quantities, should be indicated. The NFPA 704 symbols can be used to illustrate the degree of hazards. Symbols used can be color-coded to denote various items, for example: blue for utility shutoffs, green for ventilation avenues, red for fire department equipment or water supply, yellow for hazardous material locations, and orange for valuables. Diagramming is expedited in some departments by having symbols placed on rubber stamps, while others are utilizing a rub-on symbol system. There is the possibility that an area diagram showing boundaries, exposures, streets, compass direction, and water supply information will be necessary in addition to a building diagram. Noted obstructions can be marked on this diagram. A sectional view may be necessary to show floors, heights, or processing sections. Also, in some cases, separate layouts will be required for each floor to show dangerous locations.

Maps of transportation routes, such as shown in Figure 7-12, should be prepared to

PRE-INCIDENT SURVEY
FIRE PROBLEM SUMMARY

PROPERTY NAME _____

ADDRESS _____

GENERAL OCCUPANCY _____

INDEX

0	NO SPECIAL PROBLEM	FLAMMABILITY	**0** NO SPECIAL HAZARD
1	MODERATE PROBLEM	HEALTH REACTIVITY	**1** SLIGHT HAZARD
2	SEVERE PROBLEM		**2** MODERATE HAZARD
			3 SEVERE HAZARD
			4 EXTREME HAZARD

POTENTIAL PROBLEM

☐ LIFE SAFETY
☐ PROPERTY SAFETY

OPERATIONAL PROBLEMS

☐ ACCESSIBILITY and ENTRY
☐ BUILDING EQUIPMENT
☐ CONFINEMENT
☐ CONSTRUCTION
☐ EXPOSURES
☐ FUEL LOAD
☐ HAZARDOUS MATERIALS
☐ PROTECTION EQUIPMENT
☐ RESCUE
☐ SALVAGE
☐ UTILITIES
☐ VENTILATION
☐ WATER SUPPLY

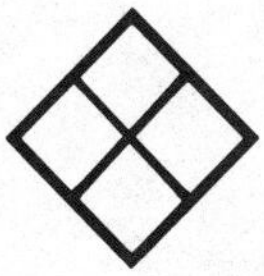

Figure 7-10. Cover sheet of pre-incident survey as a summary to indicate hazardous conditions in various areas

illustrate the possible evacuation areas along the routes, changes in grade that may require damming, the location of sewers, manholes, or open waterways that would have to be protected, the locations of water supplies, the need for special water supply operations, obstructions, and the locations of exposed target hazards. Maps of this nature will give the first-due companies considerable information for planning along the entire route. Once the exact location of an incident is known, the map information can be referred to in determining tactical operations.

The finalized diagrams can be stored and used in a number of ways in addition to being in the hands of responding companies. They can be carried in chief cars or command vehicles. Several cities are now putting the information on microfilm to be carried with other pertinent information such as water and sewer maps and hazardous materials references in a command post vehicle equipped with a microfilm reader. The information can be placed on microfilm and retrieved at the dispatching center with the information then relayed to companies at the incident location. Information can also be committed to slides with selected retrieval projectors that will bring the information to the dispatcher in a matter of seconds. Those departments converting to computerized dispatching with CRT capability in the apparatus could provide preplan information directly to responding officers.

The data can be used not only by first-responding units in their initial operations and safety considerations, but also by subsequent-responding command officers. When dealing with hazardous materials, this is important because of the possible long duration of the operation and the potential for disaster which makes imperative the accessibility of detailed, clear, concise, and accurate information for incident commanders.

Figure 7-11. Aerial photos assist during planning and in final action plan.

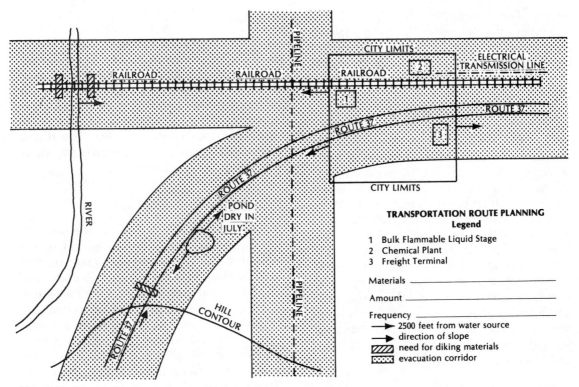

Figure 7-12. Maps of transportation routes in preplans illustrate evacuation areas, topography changes, water supplies, and other pertinent features.

TENTATIVE ACTION PLAN

Once the survey has been completed and the pertinent data collected and organized, all information must be analyzed and put into a usable format. This is the actual planning stage, in which the survey information is converted into tentative actions, operations, and resources required. The utilization and placement of the resources, will also be covered.

The planning effort should be initiated as soon as final drawings and diagrams are ready. The work will be done by the officer and team who made the pre-incident survey based on the data they collected.

The survey material must be reviewed to estimate the probable nature of an incident at the location. Thought should be given to where the fire is most apt to start and how it will spread. Numerous areas of manufacturing, processing, storing, or handling could be potentially hazardous areas. Ignition sources run the entire range from hypergolic and spontaneous to lightning and static electricity. The survey team should consider all possibilities and probabilities.

The first task of the planners will be to determine the objective they are attempting to achieve as they arrive at the scene; for example, rescue, control, or confinement. They, of course, must set the objective and do the planning on the basis of the worst possible situation. Weather is an important consideration that may modify both fireground actions and conditions. Of vital concern at this point are the resources necessary to achieve the objective. The number of fire fighters and equipment needs, especially any specialized equipment or extinguishing agents, should be determined and their availability ascertained. The tentative action plan must also include the establishing of a command post and staff to handle the incident.

With the objective well defined on the basis of the fire location and worst possible situation, the plan should outline the routing of responding apparatus through multiple alarms. The survey material must be reviewed for route obstructions such as shown in Figure 7-13. Tentative placement of apparatus based on accessibility, water supply, and safety considerations should be determined. In the placement, ample consideration must be given during the approach to the wind direction and the flow of vapors and fumes. Apparatus positioned in a vapor cloud, which may be invisible, could be an ignition source. Some vapors, of course, could kill the apparatus motor and place it out of service.

Figure 7-13. Methods of overcoming obstructions such as these railroad tracks should be considered in preplanning.

Rescue

Rescue and the saving of lives must be given first consideration. This generally will be a larger problem at transportation accidents that happen in populated areas. The rescue may have to be made from a vehicle or vehicles, from an involved structure, or throughout an entire complex. The rescue effort may extend to exposed property where evacuation will be necessary.

Evacuation

Evacuation will depend on the product, the fumes, or combustion by-products: the wind direction; the potential for tank rupture or explosion, toxicity, vaporization rate; and dispersion patterns, among other considerations. Evacuation may be critical in case of a target hazard such as a school. In preplanning, thought should be given as to how far to evacuate for the particular incident, how the evacuation is to be accomplished, where sufficient personnel will come from, the need for bilingual assistance, how and where the perimeter is to be established, and where the evacuees will be relocated. A number of recommendations for evacuation distances are currently available in different reference texts. It appears that for practical use all civilians and nonworking emergency response personnel should be evacuated for one mile (1.6 kilometers) and then up to two miles (3.2 kilometers) as time allows. Downwind evacuation from fumes or some chemicals such as acrolein could be much further depending on the size of the cloud released.

Other Tactical Operations

Numerous considerations at a hazardous materials emergency will determine initial tactical operations. The positioning of attack lines, after evaluation of the ground slope and location of the ends of horizontal tanks, will be critical. The length of preconnected lines must be considered and the time necessary to extend them if required. The hose lays necessary to support the attack lines must be planned.

A line of retreat should be determined and kept available at all times. Operations around hazardous materials should include a retreat signal that all personnel must be aware of and able to hear under the adverse conditions of the incident.

When recording details of the water supply, an estimate of the required fire flow should be made and compared to the available supplies to see if there is adequate water for the worst possible situation. If not, additional methods of providing water must be developed.

Both the type and size of lines and streams must be planned, which will assist in the laying out of the necessary water supply operations. Foam lines and elevated or master streams may be required and this would necessitate additional alarms or mutual aid. There should be no hesitation in calling assistance during an actual fire, and if given proper forethought, it will become a part of the overall tactics. Water supply requirements may bring about a need for additional or large diameter hose or special pump operations. Relay or supplementary pumping may be used in areas of weak supply; tanker shuttles in unprotected areas; or dual (tandem) pumping where ample water is available. Methods of protecting exposures may include diking or channeling a product, water

curtains, or evacuation of occupants.

During the planning an estimate of the requirement for any special extinguishing agents should be made. Items such as foam, alcohol foam, dry chemical, or dry powder may be needed. How much will be necessary and where supplies are available is important to the preplan. Foam supplies should be sufficient to coat exposed tanks or cover large spills in order to control the vapors where these operations are planned in addition to extinguishment requirements. The availibility of special extinguishing agents for water-reactive or hazardous chemicals may be difficult to determine: therefore it should be done before the incident.

As the plans take form there must be continual consideration for the unexpected and possible consequences should something unusual occur. Since there is always the possibility and probability that the situation may develop differently than planned, the tentative action plan must remain flexible and not be "carved in stone."

Plan Review

With the plan completed, it should be reviewed by the chief who would be in charge of the first-response team for suggestions for improvement and approval. The other companies and shifts who could respond should also review and comment on the plan. Once the plan has been drafted, it should enter the training stage.

TRAINING

With the plan thought through, the companies should hold a training exercise to familiarize all personnel with the plan and more importantly to see if it will work. The plan may have to be revised and if so, this should be done before it is put in final form and disseminated.

The tentative action plans and pre-

incident surveys must be periodically updated to keep them current with changes in location layout, processes, and new products. Updating also allows new emergency response and facility personnel to get to know each other and improve coordination. All shifts should go over the plans during in-station training sessions to keep the personnel mindful of the hazards involved at the particular location, and to familiarize new personnel with the problems. In most cases, full-scale drills should be held annually to maintain the adequacy of the final action plan.

DISTRIBUTION

The final approved plan that includes the pre-incident survey forms and diagrams should be disseminated to the department with the original data and information prepared and kept in a more permanent form. The preplan should be supplied to first and second alarm companies and chiefs, head of the department, any deputies or assistants who might assume command, command vehicles, fire prevention bureau, dispatching center, and the pre-incident planning master file. Copies could also be important to other emergency response departments, city agencies, or organizations who would be called upon to respond. Distribution must be determined on a local level. Additional copies can be used by the training division in its classes. At least one copy should be kept in a protected location in case others are lost or destroyed.

The importance of pre-incident surveys and planning will only be realized after they have been completed and an emergency arises. The prior knowledge and forethought will prove extremely beneficial when dealing with hazardous commodities. However, the only way to reap the benefits is to initiate such a program in your department, improve your on-going program if currently deficient, or persist in your efforts if they are now adequate and satisfactorily doing the job.

QUESTIONS FOR REVIEW AND DISCUSSION

1. Differentiate a preplan from a contingency plan. Differentiate a hazardous material study from a hazardous material survey.

2. Develop a preplan for a local railroad, highway, waterway, or pipeline.

3. What useful information can be obtained from the manager of a local hazardous material installation?

4. Will the gathering of cloud dispersion information and plans for computer model-ing assist in preplanning? Which of these will be most helpful to what portion of the preplan?

5. Why must tentative action plans be thoroughly reviewed?

6. What is a good distribution for preplans in your jurisdiction?

7. Develop a preplan for a hazardous material installation in your area. Include sketches of the structure and diagrams of tentative actions.

CASE PROBLEM

The mainline tracks of the WI and GC Railroad traverse the town of Annburg, Ohio. Because of the large amount of rail traffic, a large rail terminal has been constructed. It is a classification yard for westbound trains. The yard has a hump and 12 classification tracks.

Within the terminal is a small shop area south of the classification tracks. The shop handles minor repairs to cars and locomotives, and refueling operations.

There are four structures in the yard: the repair building of ordinary construction measuring 80 by 225 feet; a one-story structure measuring 30 by 75 feet for offices, locker rooms, and a cafeteria; a one-story frame storage/warehouse measuring 40 by 60 feet, which is adjacent to the shop and is used for parts, materials, and equipment storage; and a four-story frame switch tower for controlling the classification yard, measuring 20 by 20 feet. The first three buildings mutually expose each other.

The main through tracks run along the northside of the classification tracks. North of the through tracks is a siding which serves the PL Chemical Company, a paint, stain, and varnish producer; the Edwards Elevator, a grain dealer; and the M & M Petroleum Company bulk plant. These companies face north on a street that has an 8-inch main with hydrants spaced every 1000 feet. There are no hydrants or static water sources within the terminal. A small creek runs under the yard and several yard storm drains empty into it. A street parallels the yard about 150 feet south of the shop building. Hydrants are located on an 8-inch main along this street. Access to the yard is from the south through a parking lot east of the shop and office building. From the north, the only access is through the property of the above-noted companies. The terminal is approximately one mile in length.

The area south of the terminal is residential while that to the north is mixed industrial, commercial, and multifamily residential. The ground slope is gradual to the north and west toward the creek. Wind direction is predominantly from the south.

High voltage electrical distribution lines run along the main line tracks. Wires cross the yard to the shop area.

Diesel fuel is in an above-ground tank west of the storage building and north of the shop. Some paint and flammable liquid cylinders are stored in the storage building and used in the shop.

Several trains are made up in the classification yard each day. Tank cars, box cars, and hopper cars of hazardous materials as well as combustible materials are constantly in the yard and passing through on the main line.

1. In preplanning a rail terminal like this one, what information would need to be obtained?

2. What drawings or photographs would be beneficial to a preplan?

3. What information not provided above would be required to complete the planning and develop safe operations for an incident in the terminal?

ADDITIONAL READINGS

Alkire, Melvin G. "Viewer on Pumper Gives Prefire Data." *Fire Engineering* 129 (April 1976): 79.

Anderson, Russell C. "A Prefire Planning System." *Fire Chief* 14 (November 1970): 73.

Chamberlain, Mark M. "You Can't Get Away from Chemistry." *Fire Engineering* 123 (January 1970): 38–40.

Fire Department: Facilities, Planning and Procedures. IFSTA No. 302, 2nd ed. (Stillwater, Okla.: Oklahoma State University).

Fried, Emanuel. *Fireground Tactics.* Chicago: H. Marvin Ginn, 1972.

Hayes, Terry. "Train Wrecks—Fire and Explosion." *FDIC Proceedings*, 1970, pp. 163–169.

Hopkinson, Allen G. "Preplanning for Emergencies." *FDIC Proceedings*, 1975, p. 194.

Hulett, Allen. "A New Concept in Prefire Planning." *FDIC Proceedings*, 1970, p. 82.

Isman, Warren F. "Prepare for Incidents Involving Hazardous Materials in Transit." *Fire Engineering* 127 (April 1974): 61–62.

Kenney, L. L. "What to Note in Prefire Planning." *Fire Engineering* 123 (April 1970): 48.

Layman, Lloyd. *Fire Fighting Tactics.* Boston: National Fire Protection Association, 1953.

Link, Ralph. "Cooperation and Preplanning by Water Departments and Fire Departments." *FDIC Proceedings*, 1972, p. 142.

Lyons, Paul R. "What Do You Need to Know?" *Fire Command* 39 (January 1972): 16–19.

"Prefire Planning, Signs Can Avert Pesticide Dangers." *Fire Engineering* 126 (March 1973): 49.

Russell, Robert M. L. "The Importance of Prefire Planning." *Fire Journal* 64 (November 1970): 73.

Special Fires I. College Park: University of Maryland, 1973.

"Taped Information Played in Cab During Response." *Fire Engineering* 126 (October 1973): 52.

8 Command and communication

Command is the defining of the lines of authority at an incident. Communications refers to the directing of information and orders up and down the lines of authority and beyond the lines to other agencies involved in the incident or from whom assistance is sought. At a hazardous material incident, definite command and clear communications are vital because of the tremendous potential for death and destruction.

Command, to be successful, must be well planned and organized. It must establish leadership of personnel and control of the overall operations. This places great responsibility on the officers who must initiate the command system, and the demand increases with the magnitude of the situation.

A good incident commander will make sure that company officers are well trained and prepared to handle the possible hazardous material situations within the jurisdiction of the company. The command officer should be experienced in fireground organization and how to initiate control at the emergency scene. Finally, all command officers should be thoroughly familiar with the department procedures for establishing and operating a command post and initiating the contingency plan.

The preplanning function has been covered in Chapter 7. Ideas were presented on how to acquire information prior to an incident and how to develop tentative action plans. Preplanning should also include making plans for a sufficient command structure and designation of a command post. Adequate command personnel must be available to supervise the additional companies as extra alarms are struck, to cover internal and external exposures, to cover rescue and suppression operations, and to properly staff the command post.

In this chapter, the role of the incident commander, staff, and line officers will be discussed. The development of command through the vehicle of a command post operation will be detailed, and the subject of communications at an incident reviewed.

COMMANDER

Of major concern at hazardous materials incidents is the question of who is in charge. This is especially true when the scene of the emergency is outside an incorporated area or on a highway or railroad right-of-way. Police officers, and officials of the environmental protection agency, water resources, civil defense, and the carrier, among others, will all want to get into the act. Generally, the fire department is in charge where there is a fire and/or spill, especially if there is a threat to life or property. The highest ranking officer on the fire scene of the jurisdiction where the emergency occurs becomes the incident commander. However, law enforcement personnel may feel they are in charge of a highway accident when hazardous commodities are involved. In some states they are legally in charge. Railroad officials who own the right-of-way may want to take charge. If these groups are untrained in the dangers and precautions to take, the outcome could be unfortunate.

In advance, as plans are made for an organized operation, lines of authority should be established so it is immediately apparent who is in charge and responsible. This should be determined after legal authorities have checked federal, state, and local laws covering the subject. If necessary, an agreement should be drawn up at this time, which specifically designates the responsibility and authority of the various agencies that can be involved even at minor incidents. At the actual time of an incident, cooperation should be given to the Bureau of Explosives, Federal Railroad Administration, Federal Highway Administration, Pipeline Safety, Coast Guard, or National Transportation Safety Board personnel. This should not be construed to mean a transfer of responsibility, but a willingness to help them do their job and, in return, receive benefit from their experience by seeking their technical assistance.

Responsibilities of Commander

Essentially, the incident commander is responsible for managing all emergency scene operations. This includes a command post appropriate for the magnitude and nature of the incident, acquiring necessary technical assistance and outside resources, and directing the operational forces. This overall supervision is maintained by utilizing the line and staff officers through good emergency scene communications. The commander must coordinate fire fighting tactics with any process, valve, or pipeline shutdown, and seek advice from plant or carrier personnel with knowledge or specialized training in handling the products involved. The incident commander must oversee planning ahead logistically for personnel, equipment, and other outside assistance or support that will be needed, and must be prepared to apply the tactical operations necessary to confine and control the incident.

The incident commander has three primary means of appraising at the emergency scene. The first is by visual observation of the emergency scene by the commander. However, since this is not always possible because the command post may not be adjacent to the scene, the second method, that of reconnaissance, may be used. This requires ordering someone to make a visual check of the scene and report to the commander what is happening. The third method is the use of preplanning information that should be made available at the command post if not already there.

Transfer of Command

As the incident escalates, additional alarms are sounded, and the overall incident commander usually changes. This transfer of power must be made with as little disruption as possible. The ranking officer assumes command after communicating with the current commander and operating staff. The new incident commander should be brought up to date on the general situation status, and briefed on the effectiveness of tactical operations—so that he has all the information needed to make logical decisions. The new incident commander should become aware of the deployment and assignments of units and any resources currently needed. The condition of the hazardous materials on the scene should be addressed. Any required changes in objective should be determined and then those operations essential to achieving this objective carried out by assuming command and issuing the necessary new orders.

It should be pointed out that the arrival of a senior officer does not automatically transfer command. The assumption of command should take place only after fulfilling the procedures outlined. In some cases a junior officer who is doing a good job should be allowed to continue, with the ranking officer's support, in order to gain experience. Although the ranking officer has the responsibility, the subordinate can carry out the operations. Relieved officers should be used by the commander to best advantage.

The incident commander should take a position at the command post and remain there. If it becomes necessary for the incident commander to leave the command post, authority should be delegated to another officer and the commander should remain in constant radio contact. It may be beneficial for the commander to make an aerial reconnaissance of the scene. In doing this the incident commander should wear breathing apparatus if flying in or near the smoke cloud and maintain constant communications with the fire ground.

Many problems arise in command at hazardous material incidents because of the infrequency with which most commanders experience them and their seriousness. Among the factors that lead to problems are the lack of experience, possible unknown products, places that cannot be seen or easily reached [a train may be a mile (1.6 kilometers) or longer in length], allowing responding units to become committed before being given a definite assignment, difficulties in coordinating multiple companies, multidepartment operations or multiagencies, and hesitation in decision making. At these emergency incidents a fireground leader must come forth and make decisions; it is no time to procrastinate. If the decisions ultimately turn out to be wrong, they must be lived with. However, at the time, they were considered the best thing to do and far better than indecision.

OVERALL COMMAND STRUCTURE

The incident commander should learn to delegate authority and responsibility at any complex hazardous materials incident, because the ability to manage the entire operation effectively will soon exceed the capability of a single individual. This division of responsibility creates an effective span of control at the command level and removes details from the commander to other staff members. The commander is then free to develop the overall strategy and make the tactical decisions necessary with the use of the available resources.

The overall command structure consists of the incident commander with the command staff and operational line officers.

The operational line section of command is responsible for achieving the goal and objectives of the incident commander as they relate to the incident. The responsibility of the staff officers of the command organization is to provide technical assistance and support to the line sections in their efforts to meet the goal.

The operational line is headed by the suppression-rescue commander. The line command consists of divisional officers—generally one for each side or front of the emergency—sector commanders, and company officers. At smaller incidents one or possibly two of these levels can be eliminated. At large incidents the command staff will have three sections with officers in charge of each: planning, logistics, and administration. At smaller emergencies, separate officers would not head up a section, but perform in one of the functional positions within a section. An organization chart of the command structure is shown in Figure 8-1.

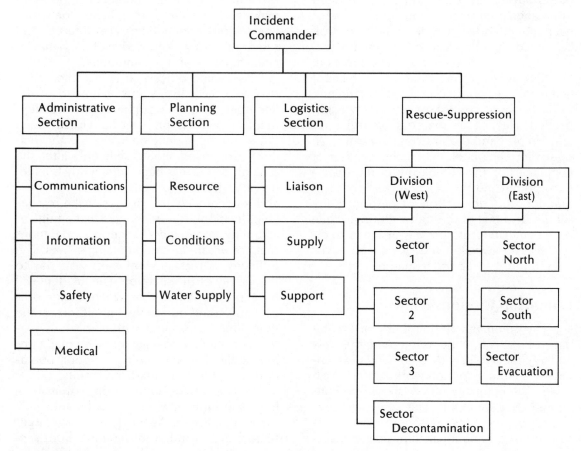

Figure 8-1.   Organization chart of the command structure

LINE ORGANIZATION

The line command should descend from the suppression-rescue commander to the lowest level, which is the company where there should still be unity of command in that the commanding officer should direct his men and be responsible for them. To maintain good command with the suppression-rescue commander in charge, there must be ample delegation of authority to various subordinate officers. The sector commanders may further delegate authority among company officers. This is all necessary because of the span of control concept, which states that one person can effectively direct only five to seven people; this concept also applies to the number of companies under the sector commander's control. As delegation of authority increases, so must communications for orders to be carried out and proper intelligence to be fed back to the commanding officer.

By dividing the line operations into divisions and sectors, the operations at a large hazardous materials incident can be divided into geographic areas of appropriate size for each commander to adequately control. Sector commanders can be assigned a section of the incident or a specific function. Division allows those within the chain of command to better control the assignment. With increased control, on-site safety is improved. Hazardous material emergencies are complex and require close control of all operating personnel. When the number of companies or the complexity of the operations go beyond the ability of a single commander, the delegation of responsibility should be initiated through sectoring. While each commander is normally assigned to direct five to seven officers, it may be advisable to resort to a smaller span of control.

Command Responsibility

The command responsibility of a company officer should be to supervise the crew and coordinate it with other engine companies so they do not work against each other. The company officer should be concerned at all times with the safety and protection of operating personnel from exposure to toxic fume inhalation, poison ingestion or absorption, pressure vessel rupture, corrosive action, or explosion. Personnel should be directed to maintain proper positions, movement, and nozzle patterns to assure safe operations by the company officer. The company must function as a team; assignments from the sector commander should be given to the company as a whole, not to individual members. The company officer will assign individual tasks to achieve the overall company objective.

All commanding officers at every level must see that personnel under their command are adequately trained to handle situ-

ations that could arise. The training that revolves around hazardous materials is extensive, including rules for personal safety, identification of hazardous materials, sizeup and evaluation, and tactical decision making. The tactical approaches for handling various commodities are different and require thorough knowledge of the behavior characteristics of the commodity.

As senior officers arrive at the scene they will have to assume command of the line operations. This can be a difficult assignment, depending on what has already been done and the impending dangers of the incident. Command should shift only after the ranking officer has been sufficiently briefed, has a working knowledge of the situation, and is adequately prepared to make reasonable decisions. It is imperative that an officer assuming command change any operations that are unsafe or could endanger lives. The difficulty of reassignment must give way to the gravity of the situation. The incident will have progressed further when senior officers arrive and due to the importance of the time limits associated with hazardous materials necessary changes must be made immediately. The continuing of poor operations could lead to disastrous results.

The rescue-suppression commander begins to function when more than one sector is established. As incidents grow, larger sectors are included in divisions. Both sectors and divisions can be identified by location as north division, east, central, and west sectors; or sectors may be named according to the function they perform, such as decontamination, or evacuation (Figure 8-1). Of course, more than three sectors can be placed under a division commander. Most situations will be handled by sectoring, although two or more divisions may occasionally be required. The rescue-suppression commander is the on-site tactical commander and has immediate responsibility for removing injured or exposed persons and limiting the fire spread. The rescue-suppression commander, who reports directly to the incident commander, must coordinate with the planning and logistics sections because he is in charge of the ongoing rescue and suppression operations. This officer must be concerned with the organization of all forces in the rescue-suppression sectors and divisions, as well as the general tactics to be employed at a hazardous materials incident. This responsibility includes decisions about (1) the type of operation: confine, control, extinguish, or withdraw; (2) resources needed by each group to carry out these operations; (3) escape routes to safe areas and appropriate retreat signals; (4) how the unexpected will be handled; and (5) how long personnel are to stay in action before rotation or being relieved.

Rescue-suppression has two major tasks: to reach the tactical objectives set by the incident commander and to report the current status of the situation. The rescue-suppression commander must report to the incident commander and the planning section on progress being made, holding, or losing ground; the inability to reach a tactical objective; the current situation in each element (sector or division); whether hazardous situations are persisting; and the occurrence of any unexpected changes.

STAFF ORGANIZATION

The incident commander must delegate to staff members the responsibility for numerous support functions. This support is supplied by three staff sections. The administrative section relieves the commander of performing detail work associated with the incident. The planning section develops alternative strategies and tactical approaches for the commander's review. The logistics section coordinates and acquires needed supplies, equipment, and personnel. These sections are particularly necessary in hazardous material incidents because too many operations are involved for a single individual to control. The incident commander as a single individual could not interface with the variety of tasks and the number of people deployed at a hazardous material emergency.

Administrative Section

The administrative section includes a communications officer, information officer, safety officer, and medical officer. The administrative section assists the incident commander in handling the details associated with an emergency incident. This section handles the media, maintains overall safety, treats the injured, and directs radio and land line communications, including those to other agencies.

Communications Officer. The *communications officer* has the responsibility of assuring that communications are coordinated and run smoothly. This begins with the early initiation of communications between the dispatching center and the command post. The communications officer must then establish communications with all units responding and on the scene. The communications officer handles all transmissions into the command post and dispatched from it, including those to outside agencies and technical information sources. It is vital that the division and sector commanders provide constant feedback that is specific, relevant, and timely to the command post. All communications in and out of the command post should be recorded and directed to the proper staff member. Recording messages assures that none are lost and can eliminate duplication or confusion. A radio log, cassette tape recorder, or telephone company recorder can be used for this. Of special importance will be information received on the products involved from CHEMTREC (discussed in Chapter 6), manufacturers, local chemists, industrial emergency response teams, or federal agencies.

The communications officer will also have to establish contact for the command post staff. Communication by members of the logistics section with other cooperating agencies such as law enforcement, street or public works department, the utilities—especially water, medical facilities, and contractors must be coordinated by the communications officer and directed to the responsible staff person. The communications officer should maintain a complete list of telephone numbers of local people and outside agencies which may have to be called. In many cases technical assistance may have to be requested from CHEMTREC, a trade association, a chemical manufacturer (TERP, HELP), or governmental agency such as EPA's OHM–TADS at the National Response Center. It should be pointed out that in many cases it is better for the incident commander to talk directly to the technical resource. This eliminates any loss during the transfer of information. The data should in any event be recorded. This reinforces the need for a telephone or telephone patch capability at the command post.

How to contact local truck terminals and railroad dispatchers in the area is also important. A temporary land line or radio telephone may have to be installed at the command post and the methods for doing this quickly must be known by the communications officer. The telephone company should be contacted for the possible installation of an emergency switchboard.

The communications officer must know local communications groups such as ham radio operators, citizen band clubs, taxis or local government vehicles that might be used as an outside communications source. The communications officer should also know where to obtain additional equipment to assist the supply officer. Such items as

power megaphones, portable radios, a mobile base, power antenna, mobile telephone, or emergency switchboard may be necessary. In smaller incidents or where sufficient staff is not available, the assignments described for the communications officer will possibly be handled by the dispatching center with the person in charge functioning as the communications officer.

Information Officer. The *information officer* is responsible for working with the news media and seeing that proper and correct information is given out. This officer frees the incident commander from the problems of interviews and disruptions during hazardous material emergency decision making. To enhance this, it is often advisable to locate the information officer away from the command post. Because many hazardous material incidents are spectacular and often involve serious or fatal injuries, accurate information must be available for the media so that erroneous or embarrassing statements are not placed on the news wires. It is best to set a policy for the press, including where they will be allowed to go. It may be advisable to hold news conferences if the emergency continues. Adequate telephone lines in addition to those for the command post should be established for the media.

Safety Officer. The *safety officer* is responsible for the life safety of everyone: emergency response personnel at the scene, the public in the area, spectators, and those in passing vehicular traffic. The safety officer should gather information on the materials involved and the danger potential of both the commodities and the overall situation. Among other duties are informing the incident commander of safety problems; assisting in strategic and tactical planning, always on the alert for factors affecting fireground safety; reviewing all sector status reports to identify operational hazards; and advising on safety matters when required. The safety officer must have the authority to stop unsafe operations immediately if deemed necessary, and then confer with those in charge as to this decision. It may be necessary for the safety officer to establish control lines and decontamination procedures, along with monitoring the condition of everyone working on the scene. This includes control of smoking and eating before decontamination; looking into the circumstances surrounding any injuries; and filling out the proper reports.

Another responsibility of the safety officer is to coordinate with law enforcement officials to block off the area, reroute traffic, and restrict access to the incident scene, the command post, and staging area. It is of utmost importance for the safety officer to make sure that full protective clothing is worn with all exposed skin covered when necessary or special protective clothing is worn where required. The positions of personnel at the scene, their exposure times to fumes, runoff, and splashes, and observations of any unusual physical abnormalities must be recorded. Lists of personnel assigned to a company may be kept by the pump operator on daily written slips. To facilitate record keeping, as emergency response personnel board the responding unit or report on the scene, their names on magnetic name tags can be placed on the apparatus. This information should then be collected for the safety officer.

Medical Officer. The *medical officer* is responsible for providing first aid to those rescued and making sure that they are promptly transported for treatment. This officer coordinates with the supply officer for medical supplies, resuscitators, oxygen, and ambulances. It may be necessary to establish an aid station to take care of victims or injuries; for major incidents, it may be necessary to set up an entire field hospital. The medical officer must be ready to initiate a triage station in the event of numerous injuries. In some cases, decontamination of victims will have to take place before first aid or transport. The medical officer should make good use of paramedics and emergency medical technicians and air support as available, and should have complete knowledge of local hospitals and notify them accordingly so that one is not overcrowded while another awaits victims. It may be helpful to have reference material on-site; for example, *Recognition and Management of Pesticide Poisonings* would be useful to carry in agricultural areas. The medical officer should be familiar with the local Poison Control Center and the proper procedures for expediently working with them. It is important to make sure that uncontaminated labels or containers are sent with poisoned victims to the hospital. The medical officer must coordinate with a coroner on identification procedures, removing bodies, and establishing a temporary morgue. In situations of lesser magnitude the duties of the safety officer and the medical officer may be combined. In all situations where fumes, vapors, or combustion by-products are a problem, one of these officers must keep records of the exposure times of operating personnel and perform a subsequent follow-up check for signs or symptoms of poisoning.

Planning Section

The planning staff consists of the resource officer, conditions officer, and water supply officer. The planning section has the responsibility for developing alternative strategies and tactical operations for the incident commander's decision making process. This requires consideration of water supply, resources, and the condition or situation status. The planning is done in coordination with the commanders of the logistics and suppression-rescue sections. All data consideration should be made as discussed in Chapter 10, "Decision Making." In addition, the planning section must consider and present alternatives on how the suppression activities can be divided into divisions and sectors; what equipment and personnel should be held in reserve; the location of the staging area; possible incident spread, safety, and special problems; how long before operational forces and decision makers should be rotated; coordination of planning with liaison, logistics, and suppression-rescue sections; and plan the safe location of the command post if this has not already been done.

Resource Officer. The *resource officer* determines the number of companies that have responded or are en route in order to establish the command post equipment and personnel status board. Other duties are recording companies arriving or at the staging area and keeping track of their assignments; supplying the incident commander with "current" resources, whenever necessary, including what companies are available for assignment. It is also the responsibility of the resource officer to maintain the command post staff organizational assignment chart and the status of command officers either responding, unassigned, handling an assignment, or assignment completed and available.

Conditions Officer. The *conditions officer* keeps records of what is happening on the scene. Progress reports will come to the conditions officer, from which a current status report of the situation is maintained for the incident commander. This profile should include the area involved, possibility and direction of spread, progress of the suppression-rescue forces, and any special factors, such as the rerouting of traffic, arrival of special extinguishing agents, or evacuation procedures. The conditions officer prepares and maintains any maps required and has responsibility for the photographic coverage of the incident. The planning section should maintain an overall tactical control chart for the incident commander, which would detail the location of companies at the scene and their assignment, what has been done, is being done, and what needs to be completed. This chart would also show the

sectioning of the fire fronts, the positioning of apparatus, and attack positions.

A variety of records must be kept at the command post during a hazardous materials situation and a record system should be preestablished. Records can be kept in a log mode by time and date. Records of all decisions should be clear, establishing who made it, why the decision was made, and the background that was the basis for the decision. Records will assist in planning for the next incident and point out areas in need of improvement. They will also serve as a justification for any monies spent during the incident.

Water Supply Officer. The *water supply officer* performs numerous functions such as determining the location, accessibility, and quantities available from all usable sources; evaluating the incident water requirement or quantity needed for the planned operations and how this compares to the supply available; informing the planning team of water supply problems; and initiating water supply operations to overcome deficiencies, such as ordering the response of a tanker task force of four tankers to shuttle water and an engine to the nearest fill site to establish a loading operation.

The water supply officer must have the ability to apply basic fireground hydraulics to the water supply situation. The officer should be familiar with the water system or have the proper maps available that indicate storage capacities, main sizes, hydrant locations, and flows available in various areas, as well as static sources and records of monthly accessibility and capability. The water supply officer will need to know apparatus capacities, locations, number of lines in operations, pressure drops, residual pressures, available hose, and discharge ports. To provide adequate supplies for the incident commander, the water supply officer needs to know how the incident commander intends to build the attack force, where the attack will be concentrated, and where additional lines can be quickly placed into service.

The water supply officer must maintain constant communications to perform efficiently. The officer must be familiar with what the incident commander's short-, mid-, and long-range plans are and know what apparatus is responding or in the staging area for replacement of an engine that has mechanical difficulties. The water supply officer may have a company assigned to be used as a special water supply company if any special water problems develop, and will take whatever action is necessary to overcome any water supply problems. This could require establishing supplemental pumping operations or relays to working companies requiring additional water or providing more water for control or exposure protection to the incident scene by dual (tandem) pumping, relays (especially if large-diameter hose is available), or by using mobile water supply apparatus. Efficiency can be improved by using pumpers to full capacity when possible, and placing uncommitted pumpers in service while maintaining an adequate reserve in the staging area. The water supply officer will have to work with the water department personnel to overcome any system supply problems. Also, all water supply information, including usage and pumping operations, should be recorded for later use in water supply and operations planning and a copy delivered to the conditions officer.

Logistics Section

The logistics section consists of the liaison officer, supply officer, and support officer. This section is responsible for providing personnel and material for the duration to control the hazardous material emergency. There must be close coordination and open communications between the incident commander, the rescue-suppression commander, and the logistics section.

Liaison Officer. The *liaison officer* coordinates all of the outside agencies who can offer assistance into an emergency response support team. The officer should know who represents the various agencies and where and/or how to contact them although this is done by the communications officer. Liaison responsibility begins as the outside agencies arrive on the scene. The liaison officer should be alert to interagency problems and prepared to solve them discreetly. Some of the agencies with whom liaison will be maintained include, but are not limited to, the following: law enforcement; disaster preparedness; Red Cross; rescue or emergency medical services; local government officials; utility company personnel, especially water, sewer, telephone, and electrical; health officials, hospitals, and ambulance services; the city or departmental lawyer for legal advice, if necessary; local, state and/or federal environmental people, with reference to both air and water; local contractors for heavy equipment; local communications organizations if supplemental radio nets are required; service groups for facilities if evacuations of large numbers are necessary; manufacturers' representatives or trade association officials who respond to provide technical assistance, such as the Bureau of Explosive or Pesticide Safety Team Network personnel; officials from governmental agencies, such as the National Transportation Safety Board, the Environmental Protection Agency, and the Department of Transportation. A major responsibility of the liaison officer will be to take charge of any evacuation operations necessary. A knowledge of all the agencies involved, from those participating in the actual evacuation to those caring for and housing the evacuees, makes the liaison officer the logical choice to direct the operation.

Supply Officer. The *supply officer* maintains the staging area; acquiring, storing, and recording all of the resources. The staging area should not be too close to the incident, probably a minimum of one-half mile (0.8 km) away. It may or may not be in the proximity of the command post, depending on the space limitations that must allow for safe and effective apparatus movement. The area needs to be large, such as the parking lot of a church or school, and well-marked. The supply officer coordinates through the logistics officer on larger situations with the planning section and incident commander. During smaller operations he would deal with the individuals directly. The supply officer works closely with the support officer and moves tools, equipment, personnel, and apparatus to the support officer on the fireground on orders of the incident commander. The supply officer must then inform the resource officer of the assignments.

The supply officer must make the location of the staging area known, especially to the fire alarm office and all responding agencies. This officer must assume a visible position so that responding units can properly report in; directions may have to be given concerning the best route to the staging area. Another duty is ensuring that apparatus and equipment are parked for immediate dispatch with no blocking.

The supply officer must keep an inventory of equipment and make sure that supplies are maintained. This information must be kept current with the resource officer at the command post. Equipment and material that might be needed during a hazardous materials situation would be: breathing apparatus; additional air tanks; a cascade or compressor unit for refilling tanks; generators and lights for nighttime operations; special protective clothing; ample supplies of extinguishing agents, especially if polar solvents or combustible metals are involved; equipment for damming and diking, such as dump trucks, front loaders, and bulldozers; extra supplies of hose may be needed for relays or reaching remote areas along railroads or transportation routes; apparatus in reserve for equipment failure or unexpected occurrences that would necessitate immediate implementation of a strike force; cranes, tow trucks, or commodity transfer equipment may be required; floating booms or absorbing materials could be necessary for oil or chemical spills; decontamination or neutralizing materials are called for with corrosives, poisons, or pesticides (these could include lime, soda ash, and chlorine bleach to name only a few); radiological and gas detectors are a necessity, as well as the normal supply of gasoline, diesel fuel, oil, and minor maintenance repairs for apparatus; and coffee, food, drinking water, sanitation facilities, and possibly dry socks and gloves for operating personnel. Extra clothing may be necessary if personnel must be decontaminated before leaving the area. The staging area may also need to establish an area for personnel to rest before returning to the fire lines.

A great deal of coordination is needed with the water supply, safety, medical, suppression, and communication officers. An aide would be advisable on larger situations to assist in keeping records of availability, assignments, and personnel and equipment moved forward.

Support Officer. Personnel and equipment moving to the incident scene report to the *Support Officer,* who is responsible for on-site logistics to the rescue-suppression officer. This position may not be necessary except on large operations; however, the duties must always be performed where necessary regardless of the title of the officer. The support officer will receive requests from the rescue-suppression commander or his staff. The support officer will use what is available and request needed items from the supply officer; the resource officer must always be advised of conditions. The support officer will need an on-site storage area and room for possibly rotating personnel; records should be kept of those reporting and their assignments. This helps to keep track of personnel in case of an unexpected tragedy. The support officer will need to keep filled air tanks, extinguishing agents, portable radios, and similar equipment available, and should have standing by a rescue strike team for any possible emergency.

COMMAND POST

The essential element of command at a hazardous materials incident is the establishment and efficient operation of a command post. The command post is the operating center from which definite control of the incident is exercised and maintained. This requires that the decision makers, those with the responsibility, be brought together under the leadership of the incident commander.

The command post staff will need data on which to act. Therefore, all intelligence, feedback, and information will be directed to this one place. This will also establish on the fireground one location to go to for assistance, to feed data and progress reports, and report unusual occurrences. It is the central communication point for the fireground with the dispatching center and possible other cooperating agencies. All division or sector commanders in charge of a section or side of the fire front must give periodic progress reports to the command post. The reports should include the current fire situation and control probability. Any rescue, evacuation, or safety concerns should be made known. The condition of exposures and any further resource requirements should be indicated, along with any special developments in the hazardous situation. This feedback is vital to the incident commander's tactical decision making and must be called for by an aide if not forthcoming. In smaller incidents where there is no staging area, incoming companies and outside assistance should report to the command post for assignment.

Initial Command Post

The initial command post should begin with the first-arriving company and officer. The company officer should provide the dispatching center with information on the nature of the emergency and the existing situation. Any necessary instructions pertinent to the dangers of the materials if known and any approach precautions should be given to other responding companies. The first-arriving pump operator becomes the command post staff, generally acting in the capacity of the water supply and communications officer (Figure 8-2). In hazardous material incidents, it is essential that a command post operation be set up immediately. A command post should be established, regardless of the size of the emergency, because of the potential of these incidents and to create a central focus point. Records or data can be kept by using a grease pencil on a highly polished fender or an apparatus window (Figure 8-3) if a clipboard or forms are not available on the first unit.

The first-arriving command officer then takes over the command post. He may build

Figure 8-2. The first-arriving unit operator as initial command post staff

Figure 8-3. Record-keeping with a grease pencil on an apparatus window

it into the overall operation usually associated with the term or it may shift to a subsequently responding chief or command vehicle.

Command Post Location

The location of the command post must be reported to the dispatching center and all units on the scene. The command post should be located back a reasonable distance from the scene in order to maintain an overall view of the situation. If possible, the command post should be upwind on higher ground than the hazardous materials and in a quiet zone protected from the weather. Ample space should be provided for the staff that must be accommodated. A large open space may be required to accommodate vehicles and operations. A shopping center parking lot would be an ideal place for establishing a command post. They are usually well lit and have accessible telephones. The physical layout size necessary grows with the size of the incident. The situation may call for renting a contractor's trailer, motel rooms, or using a school to provide adequate space for major operations. There the staff and their assistants can work together with the proper perspective and coordinate the many agencies involved in a hazardous materials incident. It is important to locate the command post so it can control the resources, the subordinate units, and the overall problem.

Figure 8-4. Traffic cone on a vehicle as a marker for the command post

Identification of Command Post Location

The command post must be properly identified so it is recognized and easily located. There are a variety of ways the command post can be marked. A brightly colored flag can be placed on a pole extending above the designated unit. Another method is to use a distinguishing colored light, such as blue or green, which may or may not be elevated. A traffic cone can be placed on the vehicle top (Figure 8-4) to denote the command vehicle. In those areas where emergency lights are turned off upon arrival, the unit with the rotating light in operation denotes the command post. If a unit is used solely for command operations it can have any one of a number of distinguishing color schemes; a checker board stripe is used on the vehicle in Figure 8-5.

Command Vehicles and Equipment

Numerous vehicles can function as a command unit. The well-equipped and suitably laid out trunk of a chief's sedan is often seen in use in southern California. Suitable reference information on local hazardous materials must be carried on the command

Figure 8-5. Command unit of the London Fire Brigade with checkerboard striping in reflective paint to distinguish it on the incident scene. *Courtesy of R. Wayne Powell*

unit; a reference file set-up is illustrated in Figure 8-6. Montgomery County, Maryland, has a number of versatile units in service. These function as a combination communications and mobile command unit. The Mid Glamorgan Fire Service in the United Kingdom is considering a command post that would be a self-contained unit on a demountable body and could be placed in service on an available chassis as needed. In addition to hazardous material references and chemical data sheets, aerial photographs or ground-level photographs illustrating

Figure 8-6. Reference source file at the command post to assist the incident commander and staff. *Courtesy of Larry R. Froebe*

important features, floor plans, and prefire planning survey information can also be carried and used with projectors or microfilm readers during an incident.

Other equipment that can be utilized on a mobile command post include a mobile television monitor, apparatus status board, weather instruments (Figure 8-7), audio tape-recording equipment, cameras with appropriate accessories and a supply of film, hazardous atmosphere sampling equipment, radiological monitoring equipment, self-contained breathing apparatus, and special protective clothing.

On clear plastic placed over aerial photographs, ongoing actions can be laid out and necessary tactics developed. Flow patterns for spills, gas dispersion patterns, evacuation difficulties for vapor clouds, exposure protection, and confinement needs become readily apparent in a situation such as in Figure 8-8. This provides a record which can later be recorded for file and reviewed in a postfire critique.

Figure 8-7. Command unit equipped with aerial photos, weather instruments, status board, microfilm readers, and radio and telephone communication equipment

Figure 8-8. Gas cloud dispersion and downwind evacuation is planned in the command post on the basis of weather information and product data. *Courtesy of Federal Railroad Administration*

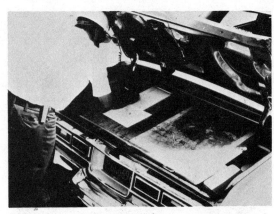

Figure 8-9. The trunk of the chief's vehicle can be made into an efficient command post. *Courtesy of Gary Girod*

Less elaborate command units can be established by improving the capabilities of chief vehicles. A modular microfilm unit can be installed to create a tactics console. A considerable amount of information can be quickly retrieved because of the small storage space necessary for microfilm. In those areas where weather conditions do not require an enclosed command post, the trunk of a sedan can be inexpensively modified. (See Figure 8-9.) Radios can be extended, a suitable shelf designed, and lights installed. There is ample room for storage of reference books; writing and record-keeping materials; maps of streets, sewers (sanitary and storm), and water supplies (main systems or static sources); pre-plan information; meters and detectors for hazardous materials; binoculars; traffic cones to protect and identify the command post; and as one chief suggests, "a bottle of Excedrin."

The larger the command unit the more versatile it can be made, since there is more room for additional storage and equipment. Preferably it would be enclosed from the many fireground elements and large enough to provide ample working room and equipment. Items not heretofore mentioned that

might be desirable include: tables, chairs, adequate lighting, a chalkboard, overlay material, necessary records and forms, emergency phone numbers, hazardous chemical data, and a command and logistics board.

A command and logistics board is a means of graphically illustrating assignments at the fire scene by means of coded magnetic symbols. These are attached to plastic-covered diagrams on a metal backing. These symbols are movable so that company locations can be changed as the situation changes. A board of this type will enable the command post to know where all units are located, determine whether all vital areas are amply covered, evaluate the need for additional assistance, and plan the deployment of responding units. (See Figure 8-10.)

Importance of Command Post Operation

In all hazardous materials emergencies, it is imperative to set up a proper chain of command and command post operation. Delay or lack of implementation will only lead to disorganization and inefficiency. There must be sufficient staff to do the job. The extensive

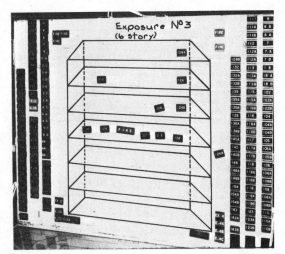

Figure 8-10. Logistics board

On many hazardous material alarms there is a lack of data to work with. Chief officers must encourage unit officers to provide and use preplanning information on hazardous material locations and routes. Contingency or disaster plans must not be forgotten; and proper communications must be established to relay information to the command. The magnitude and potential of the hazardous material incident is such that as forces are committed, the organization for managing the scene must grow also. The responsibility for various tasks must be delegated and people held responsible for their completion. This, of course, requires training officers in command post operations so they can readily implement them when needed. Assistance must be requested in time to be beneficial to the operation and situation outcome. Finally, the most important problem is the failure of the officer in charge to believe in the use of the command post concept and failure to use it in its entirety. If the concept is to be successful, it must be done correctly. Disbelievers should remember that hazardous material emergencies can be extensive and time consuming. At Roseburg, Oregon, for example, the detonation of explosives damaged over a 50-block area. The derailment in Landover, Maryland, required fire department operations for four days. At Glendora, Mississippi, purging a vinyl chloride car lasted twelve days. To deal with hazardous materials incidents properly, command posts are a MUST!

command post outlined in the preceding paragraphs is one that only larger departments could properly staff. This detailed a command operation will not be needed in every situation, but we must be cognizant of the requirements for larger disasters and prepare for the worst. On smaller operations, individuals may be able to function in more than one position. For example, the responsibilities of the safety and medical officers or the supply and support officers' responsibilities could be combined. Another alternative is to use officers from other fire departments in staff and command positions.

COMMUNICATIONS

Effective command at the hazardous commodity emergency is developed through an efficient communications network. Actually, communications are a multilevel operation. The communications chain begins at the dispatching center with the reception of the hazardous material incident report. The dispatching center then dispatches the compa-

nies and in some cases can provide additional information while they are en route. The initial fireground commander reports back to the dispatching center. In some cases the dispatching center personnel will function as the incident communications officer and handle all radio transmissions and contacting of cooperating agencies and technical assistance sources. The dispatching center then relays information back to the established command post. In larger operations the communications officer is a member of the command post staff. The communications officer confers with cooperating agencies and technical assistance sources from the command post. The command post communicates with the fireground through the suppression-rescue commander. A diagram of the communications network is shown in Figure 8-11.

In the planning stage, communications operating procedures should be developed as to who will contact CHEMTREC, activate mutual aid response if not automatic, and call informational resources and regulatory agencies. Decisions should be made about which transmissions have priority, which frequencies will be used, and how the lack of adequate equipment can be overcome. The planning should also designate who has responsibility for the various tasks, assign authority to carry them out, estimate the equipment required, and supply information on where to obtain communication assistance.

COMMUNICATIONS

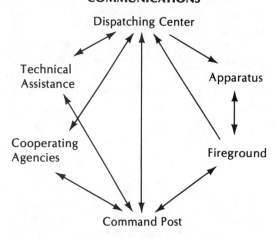

Figure 8-11. Diagram illustrating the communications network at a hazardous materials emergency

Dispatching Center

Dispatching center communications begin with securing as much information as possible on alarms involving or suspected of involving hazardous materials. The exact location and any indication as to the nature and seriousness of the incident will be especially helpful to first-responding companies.

On receipt of the incident alarm, the dispatchers should begin to ready their phone lists of information sources and search reference materials for data on the commodities, if known. This assumes that they have been trained in how to use and interpret the material in the reference books and manuals. At many incidents there will be no communications officer in an established command post. In this event, the dispatchers perform these duties. The major concern in these instances is the transfer of technical information from the resource to the incident scene. The complete, accurate transfer of all information must be stressed, because it will be vital to the incident commander. The use of a telephone patch will help to minimize the time commitment of dispatcher.

If prefire planning information is stored at the dispatching center office on slides, overhead transparencies, or microfilm, this material can be displayed and relayed to responding apparatus. This information, which would be retrieved from accurate preplans concerning hazardous materials, would be vital to fireground safety. During the operations, the dispatching center should monitor all transmissions to ensure that messages are received and acknowledged by the command post, sector commander, or unit sought.

Radio Command Vehicles

In many areas radio command vehicles respond to major emergencies. They become the command post and handle radio communications. These units are normally equipped with multiple frequencies to handle not only mutual aid departments, but also public utilities. These units are equipped with status boards to keep track of units responding to the incident. Additional information can be carried on these units and delivered to commanding officers or sector commanders as requested. The units are a suitable command location until the command post staff increases. Then they become too small and the number of people in the area interfere with communication transmissions. (See Figure 8-12.)

Fireground Communications

While on the way to the incident, responding companies should communicate with the dispatching center further information that would be of assistance. The units en route should also communicate with each other, especially if their routes cross or meet. The first-in unit should report its position to dispatching center and other units and any pertinent data on the hazardous commodities and nature of the incident. If companies are not positioned according to the preplan tentative action plan, this should be reported to other companies along with any tactical

Figure 8-12. One vehicle can service the dual purpose of radio-equipped communications unit and command unit.

instructions concerning initial rescue-suppression activities.

While en route, the company officers should also be checking the information carried on the unit that may be of assistance to them. This should include the hazardous materials study and any reference materials; for example, the *Fire Protection Guide on Hazardous Materials* or the *Emergency Action Guide*. Reference should also be made to preincident plans, which would detail response routing and barriers, the layout of the area, water supply, tentative tactical action plans, and types of hazardous materials.

Responding chief officers should keep radio transmissions to a minimum during hazardous material emergencies. However, communications vital to the safety of personnel or necessary to initiate command or tactical operations should not be delayed.

Companies responding to a staging area should report to the communications officer to receive any required instructions and then proceed to the staging area with no further radio traffic. Once at the staging area they will report to the supply officer, who will communicate from the staging area to the command post and the resource officer.

On the scene there are numerous communications from officers to personnel, from sector commanders to company officers, from the suppression-rescue commander to division or sector commanders, and from the command staff. Communication in both directions among all these elements is essential for adequate control and decision making.

Following are some general communication fundamentals that will be beneficial at the hazardous material scene:

1. Speak clearly with good enunciation and at an understandable rate.

2. Speak calmly, since emotional or excited delivery is difficult to interpret.

3. Know what you want to say and use exact terms to keep the transmission brief.

4. Set transmission priorities so that important messages are quickly dispatched.
5. Break into a transmission only in case of an emergency.
6. Since another unit may have a vital transmission, pause between messages.
7. Orders should be directed to a specific company, detailing whom to report to, where, what their task is, and the objective they are to try to achieve. The order should direct the company officer in what to do, but not how to accomplish it.

A hazardous materials problem will call for the immediate initiation of a radio communications station within the command post. Upon completion of the size-up, a progress report should be given to the dispatching center by the first-arriving unit, describing the nature of the situation, initial operations, and anticipated needs. Accurate size-up information must be supplied to the command post from inside the building, or all sides of the incident.

In many cases the lack of common radio frequencies has caused severe problems and a breakdown in intelligence. It has left command officers without sufficient information to make decisions on the unseen side of the operations. Of course, in large metropolitan areas, where a common frequency is available, then the alternative is also a problem. Multiple frequencies are necessary so that normal radio communications can be carried on one channel, while the incident can be handled on a second channel. In fact, at major disasters several frequencies will be necessary to place operations, administration, and logistics all on separate frequencies where they do not interfere with each other.

Portable radios, when available, will be of great assistance in contacting sector commanders for progress reports and issuing orders to counter doubtful situations. If radios are not available, intelligence to the command post will have to be gathered by an aide. Messengers may also be required between units with different frequencies. Whether using runners or the radio, orders must be simple, clear, and concise. This will alleviate tying up the frequency with useless chatter. Sector commanders should report to the rescue-suppression commander about every 15 minutes. Sector commanders may use nonradio communications with their companies to eliminate some radio traffic. The rescue-suppression officer will report to the incident commander at the command post. The command post should report to the dispatching center on its operations so that adequate coverage is maintained throughout the jurisdiction.

Communication Problems

There have been many communication problems in past hazardous material incidents. In smaller communities or rural areas, there has been inadequate incident scene coverage. Enough radios have not been available to cover all sectors and agencies to maintain proper coordination and deliver information to the command post. In some areas, there is a lack of common frequencies among responding fire departments, which adds to the complication. Locales with these possible problems should encourage area-wide radio nets or at a minimum, multi-channeled radios with a common mutual aid frequency, in all apparatus. In the interim, outside sources may have to be pressed into service for proper communications. The communications officer will have to know who has the capability and how to obtain their assistance. In some instances the use of a system of runners at the incident scene may be necessary.

The lack of an adequate number of portable radios has also been a problem in the past. Adequate communications are vital in these serious situations and a conscious effort needs to be put forth in those areas where inadequacies occur.

Another problem that is sometimes encountered is that of maintaining communications with other agencies involved in a major situation. The fire department normally cannot monitor or talk to them. Many radio problems can be overcome with adequate planning. Auxiliary radios can be obtained and stationed at utilities or with mutual aid apparatus on a different frequency.

Verbal communications in dealing with hazardous material must be precise. Because of the complicated names and spellings, it is easy to mispronounce a commodity and inadvertently relay erroneous information. Remember that several letters sound alike; for example, B, C, D, E, T, P, and V. Hazardous materials can have only one letter difference in the spelling, but be considerably different in makeup. An example is ethanol and ethanal. Ethanol, or ethyl alcohol, is used for alcoholic beverages and has a NFPA 704 classification of health 0, fire 3, reactivity 0; whereas ethanal, or acetaldehyde, is a reactive irritant that can polymerize and has a 704 classification of health 2; fire 4, reactivity 2. Another example, which illustrates the confusion in pronunciation and spelling, is dichlorodifluoromethane and difluoromonochloroethane. While the former is a nonflammable, nontoxic compressed refrigerant gas, the latter is a flammable compressed gas. Long product names are best pronounced by breaking them down into component parts and pronouncing each individual part. Another problem is that many products have dual names such as hydrochloric/muriatic acid and hydrocyanic/prussic acid and sulfuric acid/oil of vitriol. The solution to these problems of confusion in identification is to write down the name of the chemical and double-check by reading it back. This should ensure correct spelling. When relaying information to others, give the correct spelling, possibly using a letter-word combination such as a alpha, c charlie, etc., and then have the word spelled back to you as a check. Products with two words in their name should be indicated before spelling; for example, first word methyl, then spell it, second word bromide, then spell it. All excessive and unnecessary radio chatter must be eliminated. Whenever a change in frequency is ordered, the communications officer should notify all elements before the change is made. Fire departments that have helicopters or access to them should make ample provisions for air-to-ground communication. If this is not done, an elaborate relay system may have to be established for transmission of communications only a short distance.

Because successful operations are based on good communications, plans should be well laid out and tested for a comprehensive, well-coordinated communications network. Communication operations and procedures should be established and all personnel involved amply trained long before the hazardous material emergency occurs.

In summary, this chapter has discussed a large-incident command situation on the premise of preparing for the worst. Each individual situation will have its own requirements; the incident commander will have to tailor available resources to the needs. However, efficient hazardous material incident operations will be assured when the tasks outlined here are considered and adapted to suit the situation at hand.

QUESTIONS FOR REVIEW AND DISCUSSION

1. What are the major sections of a command staff and the responsibilities of each?

2. What are the positions in each of the major command staff sections? Who do you feel is the most important in each and why?

3. Write a procedure for positioning, mark-

ing, and reporting the location of a command post for your department.

4. What equipment and informational resources would you desire at the command post to assist in a hazardous material incident?

5. What are the major communication deficiencies in your local area and what are your recommendations to correct them?

6. Develop a command structure for the tentative action plan presented in Chapter 7 on "Preplanning for Incidents."

CASE PROBLEM

At 2:30 P.M. on a cloudy December day with the temperature at 32°F (0°C), a fire is reported on the second floor of AC Packing Company, Inc., on the southeast corner of Main Street and Fourth Avenue. The two-story frame building is 40 by 60 feet. The wind is blowing from the west at 15 mph.

Two 750 gpm pumpers and six men are the first to respond. The fire is burning out windows and an exterior second floor door on the Fourth Avenue side.

The AC Packing Company is a job packager filling many types and styles of tubes, bags, bottles, and containers. Some of the products currently stored on the second floor awaiting packaging include: flammable glue, paint, paint thinner, phosphoric acid, acetone, and sodium hypochlorite. Materials are stored in metal and fiber drums, carboys, and bags.

To the north and west of the fire building are streets, to the south at 3 feet is a one-and-a-half story frame building, 30 by 80 feet, occupied by Kirstsons Welding Supply. This former bowling alley contains stored large stocks of oxygen, acetylene, and other industrial gases. On the east side of the building are four 10,000 gallon bulk storage tanks on unprotected steel supports. The closest tank is 30 feet from the building. The tanks contain regular, unleaded, and premium gasoline,

and diesel fuel. The tanks are not diked. To the rear of the tanks, which is also the rear of the welders supply building, there are twenty-five 55-gallon drums of gasoline and diesel fuel. To the east of the tanks and drum storage is a one-and-a-half story dwelling converted into a restaurant. To the north across Main Street is a mixed commercial/residential area.

Main Street has a 6-inch main with a double hydrant on the corner of Fourth Avenue. Fourth Avenue has no water main and dead-ends at the railroad 150 feet south of Main Street.

The fire quickly spread to the roof of the welders supply and moved eastward. The building is well involved before the limited initial response had completely knocked down the original fire at AC Packing.

Mutual aid is requested from two nearby communities. Within 20 minutes two 750 gpm and a 1,000 gpm pumpers arrive. By this time, the fire has extended to the dry, oil-soaked grass around the drums and the drums are beginning to rupture. Flying brands have ignited the cafe. Fuel from the rupturing drums is running under the bulk storage tanks which now have exposure fires on three sides.

Additional assistance of two engines and a ladder company is requested. None of the mutual aid equipment has the same radio frequency as the local fire department.

Water supply is limited because of wide hydrant spacing and small mains. Four tankers are requested to initiate a shuttle and relay pumping is initiated from a static source.

1. What type of command structure should be developed?

2. Is there a need for staff officers and line sectoring?

3. What positions are necessary?

4. Should a staging area be established?

5. How should the problems of coordinating several departments, delegating responsibility, and obtaining feedback be handled by the Incident Commander?

ADDITIONAL READINGS

Chase, Richard A. "'T' Cards Provide Versatile Resource Status System." *Fire Management Notes*, Summer 1977, pp. 12–13.

"Control Unit Is Result of Three Years' Planning." *Fire*, October 1977, pp. 227–228.

"Demountable Body System." *Fire*, September 1977, p. 196.

Ellis, Don, and Joe Swindle. "Fill Your Trunk with More Than Junk." *Fire Command* 43 (1976): 31.

Fireground Command, Control, and Communications. Silver Spring, Md.: Johns Hopkins University Applied Physics Laboratory, 1974.

Firescope Incident Command System Documentation. Riverside, Ca.: Firescope Program Office, 1976.

Fried, Emanuel. *Fireground Tactics.* Chicago: H. Marvin Ginn Corp., 1972.

Gratz, David B. "Mobile Fire Safety Team." *Fire Chief* 16 (July 1972): 19–22.

_____ "Fireground Procedures: Information Systems for Decision Making." *Fire Command* 41 (October 1974): 18–20.

Halpin, Byron M., and Harry E. Hickey. *Fireground Command and Control Tactics Display Case Preliminary Report.* Silver Spring, Md.: Johns Hopkins University Applied Physics Laboratory, 1974.

_____ "New Aids for Fireground Commanders." *Fire Command* 42 (August 1975): 56–60.

Kimball, Warren. *Fire Service Communications.* Boston: National Fire Protection Association, 1972.

"Organization for High-Rise Firefighting." Orange County: Orange County Fire Department/California Division of Forestry, 1976.

Phoenix Fire Department, "Standard Operating Procedures," Phoenix, Arizona.

Special Fires I. College Park: University of Maryland, 1973.

Tactics. College Park: University of Maryland, 1971.

9
Disaster management

On January 31, 1975, the Greek tanker *Corinthos*, unloading crude oil at the BP Refinery dock facility in Marcus Hook, Pennsylvania, was struck by the chemical tanker *E. M. Q.* Twenty-six seamen were killed and damage was almost $100 million. The fire burned for 65 hours and involved 50 fire departments. The estimated cost of foam, expended not so much to extinguish the fire but to keep it from spreading, was $250,000.

On September 24, 1977, lightning ignited an eight-million-gallon tank of diesel fuel at the Union Oil Company refinery in Romeoville, Illinois. Subsequently two additional tanks containing two million and five million gallons of gasoline were ignited. The situation was brought under control after two days of fire fighting and delivery of 20,000 gallons foam concentrate from Pennsylvania. Eighteen fire departments were involved in the operations.

On September 27, 1977, a truck carrying 95 percent uranium oxide or "yellow cake" was involved in an accident and overturned near Springfield, Colorado. The commodity was in drums, several of which ruptured spilling approximately 10,000 pounds, Although radioactive, the substance was more dangerous as a heavy metal poison. The driver was seriously contaminated. An area of about 20,000 square feet and the vehicle were contaminated. The entire area, including the

truck, was covered with plastic sheeting because of the strong winds. Specialized equipment was brought to the scene and the entire spill worth $1.5 million was reclaimed with no additional injuries. The overall operation took three weeks.

The hazardous material incidents described all extended over several days and involved the cooperation of a number of agencies working together. These, along with the incidents recounted in Chapter 1, are examples of what can happen at any time in most communities with the vast amounts of hazardous materials that are currently being used and transported in our country. The possibility of a major disaster makes it imperative that large-scale planning building on that covered in Chapter 7 be carried out.

Disaster management is essentially the effective and economical utilization of all material and personnel for the greatest benefit of and protection for the people and property during a major incident. This includes the development of strategy that will be employed, preparation to handle and manage the large-scale disaster (often called contingency planning), arrangement for the required cleanup, and restoration of vital community services. In the discussion on contingency planning, additional information is supplied on emergency medical, evacuation, and logistical operations.

STRATEGY

The strategy employed in disaster management for hazardous materials is much like that in other incidents. Strategy is planning and directing forces to skillfully manage the overcoming of an adversary. Emphasis is on overall goals and objectives. The strategy must cover in detail how the available resources can be acquired and used at the incident scene. This is generally done in advance to be prepared for a major disaster when it occurs.

An important strategy element is to realize the need for and plan for larger command post operations. This includes providing physical facilities that can be expanded as additional support is requested and the situation persists. Possible sites near transportation routes, industrial areas, or known hazardous material locations should be surveyed. The following list of considerations may help in site selection:

1. Not too distant from the disaster scene.
2. Ample security can be provided.
3. Large parking facilities for staff and support vehicles or possible staging areas.
4. Adequate space for the command post out of the weather.
5. Ample electrical power, lighting, and telephone.
6. Facilities to feed, water, dewater, and quarter those involved and possibly to rest line personnel.

Several types of buildings can be pressed into service to supply these needs; e.g., schools, churches, service clubs, fire stations, shopping centers, industrial or warehouse buildings, and hangars.

Another important part of strategy is a good pre-incident planning program for specific hazardous material locations and transportation routes. This includes acquiring data sheets on the commodities generally available, reference material on possible products, and sources of technical assistance. The fire prevention bureau may be used for their expertise on hazardous commodities, reference material and technical data, and monitoring equipment. Other considerations were detailed in Chapter 7.

Strategy also includes assuring that there is adequate first alarm response to emergencies involving hazardous materials. This would include any specialized apparatus that may be necessary, such as a squad with special protective clothing, a foam or dry chemical unit, an air compressor or cascade unit (Figure 9-1a and b), a spill control vehicle, a communications and/or command unit, and lighting apparatus if at night. Larger cities will have second and greater alarms established, but should not neglect special calls that may be necessary at these emergencies. Smaller communities must plan the use of automatic aid and coordinate mutual aid for greater alarms and special equipment. Plans must be made to leave ample equipment in all areas to provide protection, although the majority is needed at the hazardous material incident. Plans should include how far this equipment or other agency assistance will have to travel and the time that it will take to arrive on the scene.

The strategy should include how local and other technical assistance will be used. The local high school or junior college chemistry professor may be an important source of information until a trade association or manufacturer's representative contacts the incident commander as a result of a call to CHEMTREC. It cannot be stressed enough that technical assistance should be sought early for hazardous material incidents.

Preparation should be made to interpret, use, and coordinate with the weather service information on cloud dispersion. Personnel must be trained in the use of charts and maps on vapor clouds and technical data similar to that in the DOT *Hazardous Materials— Emergency Action Guide*. The information

Figure 9-1.   Additional air supplies provided by a cascade, compressor, or combination unit. *Courtesy of Joel Woods*

should be obtained during contingency planning on those products common to the area so that it is available to the emergency response personnel.

It is important to determine how any neutralizing or inhibiting chemicals in a spill would be handled if required, as shown in Figure 9-2. It is essential to know what chemicals may be needed and where they are obtained, and the experts to use them. This is the area in which a local or regional environmental emergency response team, such as has been established in Montgomery County (Dayton), Ohio, to assist in spill fighting can be invaluable. These people have the chemical know-how and equipment to handle the job in a safe, efficient manner. They can also be of assistance in cleanup and decontamination operations. The use of diking or absorbent material should be considered. The large supplies of absorbing material that can be obtained from the state highway or water resources departments now must be coordinated with scene needs. Plans should be made to apply the proper quantities of foam or other special agents to extinguish a fire or blanket a spill beyond present foam capabilities.

Figure 9-2.   Neutralizing of chemicals. *Copyright by Journal of Environmental Health*

Although it is not normally the responsibility of the fire department to handle cleanup operations, the department should be familiar with the procedures for the more frequently exposed commodities in their area. Fire departments will often have to assist in cleanup, performing final overhaul, purging tanks, or supplying air and water. The actual cleanup operation should be handled by professionals trained in the business; however, the fire department will have to maintain a careful vigilance to make sure that proper safety precautions are taken, because some cleanup groups are unfamiliar with the properties of the hazardous materials or in a rush to complete the job and become careless.

CONTINGENCY PLANNING

Unfortunately, many hazardous material incidents develop into large-scale disasters. The numerous locations and frequency of emergency situations have already been discussed. All agencies involved must work together rather than each doing their own thing. Each agency doing its job well is not enough. There must be adequate coordination of operations and full communications before as well as during the incident to save maximum lives and property.

A contingency plan is the basis for conducting and coordinating operations during an incident. It provides for the management of critical resources and promotes a mutual understanding among the various participating agencies of their authority, responsibility, functions, and operations. This understanding serves as a basis for including governmental and private agencies in the emergency operations.

Many areas of the country have already developed extensive disaster plans. Where a plan has been adopted, the fire department should explore and plan for its best operation during emergencies, reviewing it to be sure that it meets all the needs for controlling a hazardous commodity emergency. It is important that the plan delineate the incident commander as defined in Chapter 8.

For those emergency response organizations which have not developed a plan for large-scale or long-duration situations, some guidelines will be presented. In hazardous material incidents the fire service will be one of the first agencies to arrive on the scene. When fires or spills occur, the fire department will generally be responsible, possibly even by statute, for organizing and coordinating the operations at the incident site. Since the fire department has this responsibility and will be in charge of resolving the situation in an organized manner, it must plan to accomplish this.

Planning

The contingency plan presented here is an expansion of the command organization established in Chapter 8. It is suggested that the following staff be augmented by working groups or committees during the planning: water supply, liaison, supply, communications, information, safety, and medical.

Representatives of agencies who will directly interface with command staff members should be brought together in a series of meetings to develop the contingency plan. The plan must include the role that would be taken by all those who could conceivably be called upon during a hazardous material emergency.

The first meeting should make clear the need for such a plan and how everyone would be called upon to meet expectant needs. Perhaps the need could best be explained by a film of a recent hazardous material incident followed by slides of local installations or transportation routes that could be involved in an emergency. The fire department command structure and operations should be clearly explained.

At this point, to gain maximum assistance, the meeting should be opened and the participants asked how they could assist during a hazardous material emergency. Some of those present will have capabilities not previously known to the fire department. They should then be requested to indicate in writing where they can best help and which other members of the group they have worked with and can jointly provide assistance with at a hazardous material situation. After the meeting, those in attendance should be divided into working committees based on the information they provided with a coordinator suggested for each command staff position.

At the following meeting, the fire department command organization should be reviewed. The committees or working groups should then endeavor to identify all agencies, groups, or organizations in the local area that could provide assistance to their command staff assignment. A list should be developed of the resources available from each party, how the resources are obtained (day and night), how they can be used, and the approximate amount of time required to become operational. The committee should send forms to representatives of the suggested parties who are not in attendance so additional resources can be added and the disaster plan inventory completed. The resource lists should be collected, edited, and compiled to ensure coordination. The section in this chapter on logistics will cover resources that may be needed during a hazardous materials incident.

At the third meeting the preparation of the contingency plan should be initiated. The committees should develop their particular section of the plan, detailing how the participants and equipment designated in the resource inventory can best be used.

The meeting should begin with a discussion of the overall goals of the plan, which are the containment, control, or extinguishment, and cleanup of a hazardous materials incident in the safest, most expeditious manner, with the least number of injuries, loss of life, and property damage. How each party contributes to this must be included; for example, spill control personnel from private agencies, an environmental protection water resource agency, or the Coast Guard will assist with spill containment operations.

The plan should include a section for each committee function, such as evacuation or safety, and outline the specific duties of each participant group and of the various divisions and units within that group. For example, the Police Group could contain municipal, county, state, and auxiliary personnel. The role of each would be designated as well as the responsibilities of specific units within the divisions, such as traffic, detective, special service, and communications.

Once each functional group has developed its section of the plan detailing responsibilities and duties, the sections must be reviewed for any areas of overlap or tasks that have been omitted. The necessary coordination of each working group within the command structure should be completed and approved by those involved. This means planning how the specific groups will assist various command staff officers or functional sector commanders as outlined in Chapter 8. Each outside agency would then know its role in the command organization and overall operations and where it would be expected to assist. Many of the parties will work into the logistics section; however, some may be assigned to function sector commanders.

Flexibility must always be maintained to overcome the eventuality of people who are not available or resources that cannot be reached. It may be advisable for the media not to broadcast reports of the incident for at least 30 minutes to keep spectators from jamming the roadways and blocking access of emergency response vehicles. However, the media may have to broadcast immediate reports in order to assist in the evacuation process. Another important point is that there must be several sources for all items. There is always the possibility that items will not be available or a large number will be required. Keep your plan flexible by providing some depth and having sufficient backups.

Part of the plan may include a system of automatic notification by the communications officer or center of specific agencies for particular locations or types of incidents. The plan should include how to contact and activate the participants (even on weekends and holidays) and who is responsible for making the contact. Details of the establishment of staging areas, and its control and record-keeping should be explained by the fire department, since this is included in the supply function. All facets of the medical needs should be carefully planned, from triage teams and on-site first aid or treatment through mass patient removal, including air evacuation (see Figure 9-3), and hospital operations for extraordinary emergency room traffic. All agencies should be familiar with their role should a temporary morgue be necessary and how mass casualties would be handled. This includes identification of bodies and protection of personal effects.

The overall record-keeping mechanism should be detailed and indicate which various command staff are responsible for different records. The larger the disaster in magnitude or time the more vital this becomes. Proper forms should be developed to expedite procedures as the situation develops into a large-scale incident.

Figure 9-3. Air evacuation for removal of victims from hazardous materials disasters. *Courtesy of Joel Woods*

Those familiar with the legal field should review Public Law 93-288, the Disaster Relief Act of 1974, and make sure that any applicable sections are included.

In review, the contingency plan must prepare for the worst possible hazardous materials incident. This will ensure an adequate margin of safety and lesser emergencies will be suitably covered. The plan must outline what will need to be done, assign the responsibility and authority to do this to the proper agency, and list the sources and locations of any special equipment or assistance that will be needed.

The contingency planning should be initiated immediately if your area does not already have a plan. This interest may lead to the creation of an overall plan for any community disaster. Procrastination will only enhance the possibility of an incident happening before you are well organized and prepared. If you have a plan, it should be reviewed periodically and up-dated annually by mailing a letter to participants requesting information on new equipment resources. Some smaller, less-populated areas may not have all the resources available to them that a metropolitan area would. This does not relieve the responsibility for disaster management, but increases it, because the smaller jurisdiction must better utilize those limited resources available.

Documentation

A formal Hazardous Materials Disaster Plan should be drawn up. The document should be as complete as possible for your area and be approved by the Chief of the department on the review and recommendation of the high-ranking rescue-suppression officers. All participants who will respond to an emergency should receive a copy of the plan. Records should be kept as to whom they are distributed so that subsequent changes and updates can be properly disseminated. Each page should bear not only the page number but also the date it was completed or revised to eliminate problems in knowing which copy is current. The plan should include a number of items, among them the following:

1. A place to record changes received and inserted.

2. An introduction covering purpose, scope, and distribution.

3. Any local ordinances denoting authorization of the plan and establishing the responsibility for it.

4. Names, addresses, and phone numbers of emergency action personnel, and the committee coordinators.

5. A disaster plan organizational chart, including who is in overall command and has authority to activate the plan.

6. A listing of known and possible hazardous material locations or routes.

7. A listing of assignments of emergency functions for the various participating organizations and where they report on the scene.

8. A listing of the resources available from each organization and how they can be obtained.

9. A listing of the size and capabilities of the various service clubs, churches, schools, motels, and hotels to handle evacuees.

10. Maps of water supply lines, sewers (storm and sanitary), buried utilities, and electrical lines, traffic patterns and routes, major transportation routes, locations of hazardous materials, sites of possible major evacuations, locations where evacuees can be moved, possible command post and staging area locations.

11. Topographical maps or aerial photos.

12. Information on how to obtain weather reports, especially temperature, humidity, and wind velocity and direction.

13. A complete listing of technical assistance sources, what they can provide, and how to use them: sources such as CHEMTREC, PSTN, OHM-TADS, the National Response Center, and local sources.

14. The system for rotation of personnel in order to overcome fatigue. This must include command as well as line personnel.

Implementation

The contingency plan should be implemented at the discretion of the incident commander. When it is ascertained that the emergency is beyond the scope of those emergency response units operating at the scene and beyond the capabilities of the departments they represent, the contingency plan should be initiated. The authority of the incident commander to do this, with no additional approval necessary, should be a part of the plan. A smaller incident that is going to extend for some duration may also necessitate the implementation of sections of the plan.

As a part of the plan the system for notification of those parties immediately involved should be outlined. Perhaps an automatic dialing telephone system with all emergency numbers on cards would be beneficial in the emergency operating center. There should be no hesitation to put the plan into operation.

Those involved in a contingency plan can benefit from the additional training even though not actually utilized during the inci-

dent and only held in a standby mode. The training will better prepare all participants for that unexpected occurrence when the plan will be tested to its limits. If the plan has not been used over the period of a year, a drill should be held to keep the involved agencies current, train new participants, and update and change the plan to keep it current with present needs.

EMERGENCY MEDICAL PLANNING

Since many fire departments are also now providing emergency medical and paramedic service, a brief treatise of their role in disaster management is included here.

Sufficient sources of medical assistance should be clearly defined. This is generally not a problem in metropolitan areas; however, large quantities of medicine or certain blood types could cause some difficulty. In smaller, more remote communities, obtaining doctors, nurses, or medicine could be a serious problem. Procedures for air lifting, possibly using military assistance, should be explored and established. The availability of multiple ambulances with trained personnel to handle burns from chemical fires, corrosive burns, poisoning, or radioactive exposure should be determined. If not available, alternative methods of transport must be found. The staffing and emergency room capabilities of local hospitals, as well as their ability to handle a large number of victims, must be obtained. The planning should include communications between the hazardous materials scene and the various hospitals in order to transport those most seriously injured to facilities where they can receive immediate treatment. This will also avoid overloading of any one hospital.

Important functions that will fall under the authority of the medical officer include victim removal, triage, establishing on-site treatment or field hospitals, transportation, temporary morgues, and victim identifica-

tion. These duties are in addition to coordinating responding medical personnel, private ambulances, if used, and hospitals.

The medical planning should include arrangements for the equipment that will be needed for multiple victims. This includes: emergency medical and paramedic equipment, blankets, disaster tags, marking pens, clipboards, triage tags, stakes for locations, body bags, "keep out" signs, wrapping or masking tape, manila envelopes, plastic freezer or garbage bags, heavy wrapping paper, bags or boxes for personal effects, twine, heavy duty stapler and staples, scissors, and various types of writing instruments and paper pads, including a large artist or flipchart pad. The miscellaneous items can then be placed in a small case and designated as a Disaster Secretary, a practice used by the Lisle (Illinois) Fire Protection District.

Arrangements should be made with area hospitals for personnel and equipment to establish a field hospital if required. Plans should also be made to establish a temporary morgue for the catastrophic hazardous materials incident. This should be coordinated with the local funeral directors' association and the coroner. Remember that bodies should not be moved unless necessary until approval has been given by the coroner. Bodies should be tagged and their locations recorded for use in identification before moving. In some cases, the fire department photo unit can be of assistance in this area.

Refrigerated vans can be pressed into use for temporary storage of the corpses.

At the hazardous material scene the medical officer will oversee the various tasks or on larger incidents delegate the responsibility for specific duties to others. When delegating tasks, the medical officer must receive frequent progress reports, preferably by runner to keep the radio channels open, to remain in control of the medical situation.

The first major responsibility of the medical officer, or the designee, if this task is delegated, will be victim removal from the immediate incident scene. In some cases, such as hazardous materials in transportation accidents, this could involve extrication. This is essentially on-site victim management. It begins with getting the victims out, then determining whether to perform triage and treatment in the vicinity of the emergency or to move victims to a quickly accessible safer area. The equipment, materials, and medical supplies must be planned so that they are available to perform these operations at the scene.

The second major responsibility is early treatment and stabilization of victims before removal to medical facilities. A large, well-marked and lighted area should be made available. On incidents with a large number of patients, the medical officer may appoint an individual to be responsible for the treatment area. Since in most cases involving hazardous commodities, it will be essential to move the victims back and away from the scene; triage, tagging, and initial treatment will not have been accomplished. This will then be the first task of the treatment area. Triage must be performed as the victims are brought into the treatment area.

A large-scale hazardous material incident could easily have more injured than can be handled by a limited emergency medical service staff. In order to care for the most seriously injured as quickly as possible, those involved should be examined and tagged or marked by the field triage team. Various triage tags, an example of which is shown in Figure 9-4, are currently available. One should be selected that meets the local needs and made available in adequate quantities. This "sorting out" will establish the priorities of those injuries or persons who will be treated first and also determine the order for removal to a hospital in order to save the maximum number of lives.

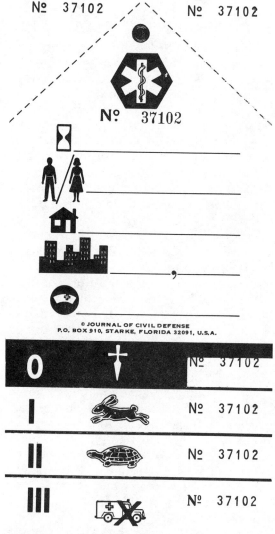

Figure 9-4a. Example of triage tag. *Courtesy of Journal of Civil Defense, P.O. Box 910, Starke, Fla. 32091*

Preferably the triage should be handled by a physician allowed for in the contingency plan; otherwise available paramedics or EMTs will have the responsibility. The injured should be grouped in categories, tagged, treated, and transported in order of need for care. The groups in decreasing need for attention are:

Classification	Treatment Assignment
1. Immediate care required, time of utmost importance	Immediate treatment area
2. Urgent, but some time available	Treatment area
3. Non-urgent, but need assistance	Assembly point
4. Dead and mortally injured	Temporary morgue or treatment area fringe

The tags should provide for a numbering system that will allow one number to be left at the locations, one with possessions, and the triage portion with the victim.

The immediate treatment area is for those victims who must be stabilized or treated to prevent death or further complications to their serious injuries. The treatment area is for those who need medical attention, but are not in imminent danger. The assembly point is for minor injuries and ambulatory victims who do not need prompt attention. Ideally patients should be placed five feet (1.5 meters) apart with their heads facing a five-foot (1.5-meters) aisle. A priority for transport from the two treatment levels should be established. Usually it would be in line with the triage priority, but could change.

The third responsibility of the medical officer is for transportation to hospitals. In addition to the normal staging area under the supply officer, it may be necessary to have an ambulance staging area near the treatment site. A loading area must be designated and marked where a good traffic flow can be maintained. If helicopter or air transportation is used, ample landing sites must be provided. The individual in charge of transportation must coordinate the dispatching of patients with the availability of hospitals. Records must be kept of the number and condition of victims moved. Depending on the tagging system used, record-keeping can be facilitated by retaining a copy or portion of the triage tag.

The contingency plan should provide for a staff to work under the medical officer.

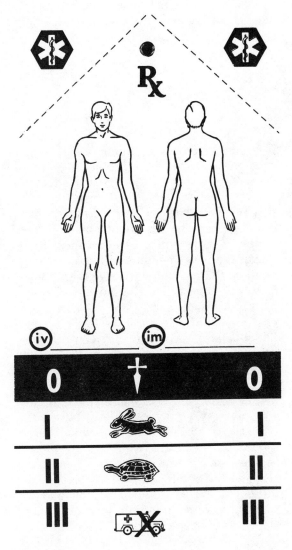

Figure 9-4b. Reverse side of triage tag. *Courtesy of Journal of Civil Defense, P.O. Box 910, Starke, Fla. 32091*

These people should have experience and training in victim removal, triage and treatment, and transportation. Although difficult to do, the medical staff must remember with mass casualties that the instinct to treat them all must be subdued and those with only a slight injury or those essentially beyond help must be left for later treatment.

The medical officer will have to communicate with the area hospitals during the emergency. A special medical communications unit may be necessary in order to operate on the ambulance-hospital medical channels. It is advisable to have a paramedic, who can relay medical messages, operate this radio. Important communications will include the hospitals' availability of emergency rooms and personnel and beds; contact with the blood bank; and alerting the hospital staff of specific problems en route. The hospital should always be notified of the number of injured on the way to avoid overcrowding and make proper preparations.

EVACUATION

Evacuation is considered to be the removal of all private citizens, including public officials, press, nonessential employees or officials, and all nonworking emergency response personnel from the immediate area and danger. Many reference texts make recommendations as to how far evacuation should extend. Vapor dispersion charts often describe how far downwind dangers can be expected. Figure 9-5 and Table 9-1, taken from the DOT *Hazardous Materials-Emergency Action Guide,* show dispersion information for Acrolein, as an example.

Evacuation may be necessary downwind for gases and vapors, downgrade for liquids or high vapor density gases, or in a circular pattern for products that polymerize, rupture, or explode. A simple rule of thumb is to initiate evacuation for at least one mile (1.6 kilometers). This did not prove to be sufficient with the bromine spill in Rockwood, Tennessee. In larger metropolitan areas evacuation in all directions for one mile (1.6 kilometers) would be a major undertaking requiring a considerable amount of time to accomplish.

Evacuation, especially when dealing with large numbers of people, immediately develops numerous difficulties and problems. Many questions must be answered in the planning stage by establishing procedures or strategies to meet these problems. Some of the questions are as follows: How to effectively alert the people quickly? How to handle persons who will not want to move unless they can see the imminent danger? Many people will be difficult to convince of the necessity and will become belligerent, almost needing to be ordered to leave their homes. How will large groups be moved? (Chicago had to evacuate 16,000 from a silicone tetrachloride cloud.) How will the traffic congestion of those leaving (if they can drive) be handled? How will persons in the area be moved if they cannot drive because of visibility? How will nonambulatory people in a nursing home or hospital be moved? How will the very young or elderly be moved, especially in adverse weather or when there is a need for medical attention, under the adversities of a hazardous material incident? How will the public be moved, if it is possible, through a vapor cloud? Both this and the routing recommended under the circumstances may need consideration before evacuation is initiated. How will the final check be made to see that everyone has left the danger area, especially at night? Where will sufficient personnel be drawn from in a minimum of

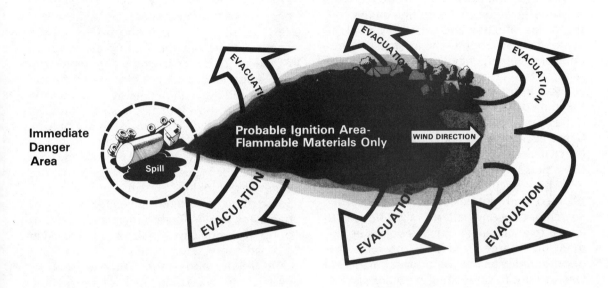

Figure 9-5. Vapor dispersion information necessary for determining evacuation areas. *Courtesy of Department of Transportation*

Table 9-1. Evacuation Distances for Acrolein—Based on Prevailing Wind of 6-12 mph (see also Figure 9-5).

Approximate Size of Spill	Distance to Evacuate From Immediate Danger Area	For Maximum Safety, Downwind Evacuation Area Should Be
200 square feet	360 yards (432 paces)	2 miles long, 1 mile wide
400 square feet	530 yards (636 paces)	3 miles long, 2 miles wide
600 square feet	650 yards (780 paces)	4 miles long, 2½ miles wide
800 square feet	760 yards (912 paces)	5 miles long, 3 miles wide

In the event of an explosion, the minimum safe distance from flying fragments is 2,000 feet in all directions.

Source: *Department of Transportation*

time to perform an adequate evacuation, especially in densely populated areas, and how will they be trained for search and evacuation?

Of essential importance is the personnel to do evacuation work. In many situations the fire service will be concentrating on the control of the situation and will only be able to complete evacuation in the immediate proximity of the emergency. Police lines should be set up at the designated perimeter and no one allowed into the area. The evacuation from the immediate scene to the perimeter will be done by the police group if they are available. Other agencies who could participate might be civil defense, fire or police cadets, Red Cross personnel, or possibly local citizens.

The evacuation should be carefully planned and not take place haphazardly. Search or dragging patterns might be used to assure total coverage. Public address equipment on emergency vehicles, power megaphones, radio or television broadcasts, Civil Defense sirens, or building-by-building visits may be used to alert the people. A considerable number of personnel could be required, depending upon the nature and populace of the area to be evacuated. Examples of difficult areas would be geriatric homes, schools, occupancies with handicapped, large sleeping areas such as a tenement district, or where people are restrained. Some method of denoting an area or building that has been evacuated will eliminate going over an area already covered. Signs can be printed that will serve this purpose.

To achieve an efficient evacuation procedure, considerable planning is needed among the cooperating agencies. All groups must work together: the police and fire departments who will essentially be alerting the people; those who may be pressed into service to transport people, such as the transit authority, charter bus lines, or the board of education for school buses; and those who may be called upon to house the evacuees.

Included in this latter group are schools, churches, fire stations, fraternal and service clubs, or any community buildings where there are sanitation and kitchen facilities, and sleeping arrangements for large groups. Often an armory or gymnasium is suitable because of the large open areas where cots can be set up. There should be ample coordination with the Red Cross, Salvation Army, or Civil Preparedness personnel who can assist in providing cots, blankets, and food. If the evacuation persists, food must be planned for: how much is needed, how long will it be needed, and how can it not only be prepared in ample quantities, but stored or kept for a suitable period of time. Along with food must be included ample supplies of pure water for drinking, cooking, and washing. Some thought should be given for those with babies or small children. Evacuation periods of up to several days should be considered in a broad contingency plan.

Of enormous importance during an evacuation is keeping track of the people moved. A registration system should be developed so that records can be kept of who is where. This will be important if families are separated or it is necessary to determine who is missing from a disaster area. It will be necessary to establish adequate communications, possibly with temporary telephones for those who will need to contact relatives.

Another facet for consideration will be personnel to guard property after the danger has subsided and until the people can re-enter their homes. In many cases guarding the property will last during cleanup operations and might require military assistance.

With the sizable number of organizations that must work together, the need for an evacuation group within the plan becomes evident. They will have considerable resources to locate, an implementation strategy to develop, and the logistics of how to move evacuees and material to the designated locations.

LOGISTICS

Logistics support the tactical operations with the supplies necessary to carry out the strategic plan. This includes providing transportation for the movement of personnel and material. In hazardous materials incidents, logistical support includes far more than the normal extended fireground operations, for which food, drinking water, and fuel are required, among other items.

The major logistical support needs of the command operation are outlined but not all items are covered. The Disaster Checklist of the International Association of Fire Chiefs' Civil Defense Committee, "Think Big," has an extremely complete list that should be checked for additional items pertinent to a local area.

Communications Support

The following support may be required by the communications officer: that of the chief dispatcher from the communications center; auxiliary communications equipment, both radio and landlines, portable radios, a portable base and antenna; the Radio Amateur Civil Emergency Service (RACES), or vehicles with a common frequency that could be assigned to cooperating agencies to establish a radio net such as could be supplied by the local taxi company, contractors, or public works; the telephone company for an emergency telephone switchboard or additional phones, including phone booths for the press and conference call capability; radio to telephone patching equipment.

Information Support

The information officer will need media support. Radio and television broadcasts may be needed to alert people to dangers, aid in evacuation, or keep spectators out of the area and emergency routes open. Accurate newspaper and wire service coverage will also be desired.

Safety Support

A vital area of logistics to the safety officer is that of protective clothing. Not only will air be necessary, but additional breathing apparatus, as illustrated by Figure 9-6, may be needed. Special protective clothing can be helpful or even a necessity. If any hazardous products are regularly used or transported through the area, the sources of special protective equipment should be contacted and proper arrangement negotiated for support during an incident. Neoprene or polyethylene gloves, boots, and facepieces should also be located for use with those substances that penetrate rubber, such as methyl bromide or Telone II.

Figure 9-6. Protective breathing apparatus for use at hazardous materials incidents

If radioactive materials are involved, the Office of Civil Preparedness or state health department may be needed. They have the training and equipment to handle these materials. Local industry or educational personnel who can lend technical assistance should be included in the contingency plan.

An area of supply not normally considered by fire service personnel is the establishment of a decontamination site. At incidents involving radioactive materials, dosimeters will have to be issued to emergency response personnel working in the area. For many commodities a decontamination control line may be required, and all personnel leaving the scene may do so only after removing their protective clothing and placing it in a plastic bag or container, removing their innerclothing for washing and showering. The logistics involved in establishing a portable shower and obtaining replacement clothing (of the right size) for a large number of personnel exposed during an incident requires that this be considered well in advance during the disaster management planning. Other items that will be needed are: plastic bags for protective equipment and clothing, soap, shampoo, towels, and blankets. Portable basins like those used in rural water supply operations (Figure 9-7) can be used for scrubbing down contaminated personnel. Proper decontamination procedures for hazardous materials which are found in the jurisdiction should be collected and recorded for references, and added to the contingency planning information. Sources of large quantities of decontamination materials should also be recorded.

In the same area of concern is preparation for equipment decontamination and control of the water and runoff from the operation. It may be necessary to channel runoff to a holding area and then remove it by vacuum truck for proper disposal. In many decontamination operations, technical assistance will be necessary. This should

be arranged for in the contingency planning stages.

Medical Support

Because a number of groups will possibly have to interface with the medical officer, the logistics of obtaining their support should be clearly outlined. Medical support includes: emergency medical service teams; triage teams; ambulance companies; the fire department physician; hospital personnel; those responsible for establishing a mobile hospital; the Poison Control Center; the coroner; city, county and/or state health departments; medical suppliers for large quantities of drugs and dressings; local trucking terminals or moving companies for vans or tractors for moving mass casualties; air support for medical evacuation, including the Civil Air Patrol who could airlift critical supplies also; funeral directors for assistance in a temporary morgue; availability of a refrigerated van to serve as a temporary morgue; and the clergy to administer to the injured, dying, or deceased.

Planning Support

The resource officer in the planning section may want support from a local government administrative officer, a financial officer

Figure 9-7. Portable water-supply basin used for initial washing of contaminated personnel

regarding emergency expenditures, a legal representative or city attorney to answer any questions of law, the engineering department and possibly the planning department for physical and environmental details.

Water Supply

The water supply officer should have access to water department assistance. Information will also be needed on auxiliary suppliers such as street flushers, milk trucks, and concrete mixers to haul water (see Figure 9-8). The logistics for an incident involving hazardous materials may require huge quantities of extinguishing agents as were used in the *Corinthos* disaster in Marcus Hook, Pennsylvania. These supplies may have to be moved over considerable distances as they were in the Union 76 tank farm fire in Romeoville, Illinois, where additional foam was supplied from Pennsylvania. The problem may not only be securing a sufficient quantity, but also where to obtain the special agents needed for combustible metals or water-reactive chemicals. Sizable quantities of alcohol foam for polar solvents or low-temperature foam may also be difficult to find. If there is a need for these products it should be recognized and the

Figure 9-8. Concrete mixers used as auxiliary water supply units

proper arrangements detailed into the logistical portion of the contingency plan.

Liaison Support

The liaison officer will need the majority of his support in evacuation assistance. This could include police departments, city, county, or state to aid in removal and establishing safe perimeters; sources of public address equipment; the National Guard for transportation vehicles, field kitchens, or guarding property; the Salvation Army for canteen services; the Red Cross to assist in relocating those moved; tanks for safe drinking water; food suppliers for meals for workers and food stuff for evacuees; methods of transporting large groups, which would include the use of school buses, charter bus services, and transit authorities; places where evacuees can be housed, such as motel and hotel operators or their associations; service clubs, churches, and Boards of Education for schools; also portable sanitation facilities for on-site needs and those evacuated. Evacuation operations will require strong logistical support in populated areas.

Rescue-Suppression Support

The line officers should also have their direct logistical needs outlined in the contingency plan.

The rescue-suppression officer will need information on mutual aid fire departments to call on for additional hose requirements, air supplies, and numerous other equipment and personnel needs. Other special requirements are: hazardous material monitoring equipment for flammable, combustible, and radioactive materials (see Figure 9-9); air support for reconnaissance; scuba teams for waterfront incidents; emergency lighting units; and snowmobiles in northern climates to reach remote locations. Fuel suppliers are necessary at extended operations, and someone to provide tire repair and mechanical service for apparatus breakdowns.

Figure 9-10. New devices using polyurethane foam for plugging leaks in pipes and tanks

Figure 9-9. Monitoring equipment for spills or leaks

Containment Support

Major logistical support is required for the containment and control of large leaks, spills, or runoff. During the contingency planning there would be close coordination between the operational and supply people to ascertain what specialized equipment and material might be required, such as vacuum equipment, containment booms, and plugging devices as shown in use in Figure 9-10. Absorbing materials or dispersants in large supplies, inhibiting agents, or neutralizers may also be needed. Where they can be ob-

tained and how they will be supplied to the scene must be a part of the contingency plan. This will require working with public works or street departments for barricades, trucks, and sand or dirt; local contractors for trucks, bulldozers, and front loaders for diking; cranes for debris removal; and skilled heavy equipment operators.

Cleanup Support

Logistics support for the cleanup procedures will entail information on cleanup and recovery companies with the equipment noted in the preceding paragraph. Septic tank cleaners can be used where vacuum trucks are not available; wreckage removal companies or towing services will be needed to clear transportation routes; commodity transfer units may be required for product removal (see Figure 9-11); and local truck terminals can provide tanks and tractors for moving the transferred products.

If site restoration activities are enjoined, the suppression-rescue officer will also work with the gas, electric, sewer, and sewerage treatment utilities.

Special Support

The incident commander may find it necessary to contact and seek assistance from government or industry officials during a hazardous material incident. Preparation for this should be made and documented in the contingency plan. Numerous agencies have been mentioned earlier. Other agencies and individuals who may be called upon are harbor authorities and the Coast Guard along waterways, airport and FAA officials for aircraft incidents, federal highway and trucking officials, or federal railroad and rail company authorities. Environmental and water resource people at all levels may also have to be notified.

Supply Officer Function

The supply officer will be responsible for coordinating and providing most of the logistical needs outlined. Many supplies will funnel through the supply officer to the support officer, while others will go to functional needs. The need for transporting many goods will require working with trucking and moving firms for vans and tractors to provide resources and supplies. The supply officer may also need to provide lighting units for the staging area and traffic control.

Figure 9-11. Product removal required before wreckage can be cleared. *Courtesy of Robert H. Shimer*

CLEANUP

Cleanup operations after a hazardous materials incident are generally not the responsibility of the fire service. However, in their overall responsibility to protect lives and property, it may be necessary that they use their authority to see that proper procedures are followed. In some instances fire department personnel may be the only people readily available who have experience or training in the area and who can provide assistance on what actions should be initiated. Personnel from state highway departments, railroad section gangs, or towing services often have little if any training or knowledge of the products involved and the potential dangers. Even chemical, biological, or health officials from government agencies, who have little if any protective equipment, fail to recognize their own possible exposure.

Of course, the cleanup operation will vary depending on the commodity involved. Cleanup can include not only picking up spilled product after containment, but also removing any contaminated debris, runoff, or soil. It may further extend to decontamination of emergency response personnel and equipment and those involved in the cleanup operations. In some cases, there may even be a need to consider the fallout area downwind, or the water table several feet below ground.

Since there are such variations in operations and techniques, it will generally be necessary to obtain outside technical assistance from manufacturers, trade associations, or professional spill removal organizations. The first place to look for information on small containers, cartons, or cylinders is the label.

A number of specialized pieces of equipment may be required. This can include the following: booms, boats, skimmers, winches, sidewinder equipped tractors, earth moving equipment, vacuum trucks, covered dump trucks preferably with steel bodies, fiberglass pumps, stainless steel pumps, plugging devices and tools, compressed gas pumps and compressors, tapping devices, steam generators, neutralizers, absorbents, belt mops, dispersants, coagulants, decontaminants, and proper protective clothing for the hazards encountered.

Considering that spills include gases, liquids, and solids in both vertical and horizontal spills, it is easy to understand that problems can be encountered. In most cases, spills or runoff of liquids or gases traveling horizontally along the ground are easier to contain than those moving vertically either into the atmosphere above or the earth below.

Some general safety precautions that apply to most hazardous materials during cleanup are: avoid any skin contact; do not breathe dust, vapors, fumes or smoke; and before smoking, eating, or leaving the contaminated area, wash thoroughly with the prescribed solution.

For small spills the procedure to follow is to absorb the material; dust or powders can be picked up with a dampened absorbent, and then sweeping up the material and placing it in a plastic bag or bags of at least four mils equivalent thickness. The spill area should then be decontaminated and the solution picked up with an absorbent and placed in a sealed container. This procedure may have to be repeated a second and third time. Absorbents locally available could include lime, sawdust, straw, commercial clay, fuller's earth, kitty litter, or high clay content soil. Decontamination solutions are not recommended in this text because of the wide range of products and the differing needs that can arise. In all cases, flushing of spilled material should be discouraged unless there is an immediate threat to lives and safety. Even then runoff must be controlled and in many cases subsequent cleanup and decontamination performed. Spills of solids, powders, or dusts may need to be kept from spreading by covering with nonporous tarps, plastic sheets, or a very light sprinkling of water.

The cleanup of an etiologic agent may occur after a fire in a post office, biological or medical laboratory, in transportation or at a terminal, especially air freight. The material should be isolated immediately by placing it in a plastic bag, using rubber gloves or a heavy cloth towel. Notify the Center for Disease Control in Atlanta at (404) 633-5313 and give them whatever information is available on the package contents, degree of damage, where the shipment originated, and where it is going. The center will provide additional information on disposition and decontamination. Fortunately, etiologic agents can all be controlled by the application of household bleach, sodium hypochlorite. A spill area can be decontaminated by mopping with large quantities of bleach. If the surface or package is not conducive to mopping, it can be covered with a bath towel, blanket, or some similar material, and saturated with bleach. When using bleach, the normal skin and respiratory protection should be used.

Spills around waterways can be a real problem in cleanup. The Federal Water Pollution Control Act of 1972 provides for oil or hazardous substance cleanup. It calls for the immediate notification of the appropriate government agency, and the cleanup must be performed to the satisfaction of the United States government. An oil spill is defined as any discharge that causes a film or sheen upon or discoloration of the water or adjoining shoreline, or causes a sludge or emulsion to be deposited beneath the surface of the water or upon adjoining shorelines. With this specific definition for oil spills, it is easy to see how it would be applied to other hazardous substances. For spills that enter navigable waterways or flow into them, the Coast Guard should be contacted at the National Response Center, (800) 424-8802.

RESTORATION OF SERVICES

Another phase of disaster management that the incident commander may be required to coordinate, especially in smaller communities or rural areas, is the restoration of services. A number of vital services could be impaired by the incident and essentially create minor emergencies of their own. These situations would be beyond the fire department jurisdiction, but still could necessitate direction by the incident commander. Examples of this might be the contamination of the ground water table supplying wells or the water source for a community's water filtration plant. Auxiliary water supplies will have to be provided for the population, which could possibly involve fire department operations. Another example would be the restoration of electrical power. Although this is done by the power company, in the interim the fire department may have to supply emergency lights or power for vital operations. Numerous other services may be required that will need fire department assistance and incident commander coordination to bring the incident to a successful completion.

In summary, a great deal of effort may have to be expended to provide the disaster management system required for potential hazardous material emergencies. Much cooperation and coordination is required for an efficient system, even in smaller communities where the resources are limited. Determine in advance what is needed, and make plans to overcome any deficiencies. The time to make mistakes is on the planning board before the hazardous material incident occurs, for failing to plan is planning to fail.

QUESTIONS FOR REVIEW AND DISCUSSION

1. What are good criteria for a command location? Are there suitable locations in your jurisdiction and have arrangements been made for their utilization?

2. Has a procedure been developed to use local people for technical assistance at a hazardous material incident? If so, describe the preplanned procedure. If not, outline a procedure for your community identifying possible resources.

3. Develop standard operating procedures for evacuation in the immediate area and secondary evacuation area for an incident in a populated residential/commercial district.

4. Review the current Emergency Medical Services disaster plan and its coordination with fire department operations. Record any recommendations you have in relation to a hazardous material incident and the current disaster plan.

5. Develop a list of logistical needs to handle adequately the commodities stored or used in your immediate response area or transported through it.

CASE PROBLEM

On Sunday morning June 18, a Falcon II jet plane left a suburban Los Angeles airport. After climbing to about 1500 feet, the plane lost power and started to descend out of control. The pilot, attempting to land on a street, missed his target, striking an apartment building, shearing off a wing, engine, and fuel tank. The eight unit 2½-story ordinary apartment building had a major portion of the roof ignited by the fuel in the portion which hit it. The apartment building had similar buildings to the south at 50 feet and to the east at 40 feet.

The plane, after striking the apartment building, veered 90°. Careening across the street, the plane sent burning jet fuel into the storm sewer, took down an electrical pole and telephone lines, and sheared off the natural gas regulator before crashing into the south (front) wall of a church during the 11:00 service. The church was ignited and the wall forced in on the congregation by the skidding plane. The church was moderate in size, of hollow concrete and brick walls, with an open nave having a laminated beam supported roof of tongue-and-groove planking. The steeple was above the area where the plane entered the building. It did not fall but was leaning toward the street in front. The church sat on a corner and had a parking lot in the rear. The only exposure was a one-story commercial building 75 feet to the east.

The plane was en route to Stanford University carrying a mixed load of radioactive materials including Cobalt 60 and Sodium 24. The temperature was 75°F with the wind at 10 mph from the west. Ample water supply was available.

1. Should the disaster plan be implemented? How would it assist in this incident? What agencies would be involved?

2. What is the priority of concerns and who is responsible for each of these actions?

ADDITIONAL READINGS

Aircraft Fire Protection and Rescue Procedures, IFSTA No. 206, 2nd ed. Stillwater, Okla.: Oklahoma State University.

Bahme, Charles. *Fire Officers Guide to Emergency Actions*. Boston: National Fire Protection Association, 1974.

Emergency Operations Reference Manual. Elk Grove Village, Ill., December 1973.

"Emergency Organization Plan." Fire Department, City of Huntington Beach, California, June 1974.

Fire Service Rescue and Protective Breathing Practices. IFSTA No. 108, 4th ed. Stillwater, Okla.: Oklahoma State University.

"Guidance for the Development of a Local Emergency Plan." Sacramento: Office of Emergency Services, 1974.

Harms, George. "The Fire Administrator's Role During Disasters." In *FDIC Proceedings*, 1972, p. 72.

Katz, William B. "Handling Flammable Product Spills Without Doing Environment Damage." *Fire Engineering*, August 1977.

O'Hagn, John T. *Fire Fighting During Civil Disorders*. Washington, D.C.: International Association of Fire Chiefs, 1968.

Page, James O. *Emergency Medical Services for Fire Departments*. Boston: National Fire Protection Association, 1975.

Shamer, Lawrence C. "Triage: The Treatment of Mass Casualties by Fire/Rescue Personnel." *The International Fire Chief*, 43 (October 1977).

"Think Big." Civil Defense Committee, International Association of Fire Chiefs, 1977.

Troeger, John L. "How to Develop a Disaster Plan." *Fire Chief*, December 1975.

10 Decision making

Decision making is the art of skillfully selecting a course of action and putting it into effect. Decisions should be made after a careful study and evaluation of the situation, on the basis of the experience of what has happened in previous similar situations, or decisions may be made haphazardly without forethought.

Fireground officers have the responsibility of making decisions. Since decision making is a mental process, officers must be trained to think situations through in an orderly manner in order to make better decisions. This is especially applicable to hazardous material incidents, which can extend far beyond the normal parameters in a short time. When a situation accelerates beyond the size and scope of an incident commander's experience, it is far too easy to abandon logical, consistent thought patterns and begin making irrational decisions, which, like the situation, are soon out of control. Also, in a large emergency situation, fireground commanders must remember to react to the overall situation, and not become bogged down with the obvious, or place a great amount of emphasis on insignificant detail. This requires training to accurately assess the situation, determine priorities, and make decisions for action.

The method of problem solving recommended in this chapter is a systems approach. It will enable the incident commander to make decisions that will be more effective in bringing the situation to a safe conclusion. When used consistently, a systems approach will train the commander to think logically and avoid many of the pitfalls of decisions based only on hunches, intuition, or "gut feelings." A clear description of systems analysis is presented, followed by application of this principle to the hazardous material scene. Once the technique is understood, it will be beneficial to go back to Chapter 1 and its synopsis of actual hazardous material incidents, perhaps delving into them in greater detail, and making your own decisions.

SYSTEMS ANALYSIS

Most of the problem analysis techniques recommended today follow a systems analysis procedure. This essentially means nothing more than placing what we know in an orderly manner for analytic study to determine the best course of action from the possible alternatives. The following is a six-step systems approach that can be used anywhere in problem analysis on the fireground, in administrative work, or during a hazardous material emergency:

1. Identify the problem.

2. Examine the conditions surrounding the problem.
3. Develop possible alternative solutions.
4. Evaluate the alternatives.
5. Make a decision based on the best alternative.
6. Implement the recommended course of action.

To illustrate the systems analysis approach, a simple dwelling fire example will be used. This fire occurs on a Tuesday night in early spring with mild temperatures and a slight breeze of 5 miles (8 kilometers) per hour from the northwest. The structure is a 1½-story wood frame bungalow 30 by 50 feet (9 by 15 meters). The first floor has a living room, dining room, kitchen, bathroom, and storage areas with a rear porch. The second floor has two bedrooms, a bathroom, and large storage areas along the eaves. The basement is finished with a third bedroom, recreation room, and work area. The fire originates in a living room sofa from a dropped cigarette and has secured a good start in the home (no smoke detector). The fire is extending horizontally with smoke traversing all vertical avenues. The family of five are at home and retired for the evening. Exposures are similar bungalows 20 feet (6 meters) to either side and an attached two-car garage at the rear.

Identify the Problem

The exact nature of the problem should be determined. Be careful not to confuse symptoms or apparent problems with the real problem. For example, in the house fire, a great deal of smoke is an apparent problem. Actually, the real problem is the fire, which must be extinguished before it spreads further. An associated difficulty may be the toxicity of the products of combustion or the high heat buildup. It must be pointed out that

in some cases there can be more than one problem. The dwelling example could entail the endangered lives of the occupants as another major problem. Although there is an interrelationship between the two problems, one will require rescue and the other extinguishment.

Examine the Conditions Surrounding the Problem

This is the data collection phase or size-up in which all the details affecting the situation must be gathered. Consider the internal and external environment and the resources available. For example, in the house fire, this could refer to the exposures within the dwelling such as adjacent rooms, the floor above, or those without, such as the attached garage or nearby structures. The fire department capabilities or resources to handle the situation would include apparatus, personnel, training, and water supply.

Develop Possible Alternative Solutions

Determine from the known information what various plans of action might be initiated. Our example possibilities might be a vigorous interior attack with small lines, a quick knockdown with a 2½-inch heavy attack line, or an approach from the rear with a 1½-inch line to protect internal exposures and control the fire. Remember that doing nothing is one alternative that should be considered.

Evaluate the Alternatives

Identify the considerations to be used in evaluating the alternatives. Criteria should be laid out by which to judge the alternatives and select the best one. It may be advisable also to assign weights or priorities to these considerations. In our example, the saving of lives has a higher priority than extinguishing the fire. The criteria may be complementary in any given situation. Extinguishing the fire may eliminate the need for evacuating adjacent homes.

Organize and interpret the facts and how they bear on the choices. Determine how each piece of information collected affects each of the proposed alternatives. Explain clearly any assumptions that have been made and what information they were based on.

Consider the implications of each alternative. If an alternative has inherent difficulties or problems, these must be stated; for example, a large-line attack alternative has the implication that severe water damage will occur if the nozzle is not immediately closed after knockdown of the fire. Each alternative must achieve the outcome desired.

Make a Decision Based on the Best Alternative

The decisions made at the incident must be based on the best alternative. They should not have any possible adverse effects that would result in the further loss of lives or property. The solution should be one that will control the problem, prevent any recurrence, and by no means initiate new problems. These parameters should be based on the criteria specified to determine which alternative will accomplish the desired outcome. In the house fire, this would be the 1½-inch line attack from the rear to confine the fire and protect internal vertical avenues of spread. The best alternative may not be obvious. It is in these cases that systems analysis will be of greatest help to ensure that all risks and conditions have been identified and the problem logically thought through.

Implement the Recommended Course of Action

This is the deployment of the tactical action plan to initiate the alternative selected to attain the desired outcome. It is the initiation of the attack on the dwelling fire. In our example, for many departments this would simply be the standard operating procedure for a house fire. The course of action should include all the necessary considerations such as the resources necessary, safety of personnel, and possible outcomes.

DECISION MAKING AT HAZARDOUS MATERIAL INCIDENTS

"Too frequently, the impulse to act immediately overwhelms reason in crisis situations; risk-taking becomes excessive, and losses escalate rather than decline."[1] At a hazardous materials emergency, the incident commander should apply a systems approach to prohibit this.

Numerous decision making processes have been developed. These methods must be familiar to the officer in charge long before the fact so that they can be put into instantaneous use when the need arises. Three fire service oriented decision making methods are those contained in *Fire Fighting Tactics* by Lloyd Layman,[2] "Decision Making" by James E. Heller,[3] and "D.E.C.I.D.E. in Hazardous Materials Emergencies" by Ludwig Benner, Jr.[4] A hazardous materials situation is no time for indecision in developing and implementing a plan of action.

It should be pointed out that no decision making guide will be a panacea for all the problems that might be encountered. Training, experience, judgment, proper preplanning, good command post operations, and improved data collection all have a part in successful decision making. While the decision making guide presented here can be used at any emergency, it is primarily aimed at decision making at hazardous material emergencies. Following are the steps in the process:

1. State an objective.
2. Collect and evaluate pertinent data by size-up.
3. Develop alternatives.
4. Anticipate the unexpected.
5. Select the safest feasible alternative.
6. Implement the decision in a plan of action.
7. Monitor the progress.
8. Take corrective action if necessary.

State an Objective

Since the overall goal of the fire service is to protect lives and property, the decision making starts with setting the objective necessary to achieve this at a hazardous materials incident. All of the alternatives developed must be consistent with the objective. The objective for any hazardous materials emergency is the containment, control, or extinguishment in the safest, most expeditious manner, with the least number of injuries and losses of life and a minimum of property damage. It is most important to keep this objective constantly in mind. Because of the possible outcome during hazardous material incidents, the best decision may be less than an all-out attack; in fact it may even be no attack at all.

Collect and Evaluate Pertinent Data by Size-Up

Size-up is the determination of the problem, its nature, and the conditions that exist in the immediate area. The size-up serves as a data base for developing and evaluating the various tactical alternatives.

The size-up of a fireground situation has been taught in officer training and tactics courses for some time. Therefore, a lengthy

detailed procedure will not be described here. A brief review, however, will be helpful.

Size-up essentially covers those items that were considered in preplanning. The size-up will reaffirm and update the preplan information when possible. However, because of restraints of time and conditions, the preplan data may have to be heavily relied upon. Preplanning has the distinct advantage of being done in a relaxed atmosphere without the pressure of an emergency. It also allows the discussion of all important considerations and the determination of possible problem areas that could easily be overlooked during size-up. Preplanning enables an exact look at the inside with facility people, which cannot be done during size-up.

In hazardous material situations, there must be an evaluation for approach. This should begin immediately with receipt of the alarm, and includes the address, and any additional information on the emergency received from alarm headquarters. En route, the approach evaluation should weigh the use of alternative routes because of traffic or road conditions, the location and products involved, the accessibility to and/or into the incident, wind and weather conditions, and responding personnel and equipment. In anticipation of a hazardous material problem, a check should be made with the dispatching center to see if further information has been received. The preplan information should also be reviewed to thoroughly familiarize the responding officers with the site.

The approach evaluation is essential to keep emergency response personnel out of the hazardous area until identification of the materials can be made and initial assessment completed. This is vital because emergency response crews are reaction-oriented and must be taught to approach hazardous materials cautiously before they become too deeply involved and find themselves a part of the problem rather than the solution. Care must be taken to approach and place apparatus upwind and upgrade from the problem. The vapors may not be readily seen and could be ignited by responding units. This recently happened in Vicksburg, Mississippi, when a pumper responding with four fire fighters to a leak emergency at an LPG bulk plant drove into the vapors and ignited the escaping gas. Three fire fighters were killed in the resulting flash fire. An explosive meter should be available to all fire departments to check for hazardous vapors that are not visible.

Topography, wind direction and velocity, vapor dispersion patterns, and water supply, the position of hazardous materials, and exposures must all be considered in placement of apparatus. All apparatus must be positioned for a safe tactical size-up. Before the tactical size-up is initiated, an escape route should be determined and all emergency response personnel informed of it. Tactical size-up will include problem identification, life safety considerations, exposures, location, position, construction details, time parameters, water supply, weather, obstructions, attack position, personnel, apparatus and equipment, private fire protection, additional help available, and action already taken.

Once the size-up data have been collected either by the incident commander touring the area, receiving reports from sector commanders, referring to preplanning information, or a combination of these sources, the size-up data must be evaluated.

The following items are not listed in priority order and some may be of greater importance than others in a particular situation. Many items when evaluated will be dependent on other data collected. The material collected during size-up and how it was evaluated will be a crucial influencing factor in developing alternative solutions.

Problem Identification. The immediate problem must be identified. This information may be given to responding units by the dispatcher; however, it could be incomplete or in error. As the first unit arrives, it should give a report on the immediate situation. Remember that additional information obtained during the size-up may change the problem or identify a more serious problem and that symptoms should not be confused with the problem itself. The Phoenix Fire Department dispatched companies to a car fire. When the companies pulled onto the apron, they found the car fire was under a fully involved gasoline transport.

Of major concern when hazardous materials are involved is the nature of the problem: whether it is a spill, leak, or fire. What products are involved and their chemical and physical properties (flash point, flammable limits, ignition temperature, specific gravity, and vapor density) and makeup will be of immense importance. Is the commodity toxic, flammable, combustible, corrosive, radioactive, unstable, or water-reactive? The amount involved in the incident will have a bearing on the operations; a broken 1-gallon jug, a 55-gallon drum, or a multithousand-gallon tank all present different considerations. The rate of flow or vapor release should be observed, as well as whether the involvement is limited, partial, or total.

To obtain this information and possibly assist in identification, binoculars may be necessary. A flammable gas detector will also be required to detect hazardous, yet invisible concentrations. It is pertinent to life safety that vapors and harmful concentrations be recognized before personnel are needlessly exposed or ignition sources are inadvertently introduced into the area.

Identification of the material can often be difficult. This subject was discussed in Chapter 6, but its importance in size-up cannot be overlooked. The type of container is important and may help in identification. Since many package labels and vehicle (highway or rail) placards give only a warning, alternative methods of identifying the materials must be learned and exercised. These include the use of shipping papers (both highway and rail), trip tickets, manifests, switch tickets, tractor or trailer numbers and the company name, rail car numbers and company names, plant or warehouse inventories, and key personnel from facilities handling these products.

Evaluation. When the hazardous materials and the problem have been identified, it is important to evaluate whether the situation is decreasing or growing in intensity. Is it serious, dangerous, controllable, minor, or of little significance? How will the products react under the present circumstances? Has there been a mixing of products or additional stress added that change the normal procedures for handling the product? Is sufficient information available from on-site references, responding personnel, or CHEMTREC which indicates possible changes in the material and the indications or warnings of such changes.

How must you react to control the spill, leak, or fire? What immediate actions must be initiated with this particular material to reduce any possible destructive outcome? Is the hazard major, moderate, or low? Will a disaster plan have to be activated? Are any special considerations necessary because of the nature of the problem?

Life Safety—Of primary concern during size-up are the life safety considerations. Necessary information includes how many people are involved and where they are located. This includes emergency response personnel; any victims involved in the incident whether in a structure, a vehicle carrying hazardous materials, or other involved vehicle; employees; people passing by; residents; or spectators. Any high life concentrations (target hazards) that could add to rescue or evacuation difficulties must be quickly identified. Large hospitals can present severe life safety problems. They also have hazardous materials on the premises. Where people are nonambulatory, the situation intensifies.

Life safety must concern itself with all the possible ways in which people could be injured or exposed to toxic substances. This in-

teracts with a number of other size-up considerations, such as nature of the problem, exposures, and position.

Evaluation. The safety and welfare of those people involved or exposed and what must be done to protect them must be evaluated. Are rescue, evacuation, or withdrawal operations required? Has the period of flame impingement been such that a rupture could occur? Will the materials possibly polymerize or explode? Are all exposed persons in the immediate area or will the flow of vapors or liquid call for actions remote from the site? How will those operations be initiated and carried out? Are sufficient personnel available to do this and maintain other operations, or must life safety be given priority?

Exposures. All exposures must be recognized and identified during size-up. Exposures fall into three broad categories: life, property, and other hazardous materials. This can include exposure to vapors, fumes, or products of combustion. Smoke and the fall-out it includes can add to the problem by endangering nearby homes and other structures. After the fire in a pesticide-formulating plant in Ennis, Texas, several homes in a downwind subdivision required hosing down to remove the fallout from the fire. There may be severe fire exposure by radiation, conduction, or convection to other nearby buildings or tanks. The liquids being released must be checked, as well as runoff before it endangers an exposure. Is there any human exposure to inhalation or absorption of hazardous materials, or danger if a violent rupture or explosion occurs? There could also be the chance of internal exposures within a large manufacturing or warehouse complex.

Exposures in hazardous material situations are much broader than in the normal consideration of fire spread. Furthermore, the exposures can be at a great distance where wind conditions move a vapor cloud. Also, runoff getting into a waterway can develop distance exposures. Even wells and the ground water table can become an exposure.

Evaluation. The main concern with exposures will be how to protect them. This may require normal fire fighting operations to protect structures, unmanned streams or foam to protect exposed tanks, or special considerations for the people problem. Exposure protection will depend on the density of people or structures in the area. Other points to evaluate are what operations will be necessary and what equipment must be used to protect the various types of exposures encountered and how the size or quantity of exposure affect the situation. For example, is enough high-expansion foam available to cover a fuming acid spill and thus reduce the vapor exposure problem? Will evacuation of exposed buildings be necessary, and if so how will it be accomplished and for how far? Evaluation of exposures revolves around knowing what there is to protect and determining how this will be accomplished.

Position and Construction. The exact location of the incident must be ascertained in the size-up and the location of the hazardous commodities within the situation area. The location will be important to rescue operations, evacuation, exposures, confinement, runoff control, positioning of attack personnel, and the control operation selected. Of significance is whether the incident is located in a built-up or remote area. Those located in congested metropolitan areas will add astronomical considerations. Also of importance is the position of the products or containers. The position of a spill of Telon II, which penetrates rubber, would require more consideration than a drum of flammable liquid that could be uprighted to stop the flow. The position of a liquefied petroleum gas cylinder, which could be moved away from a burning trailer, is important, but a tank car of vinyl chloride under flame impingement, which can only be approached from the end, is more significant.

Information on construction details must be acquired during the size-up. This includes not only the pertinent features normally sought during a structural fire when the hazardous materials are in a building, but also any special information on how they are stored, as, for example, in a special locker or vault. For transportation incidents, information on the important features of the truck, rail car, ship, or barge will have an effect on the situation. An important detail during accidents will be the condition of the container storing hazardous materials. Has the radioactive material "pig" been opened, have tank cars received mechanical damage that increases the physical stress and possibility of further problems, or have safety devices been damaged or negated? The construction features of small containers such as drums, carboys, and pails is also important, and should have been acquired during preplanning.

Data on transportation vehicles should be learned during training sessions and in visits to local terminals. It is often difficult during the confusion of an incident to obtain all the details, especially when considering the possibility of its happening at night, being obscured by smoke, or involving hazardous vapors and products of combustion. Details on transportation vehicles are given in following chapters. Also worthy of note are any safety features for the control of hazardous materials, such as explosion venting or special drains for spills or runoff. The safety features and the manner in which they operate will help in evaluating the situation. Of importance also is the condition of any containers and the relevance of this to the situation.

Evaluation. What effect does the location of the incident and the position of the hazardous commodities have on operations? Are the preconnected hose lines on responding apparatus long enough to reach the problem? Will additional equipment or personnel be necessary to stretch lines by hand or to reach hard-to-get-to locations as in a railroad classification yard? How does the location or position affect the attack positions? Because of the location, must attack lines be placed to protect exposures?

Construction details must be evaluated. Signs of structural weakness or failure are extremely important. How must the situation be handled based on the condition of any tanks involved and the reaction of safety devices? All stress factors influencing the container must be carefully evaluated. This includes mechanical or physical damage, chemical stress, pressure differences due to temperature changes, and the effects of thermal stress. If these stresses are evident, all safety precautions must be taken.

Time. An estimate should be made of the time lapse in the duration of the incident. This must include the time expended in responding, the time of flame impingement if closed containers are involved or the products are heat sensitive, and the time if containers are exposed to corrosives. The time required to amass the needed resources, implement operations, and achieve results must also be included. Exposure time for involved persons and emergency response personnel should be determined and recorded.

Evaluation. Time is crucial at a hazardous materials incident. The amount of time that has elapsed can be useful in determining the amount of fuel remaining, the need for and quantity of extinguishing agents, or the amount of time available for withdrawal and evacuation operations. Where harmful commodities are released, the length of time personnel have been exposed is vital. The evaluation must consider how much time has been expended and how much safe time remains.

Where there is severe fire exposure to enclosed tanks, the time it will take to apply

Table 10-1. Closed Container Explosions

Date	Location	Contents	Time Lapse Between Initial Incident and Explosion
Jan. 1968	Dunreith, Indiana	Ethylene Oxide	45 minutes
Jan. 1969	Laurel, Mississippi	Liquefied Petroleum Gas	Immediate
Jun. 1970	Crescent City, Illinois	Liquefied Petroleum Gas	1 hour to 4½ hours
Jul. 1973	Kingman, Arizona	Liquefied Petroleum Gas	19 minutes
Jan. 1974	West St. Paul, Minnesota	Liquefied Petroleum Gas	17 minutes
Feb. 1974	Oneonta, New York	Liquefied Petroleum Gas	33 minutes
Test	White Sands, New Mexico	Liquefied Petroleum Gas	24 minutes

adequate quantities of water for cooling is extremely important. Was there a delayed alarm, how long did it take to arrive on the scene, and the essential question, what can be done in the time you have available must all be considered. Although no one knows how long before a violent rupture, polymerization, or an explosion may occur, an estimate may need to be made on when it is possible or probable. The time lapse from past incidents as shown in Table 10-1 may give some insight into what can be expected.

Water Supply. Detailed information on the water supply will be required in most incidents, since water may be needed for confining, controlling, cooling, or extinguishing. The command post will need to know how much water will be required, the number and size of available sources, and whether any special pumping operations are necessary. At least two sources should be used, when available, to increase reliability. It should be determined early whether relay, supplemental, dual or tandem pumping (Figure 10-1), shuttles with mobile water supply apparatus, or special railroad equipment are going to be needed.

The application rate that must be provided and the time to deliver it can be critical. It is suggested that operations be developed to apply a minimum of 500 gallons (1,900 liters) per minute for any hazardous materials incident within 15 minutes from its inception. Although this amount may not be needed for small containers or quantities, each large tank can require this amount. The amount of water required would increase as the number of tanks involved, their size, or the size of the structure increased. What application equipment is available to deliver the necessary rate is important. A final important question on water is the length of time the flow will be necessary. Hazardous materials incidents have been known to last several days and could easily overtax many small water systems or supplies.

Figure 10-1. Dual pumping operations

Evaluation. In evaluating the water supply, it is important to know whether it is adequate, can be applied where it is needed in the proper amounts, and whether it will be effective. For example, a deluge gun is not needed for the expanding gasoline from an auto fuel tank on a hot summer afternoon, nor are booster lines effective on ammonium nitrate. Are the number and size of hose lines and the master stream devices required readily available? Will the streams reach the problem area from a safe or protected location? Can the water be placed where it is needed or are there obstructions such as fences, piled up rail cars, toxic vapors, or intense heat? Enormous quantities of water will do little good unless they can reach points of impingement, aid in cooling or extinguishment. Whenever possible preplans should develop provisions to supply maximum quantities with 500 gallons (1,900 liters) per minute established as a starting point for each large container. If the product is water-soluble or water-reactive, how does this affect operations and what can be done to achieve control?

Weather. The weather conditions can be picked up en route, but any changes must be noted during size-up. Wind direction, temperature, humidity, precipitation, and inversions can all be important and must be constantly monitored for changes.

Evaluation. How do the current (or expected) weather conditions affect the situation and what must be done to overcome the problems imposed by nature? Will strong winds readily dissipate toxic fumes and vapors, or will they create the necessity for evacuation of a large area? Strong winds will make it difficult to use high-expansion foam and make regular foam more desirable for covering a hazardous spill. Extremes of temperature will bring about the need to rotate personnel faster and may cause higher consumption of air for breathing apparatus. High humidity or an inversion may bring about a need for increased ventilation equipment. High temperatures can cause the autoignition of some exotic materials (diborane, for example), which calls for evaluating the weather conditions as soon as possible.

Obstructions. Obstructions into the property such as fences, locked gates, parked equipment, or critical storage must be noted. Storage may be high-piled stock that could fall, drums, bags, or containers that could burst or break and release dangerous commodities, or only a small amount of etiologic agent or toxicant, which if spilled could impede progress. Construction details that are dangerous should also be noted; signs or overhead piping, for example, which could fall. All types of obstructions must be watched for during size-up because they affect tactical operations.

Evaluation. Obstructions may necessitate longer hose lays, extending lines, and numerous operations that could cause delays or even endanger lives. Obstructions may also hinder getting to the dangerous commodities

or have a critical effect on line and manpower placement. The time lost in coping with obstructions, the additional equipment and manpower required, and the effect of this on the situation should be estimated and evaluated.

Attack Positions. All possible attack positions should be determined during size-up. This will enable the incident commander to develop a plan of attack, discarding those positions that are not safe and using those that can be quickly put into service, have strong water resources, or are from protected locations.

Evaluation. Attack positions must be evaluated as to their adequacy and safety. The number of positions must be sufficient to deliver the required quantities of control or extinguishing agents. The lives of emergency response personnel must not be jeopardized haphazardly by poor position selection. There will be times when unsafe attack positions may be temporarily necessary during a holding action while exposed structures are evacuated; however, personnel in these positions should be covered with protective streams and replaced by unmanned streams as quickly as possible. Unsafe attack positions should never be taken unless there is a severe exposure to lives requiring a holding action. Downhill, downwind, above or over, at the ends of tanks or in a circular area around explosives, unstable materials, or containers that may rupture are all dangerous attack positions. Tactical operations must be developed to alleviate these.

Personnel. The personnel available on the scene, en route, or available through subsequent alarms, recall, or mutual aid will have to be considered. This should include all emergency response personnel who may be called in from other agencies or departments.

Evaluation. Personnel availability must be evaluated to determine whether it is sufficient to perform all the tasks required of the situation. This will include everything from stretching lines to providing backup protection. How long will it be before additional help arrives? How long will the present forces be able to maintain the battle? Call for assistance early enough to be of help. Do not needlessly expose fire fighters to hazardous materials to impress news cameramen. Use only those forces necessary to do the job. Keep all remaining personnel in a remote, safe, protected staging area.

Apparatus and Equipment. Along with personnel, apparatus and equipment needs must be appraised. This is especially pertinent if special agents or devices in large quantities are necessary; for example, alcohol foam, or specialized equipment, such as floating booms, inhibitors, or neutralizers for spill control. Recent chlorine problems have required large quantities of caustic soda and special operations for neutralization. Size-up must include the determination of needs based on the situation, such as how much hose will be required.

Evaluation. The evaluation of apparatus and equipment must determine if what is available (on the scene, en route, or from mutual aid) is capable of dealing with the situation. If additional resources are needed, what, how much, and how will they be utilized? It is not necessary to amass far more equipment than can be used and possibly leave another area with minimal protection, if not unprotected, because of a large hazardous materials incident that is better handled by controlled burning. The evaluation should include the need for special agents or devices such as a vacuum truck, soda ash or lime, or additional foam eductors as required. In a situation of large magnitude, some backup apparatus should always be held in reserve in a staging area.

Private Fire Protection. Any private protection devices that can provide assistance and their locations are of special interest. Automatic extinguishing systems (Figure 10-2) or private water systems will be most common. Any knowledge required for their operations will have to be acquired quickly if this

Figure 10-2. Private protection such as automatic sprinkler systems assessed in size-up

has not been preplanned.

Evaluation. It must quickly be determined if the private protection can be of assistance or whether it might cause additional difficulties. How can it best be utilized to better the incident? In some cases, the private protection will be damaged or inoperable due to the fire, and time or water supply may be wasted if the size-up and evaluation are not accurate as to its utility.

Additional Help Available. For additional help, all persons and agencies covered in Chapter 9 on disaster management should be contacted. All additional help that is readily available or could be subsequently needed should be determined early. This provides the maximum amount of time for notification, information to be received, or for direct response. Who to call must be determined before technical assistance can be summoned.

Evaluation. Do not overlook specialists in the field who are available or outside agencies who can provide assistance. The critical factor is to determine how these resources can be used and to have an assignment ready when they arrive. Adequate disaster planning before the incident will be invaluable. It is essential also to know how long before help will arrive.

Action Already Taken. A final important item to check during size-up is the action already taken. Have personnel on the scene started notifying other agencies or summoning technical assistance? Have the first-responding emergency response personnel initiated good, safe procedures or have they exposed themselves to harmful smoke, vapors, or solutions because of inadequate or a lack of protective clothing and breathing apparatus while making a small-stream attack? What has been done may very well determine where the starting point of your tactics will be. Have the first-arriving units initiated control efforts that place them in danger? One of the most important life safety considerations has to be for your own personnel, who may have to be ordered to withdraw from a life-threatening situation.

Evaluation. The operations initiated by the first-arriving emergency response personnel must be evaluated. If initial operations are proceeding satisfactorily, such as multiple master streams on a derailment of fuel oil tank cars, no change may be required; however, if the runoff has not been diked and is running into a major waterway, a change in operations may be required. The presence of water soluble, reactive, or poisonous materials can call for reevaluation. The most important evaluation of the action already taken is whether the present action is endangering lives unnecessarily.

Develop Alternatives

All operational alternatives must be formulated and reviewed. Alternatives must designate both what tactical operation is to be performed and how the operation is to be carried out in a plan of action. Any legal requirements that pertain to the alternatives must not be overlooked.

Essentially there are several tactical operations or combinations thereof that can be selected: rescue, evacuate, confine, control, or extinguish. In all cases, rescue of victims or trapped personnel must take precedence. A rescue operation is immediately followed by the necessity of evacuation. Evacuation is defined as the removal of all nonoperating emergency response personnel and the entire public, including officials and law enforcement officers, for a minimum distance of one mile (1.6 kilometers).

The operational alternatives are confine, control, extinguish, or withdraw. Confinement is used for fires, liquid and solid spills, and in a few cases, gases. Control tactics can be used on fires, and liquid, solid, or gas spills and leaks. Control can be extended to limiting ignition of spills and leaks. Extinguishment is accomplished in many cases only after confinement, control of the fuel source, or at times, during and to enhance rescue so that there is a definite interrelationship between the various options. A withdrawal operation generally includes rescue and evacuation during a holding action and then removal of all emergency response personnel for a distance of at least one-half mile (0.8 kilometer).

An important evaluation to make at all hazardous materials incidents where there is any possibility of endangering emergency response personnel or others is what will happen if nothing is done. In many cases, the end result cannot be altered by either successful or unsuccessful fire department operations; in these instances it is more desirable and less dangerous to do nothing, except perform the necessary evacuation effort and thus reduce the severe life hazard, and withdraw from the area. At times emergency response personnel will be called upon to risk their lives to protect others until evacuation can be executed. During these times all available safety measures should be taken, including unmanned hose holders or master streams using protected locations or lines tied in place. Personnel should wear full protective clothing, including breathing apparatus, and protective streams should be played on them as well as backup lines provided.

The alternative of safely allowing the fire to burn out may be the most practical. Salvaged materials from an extinguished fire will have to be transferred from the site and disposed of in some cases by burning. Transfer operations can also be dangerous and will probably require fire department standby operations.

It must be kept in mind that for any alternative there may be many methods of achievement. For example, a spill might be covered with foam to reduce vapors, dammed to reduce spread, or vacuumed up; all of which would lead to control. The alternatives formulated must include a plan of action.

It will next be necessary to evaluate the data collected during size-up and relate it to the alternatives available and the desired objective. A major problem will call for a large-area evacuation with confinement improbable and control impossible. A moderate situation will demand evacuation, but can be confined or controlled and/or extinguished. A small incident will not require evacuation and generally can be easily handled, but all safeguards and precautions must be taken.

Anticipate the Unexpected

At all times there must be anticipation of what could unexpectedly happen. Hazardous material fires differ considerably from the routine structural, vehicle, or brush fires and the fireground officers must be ever alert for the signs of forthcoming danger. Anytime a problem is diagnosed that is beyond the capabilities of responding personnel, seek additional help and technical assistance immediately!

Two incidents occurring early in 1978 illustrate the need to be ever anticipating the unexpected. At Waverly, Tennessee, a tank car of liquefied petroleum gas was derailed two days before rupturing unexpectedly and killing 15 people. This is the first time a tank of this type, stationary or in transportation, ruptured without fire exposure. At Sidney, Nebraska, a car containing white phosphorous ruptured after several hours of fire exposure. This was not anticipated and six railroad workers in the immediate area were injured. The car had a double-shell insulated tank and was the first car of this type known to rupture

under fire exposure.

Select the Safest Feasible Alternative

Finally, a decision must be made. "The buck stops here" would be an appropriate sign for the command post. Of the alternatives available, the one that will provide the safest, most feasible, and earliest control should be selected. The overall goal of protecting lives and property must never be forgotten.

Unfortunately, decision making must be done very quickly for there is little time for deliberation with hazardous materials and no place for procrastination. The time available during a situation to perform the size-up and evaluation, and for the decision making process is not only critical but extremely limited. The maximum time for this would be well under five minutes.

Implement the Decision in a Plan of Action

Once a course of action has been selected, it must be carried out immediately. Implementation has two dimensions: the present situation and what could happen. The details of these have been covered in previous sections. The implementation of the tactical operations must include coordination of all elements, from the timing of the attack until the removal of dangerous products and decontamination of the area and all personnel involved.

Monitor the Progress

The progress of the plan of action and its results must be monitored. Sector commanders must report developments continuously so that progress or lack of it can be taken into account in subsequent tactical decisions. Size-up becomes a continuing operation; sector commanders must be trained to constantly be watching for danger signs and report them immediately to the command post. Subsequent decisions can only be made after reassessment of what should be done and the effectiveness of how to achieve it.

Take Corrective Action If Necessary

Whenever necessary, take corrective action to bring the operations to the level required to achieve the objective. Continual changes may be required to alter or halt the progress of the incident. These changes must be made without delay or hesitation. Continuing to use up resources in an unsucccessful course of action may make a good show for the public, but it will not achieve the level of efficient fireground operations that will ensure an early and safe conclusion to the incident.

FIREGROUND DECISION MAKING FOR CLOSED CONTAINERS

A hazardous materials emergency involving a fire with flame impingement on a pressurized tank or a closed nonpressurized tank, generally of steel construction, is extremely dangerous and requires prompt action. For this reason, special consideration is given here to the decision making process. The systems approach is used, but has been simplified to pinpoint the specific conditions that apply in this particular situation.

The officer in charge of the first-arriving apparatus at the scene must make the initial decision, based on size-up and evaluation, as to whether to proceed with operations or to withdraw and evacuate. The decision to proceed with operations should only be made when the following six questions can be properly answered:

1. What material is involved?
2. How long has the fire been burning with flame impingement on the container?
3. What are the exposures?
4. What is the water supply?
5. Can water be applied to the point of flame impingement?
6. How quickly can water be applied?

If any one of the above questions as illustrated in Figure 10-3 cannot be properly answered, withdrawal and evacuation operations should be started immediately.

What Material Is Involved?

An immediate determination of the hazardous materials involved will be necessary. If the commodities are classified as explosives A, poisons A, oxidizers, organic peroxides, or are toxic unstable or reactive such as nitromethane, or polymerizers such as ethylene oxide, there should be immediate withdrawal and evacuation.

If the material does not fall into one of these categories, the next question should be considered.

How Long Has the Fire Been Burning with Flame Impingement on the Container?

If the fire has been burning with direct flame impingement on a closed tank shell for over 10 minutes, even with the relief valve operating, there should be immediate withdrawal and evacuation. The time expended in reporting the alarm and responding is a vital consideration in calculating the length of time the fire has been in progress. Tanks have violently ruptured with disastrous results in 17 or 19 or even less minutes of fire exposure (see Table 10-1). It is possible that 10 minutes

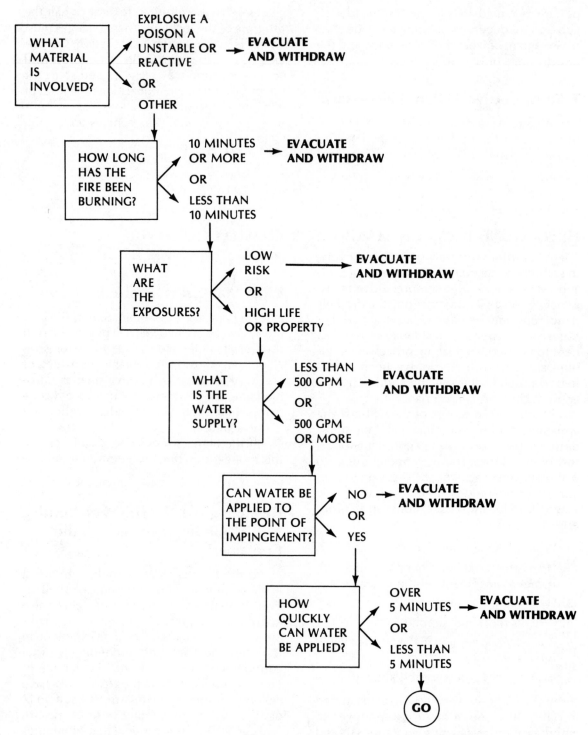

Figure 10-3. Parameters for decision judgments at a closed tank incident. *Copyright © 1975 by Gene Carlson*

could be exhausted by the time apparatus arrives on the scene.

If the direct flame impingement has been of *less* than 10 minutes duration, the next question should be considered.

What Are The Exposures?

If there are no high life potential or large property exposures, withdraw and evacuate the area. However, when exposures involve a large number of lives or exceptionally high value property, consideration must be given to controlling the fire to avoid tank rupture and protecting these exposures. Remember, is any building worth a fire fighter's life? In cases such as hospitals, nursing homes, occupied schools, large tenements, or numerous large, well-occupied buildings, control should be attempted at least during the evacuation procedures. While placing lines in position fire fighters can be protected by fog streams. Full protective clothing, including positive pressure self-contained breathing apparatus, must be used in many cases. A minimum number of personnel should be assigned to these operations because of the extreme risk involved and the need for all available help in the evacuation operation. Use unmanned streams, if possible.

If there are large life or property exposures, the next question should be considered.

What Is the Water Supply?

In order to control tank fires, large volumes of water will be necessary. A minimum consideration must be 500 gallons (1,900 liters) per minute for each large-capacity exposed tank as found in transportation and the smaller tanks in storage areas. Ten percent of the capacity of a tank may be a much more realistic figure as to the amount required for adequate cooling. If quantities in this range are not readily available, it is extremely unlikely that the fire can be controlled, and

withdrawal and evacuation should begin immediately. In the event of a high life risk, it will be necessary to use whatever water supplies are available in a holding action during the evacuation, preferably with unmanned streams or from protected locations. After evacuation of the high life risk, total withdrawal is essential. Fire department operations for incidents of this nature should be preplanned to provide water capabilities of this magnitude and larger.

If 500 gallons (1,900 liters) per minute per tank are available, the next question should be considered.

Can Water Be Applied to the Point of Flame Impingement?

If the tank or tanks are arranged in such a configuration that water applied will not reach the point of impingement, water application will not be effective. This can be a common occurrence in transportation incidents. If water cannot be applied to the point of impingement, withdrawal and evacuation must be initiated.

If water can be effectively applied, the last question should be considered.

How Quickly Can Water Be Applied?

There is little time to initiate strong fireground operations. If effective streams cannot be applied before 15 minutes of flame impingement, the fire crews should be withdrawn and evacuation carried out. If effective fire operations can be started within 15 minutes, they should be set up under the cover of protective streams and then unmanned hose holders or master stream devices used. Whenever possible, fire crews should operate only from protected locations. All nonworking fire fighters, press, or other public personnel, and staff officers should then be withdrawn from the immediate area.

If the above six questions can all be answered with a positive response, fire fighting operations should be carried on with all due precautions. In most cases it will be safe to do so; however, anticipate the unexpected.

RECAPPING DECISION MAKING

Dealing with hazardous materials is not the same as structural fire fighting and a mental adjustment must be made by everyone on the scene. Fire officers have to think differently when considering the basic procedures. Rescue is more than removing someone from a superheated smoke-filled room; rescue may be from corrosive or flammable vapors, from a toxic liquid spill, or from burning pesticides. Exposures may be much further away, miles downstream for escaping liquids, far downwind such as the bromine spill in Rockwood, Tennessee, or in all directions for explosives. The evacuation problems in exposed areas can be tremendous. Confinement is no longer to the room of origin, but may require foam to hold fuming acid clouds, heavy equipment and sand to dike flowing liquids, or special booms to stop an oil flow. Extinguishment is not simply cooling down the ordinary combustibles, but must consider whether extinguishment should be attempted, whether sufficient quantities of the proper agents (dry powder or alcohol foam) are available, or whether it is better to let the fire burn. Overhaul goes further than total extinguishment and may include purging tanks and removal of residue or runoff. Decontamination procedures may be required. Salvage can require moving exposed rail cars or the truck tractor or even commodity transfer to another vehicle. Special problems may result now that the material has been disturbed, mixed, shocked, or exposed to heat.

Decision making at hazardous materials incidents requires swift size-up and accurate evaluation of the situation in light of the possible alternatives, selection and implementation of a well-defined plan of action, and continual monitoring of the progress to bring about a safe conclusion.

If the situation is uncontrollable, and cannot be confined or extinguished *with the resources that are available* at the situation *and within an acceptable safe time frame*, then RESCUE, EVACUATE, and WITHDRAW.

Decision making at a hazardous materials incident is an awesome responsibility for the incident commander. Decisions must be made with only the immediately available information and in a matter of seconds. Be prepared. Make sure your decision is a wise one.

QUESTIONS FOR REVIEW AND DISCUSSION

1. After reading the case problem, list what should be covered during size-up of the situation.

2. Apply the systems analysis method described in Fireground Decision Making for Closed Containers to a recent incident in your area.

3. Apply the systematic decision making process to the same or another local hazardous material situation.

4. Describe when taking no action can be a good alternative in your decision making process.

5. List eight unanticipated developments that could take place at a hazardous materials incident.

6. Why is it important to evaluate the action already taken when you arrive on the scene?

CASE PROBLEM

The fire occurred on a Wednesday afternoon in early June near a midwestern county seat town of 12,000 population. The occupancy is a farm and garden supply center approximately 2½ miles (4 kilometers) north of town. The weather is unseasonably warm with the temperature about 85°F. The wind is light and variable from the northwest. The humidity is high and there is a chance for a late afternoon thundershower. This is the first day of a four-day sale on "your needs for spring planting and gardening" and customer traffic has been heavy.

The diagram in Figure 10-4 shows the location of the various buildings. Building 1 is a one-story concrete block sales and office with display shelves and ordinary shelf stock of garden fertilizers, weed killers, insect and bug sprays, etc. Building 2 is approximately 60 feet (18 meters) southeast and is a recent metal-on-metal building with exposed polyurethane insulation. It is 40 feet by 40 feet and 20 feet to the eaves. It is used for the storage of bagged fertilizers including ammonium nitrate. Building 3 is the original building and is 2½ story wood frame. It is 40 feet (12 meters) by 30 feet (9 meters) and sits 30 feet (9 meters) directly east of Building 2. It is used for the storage of pesticides on the first floor. The second floor was formerly offices and is now used mainly for some storage of shelf stock for the sales area. Forty feet (12 meters) to the east across the roadway is a large bulk anhydrous ammonia tank (number 4 on the diagram in Figure 10-4). Delivery vehicles, nurse tanks, and applicators are parked north and east of the storage tank.

The only water supply for fire fighting would be from a small lake formed by a dam north of Building 3. Fire apparatus could draft from the dam and sufficient water is available. The creek is crescent-shaped around the property with all surface drainage flowing toward the creek south of the property. The creek flows into the municipal water supply reservoir.

The fire department response to this area would be a light attack unit, a 1,000-gallon-per-minute (3,785-liter-per-minute) pumper, and a 1,500-gallon (5,677.5-liter) tanker. Personnel would total 10 on the first alarm. Adequate mutual aid and emergency medical service are available on request.

The fire has started in a second-floor room of Building 3 used for miscellaneous old furniture and records. The alarm is received at 2:00 P.M. and the first unit arrives at 2:05 P.M. First-floor contents include: chlordane, lindane, Sevin, Paraquat, parathion, malathion, and thimet in containers from one gallon to 55-gallons (3.8 to 208 liters). Many similar products are in household size containers in other rooms on the second floor.

The following illustrates how the decision making process could have improved the actual results in this example.

What Could Have Been Done	What Actually Was Done
1. Establish an objective of containing the fire and protecting all exposures—human, property, air, water and soil—from any further damage.	1. Initial objective was to protect remainder of building; fire was considered as ordinary structural fire with no regard for hazardous materials.

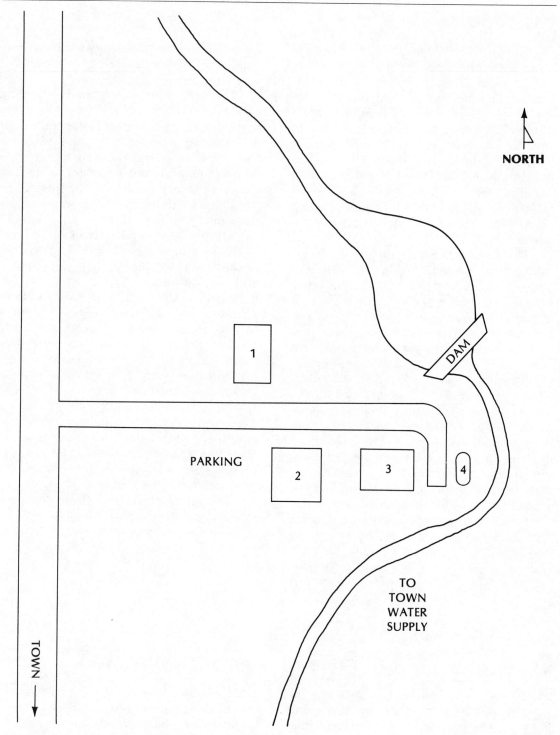

NORTH

DAM

PARKING

1

2

3

4

TO
TOWN
WATER
SUPPLY

TOWN

Figure 10-4. Diagram of building layout and surroundings

What Could Have Been Done	*What Actually Was Done*
2. Collect all data on the fire, including downwind exposure, surface drainage, contents of the building, and the potential for harm to any exposures. Determine equipment and assistance required to achieve objective.	**2.** Little size-up was done as personnel committed themselves to attack. Manager was not contacted for assistance nor was preplan used. No assistance was deemed necessary. The environmental and building exposures were not considered in the haste of the attack.
3. Develop the alternatives that could be used to control the situation: **a.** Attempt extinguishment of the fire based on the preplan. **b.** Let the building burn, withdrawing from the immediate area and using unmanned streams to protect building 2 and tank 4; evacuate downwind. **c.** Protect exposures with handlines and use water fog to control smoke and products of combustion containing the runoff water.	**3.** Rather than considering the alternatives, immediate fire attack was initiated with the limited water supply of the initial pumper. The brush unit was sent to the rear to contain a grass fire ignited by burning material falling from a burnt out window. The tanker was to supply the brush unit.
4. Evaluate all size-up data relating to personnel protection required, exposures, water supply, downwind evacuations, medical needs, additional assistance, and nature of the problem caused by the building contents after obtaining technical assistance.	**4.** With the lack of adequate size-up, little evaluation was done. The major result was the failure to use the small lake for a water supply. The initial engine depleted its tank and the fire began to spread quickly through the old frame building.
5. Anticipate the unexpected occurrences such as container rupture, injuries to fire fighters, or human errors.	**5.** With the lack of water and the fire spreading, a line was brought from the tanker to the engine and a second assault made on the fire; however, the delay had increased the fire to the extent that containers were beginning to rupture on the second floor. Mutual aid was requested for water supply.
6. Select the safest feasible alternative for all involved, which would be to allow the building to burn. This would destroy most of the pesticides and not create a water or soil pollution problem. Withdraw personnel, set up unmanned exposure protection streams, and evacuate downwind to control any runoff.	**6.** With the fire spreading unchecked, it was decided to proceed with an exterior attack. Because of the heat, many personnel removed their protective clothing.

What Could Have Been Done	*What Actually Was Done*
7. Implement this decision by coordinating the withdrawal and evacuation with police and the fire alarm headquarters. Protect exposure, using unmanned streams or hose holders. Give orders to carry out the plan of action, including completion of evacuation.	7. With the limited water application, the fire quickly extended. Soon the 55-gallon (208-liters) drums on the first floor began to rupture. Several of the fire fighters involved began to feel nausea and effects similar to smoke inhalation.
8. Continually monitor the progress of the operation. Use the chain of command that has been established to obtain information for subsequent decision making.	8. Subsequently the Chief arrived to find several fire fighters down, a limited feeble attack being made, and mutual aid units initiating a water supply operation.
9. Alter the operations when necessary to achieve the original objective of limited damage to personnel and property.	9. All fire fighting was discontinued. Medical personnel were called, triage and transportation initiated. CHEMTREC was contacted, a dike built, and necessary authorities notified, while the pesticides were allowed to burn and decompose.

The end result of the fire was a need for triage, additional ambulances, initiation of hospital disaster plans, the treatment of several scores of people with one serious pesticide poisoning, test of drainage for contamination, and several thousand dollars spent for removal of runoff water. All apparatus, protective equipment, and personnel had to be decontaminated. Had a good decision making process been applied, there would have been little if any need for most of the extra measures called for in the actual incident. A safe, effective plan of action would have been developed to bring the situation to an early, uneventful conclusion.

NOTES

1. Ludwig Benner, Jr., "D.E.C.I.D.E. in Hazardous Materials Emergencies," *Fire Journal* 69 (July 1975): 13.

2. Lloyd Layman, *Fire Fighting Tactics* (Boston: National Fire Protection Association, 1953).

3. James E. Heller, "Decision Making," *University of Maryland Fire Service Bulletin,* 1972.

4. Benner, "D.E.C.I.D.E. in Hazardous Materials Emergencies."

ADDITIONAL READINGS

Benner, Jr., Ludwig, "D.E.C.I.D.E. in Hazardous Materials Emergencies." *Fire Journal* 69 (July 1975): 13.

Carlson, Gene P. "Making Decisions at Tank Fires." *Fire Engineering* 128 (April 1975): 26.

"Command Decisions." *Fire Command* 39 (February 1972): 19.

Heller, James E. "Decision Making." *University of Maryland Fire Service Bulletin,* 1972.

"In the Line of Duty: Fire Company Decisions and Action." *Fire Command* 37 (July 1970): 32.

Layman, Lloyd. *Fire Fighting Tactics.* Boston: National Fire Protection Association, 1953.

"The Size of Your Size-Up." *Fire Chief* 14 (August 1970): 25.

11 Railroad transportation

In this chapter the problems pertaining to the more spectacular hazardous materials incidents will be addressed: those happening along our nation's railroads. The causes of the incidents and how rail cars react under stress conditions will be described. Information on the preplanning before and the tactics necessary during an incident will be provided. Fire fighting considerations and the handling of special problems will also be discussed.

CAUSES AND REACTIONS

Of the numerous train derailments occurring, approximately 10 percent involve hazardous materials. However, these often are so spectacular and devastating that great concern is warranted. The number of hazardous materials rail incidents has been steadily increasing, as charted in Figure 11-1. More than two occur each day, which shows the significance of the problem.

The magnitude of these hazardous materials incidents is related to the large quantities that can be hauled by rail. Tank cars can range in size up to 34,500 gallons (130 cubic meters) water capacity (the regulated maximum size) and box cars may carry up to 100 tons (90 metric tons). Along with these sizable single-car capacities is the fact that multiple carloads of hazardous materials are normally in the same train and frequently placed together. Thus, an incident usually will have several cars involved, which can increase the

severity of the problem by adding to the fuel load, exposing each other during fires and ruptures, or creating flame impingement problems.

Hazardous material incidents on the railroad do not always stem from a derailment. Fires can start in cars en route and involve a hazardous lading. A grade crossing accident with a vehicle may not derail the train, but hazardous materials in either the train or the other vehicle could be involved. Leaks may arise from the various openings in the cars. A number of these and other situations will be described throughout the chapter.

The causes of train emergencies are numerous but can be classified into four general categories: track deterioration, equipment failures, human error, and other causes. Table 11-1 gives the breakdown by percent in 1976 accidents.

Track deterioration is the major cause of

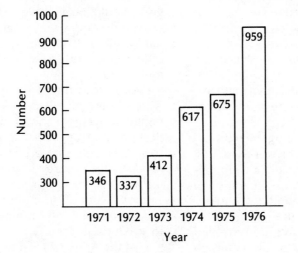

Figure 11-1. Rail incidents involving hazardous materials. *Courtesy of Federal Railroad Administration*

Table 11-1. Railroad Accident Catetories

Railroad	Accidents	Accidents per million train miles	Damage per million train miles	Percent accidents caused by		
				Track	Equipment	Humans
Baltimore & Ohio and Chesapeake (Ohio Railroads)	655	15.4	$349,000	40.5%	25.49%	18.32%
Consolidated Rail Corp.	1,082	12.7	191,000	45.4	19.4	21.2
Norfolk & Western Railway	221	7.7	194,000	33.5	26.6	19.0
Louisville & Nashville Railroad	700	25.3	455,000	47.6	21.7	21.3
Illinois Central Gulf Railroad	601	23.2	551,000	56.6	11.9	15.5
Atchison, Topeka and Santa Fe Rwy.	239	4.9	126,000	23.8	28.8	35.6
Burlington Northern Railroad	783	12.5	310,000	34.6	24.4	27.9
Chicago & North Western Transportation Co.	778	35.8	650,000	50.5	14.1	16.6
Chicago, Milwaukee, St. Paul and Pacific Railroad	525	27.8	621,000	48.8	16.8	19.2
Chicago, Rock Island & Pacific Railroad	509	37.4	548,000	67.0	17.5	9.8
Missouri Pacific Railroad	165	6.0	197,000	32.7	29.0	26.0
National average	10,248	13.2	293,000	41.6	21.0	23.0

Source: 1976 figures from the Federal Railroad Administration.

problems and stems from the poor economic condition of many railroads. The cost of proper preventive maintenance and constant upkeep has become too costly today. Major competitors—trucks and barges—travel on federally funded highways and waterways, while the railroads own their own right-of-way and tracks and thus are responsible for the maintenance.

Some of the problems along the right-of-way include broken track, broken switches, separating joints or occasionally a weld, deterioration of roadbed and ties, separation of rails, generally poor track conditions, and objects on the tracks. Most of the new track today is ribbon rail which is continuous welded rail up to one-half mile (0.8 kilometer) in length. This improved type of construction will help to reduce track problems. However, it will be a long time before the nation's rails are all of this type.

The major equipment failures have been the disintegration of wheels, broken wheels, truck bolster problems, broken axles, coupler failure, broken center sills, broken switches, and journal failures. Once again many of these problems are the result of inadequate or improper maintenance.

To alleviate many of these problems, the railroad industry in some areas is using infrared hot journal detectors, detectors for hanging or dragging pieces, roller bearing axles, interlocking couplers, and rail flaw detection equipment. The Federal Railroad Administration has recently mandated the retrofitting of all single-shell pressurized cars to include interlocking couplers; also head sheet protection and insulation will be required on many, depending on the commodity carried. High carbon wheels had to be removed from tank cars by April of 1978, from hazardous material trains by July, and from all cars by the end of 1978.

Human error difficulties stem mainly from the use of poor judgment and attempting to take shortcuts to cut costs or reduce the time necessary to complete a job. In this category would be included the flying switch; shoving hazardous material cars together in a flat switchyard; careless or unsafe clean-up operations; and personnel working while inebriated.

The Department of Transportation has issued Emergency Order Number 5 concerning the movement and striking of DOT specification 112A and 114A single-shell pressurized tank cars that are not equipped with head shields and are carrying a liquefied flammable gas. The order states that these cars cannot be cut off in motion, nor can any car moving under its own momentum be allowed to strike these cars.

At the April 1978 National Transportation Safety Board hearings, representatives from Kentucky testified that 493 accidents occurred at only 292 different locations in their state. When there is such a large number of recurrent incidents, questions are raised about the quality of repairs made after a derailment.

Other problems fall into the areas of vandalism and sabotage. These are caused by children playing on the right-of-way, pranksters, and disgruntled workers or patrons.

If a derailment occurs, the release of the commodities is a major concern. This can happen as a result of the physical damage incurred. The mechanical forces involved during the derailing of rail cars are tremendous. Consider that a freight train can easily consist of a hundred cars or more, weighing 100 tons (90 metric tons) or more each, and moving at a speed of over 40 miles (64 kilometers) per hour. This creates a kinetic force of sizable magnitude. These forces will crush box cars like toys and toss tank cars about like toothpicks. Tank cars can be crushed, punctured by couplings, have the skin torn, have the loading/unloading valves in the dome or bottom torn off, and insulated cars may have the outer shell ripped and the inner shell punctured.

In a derailment, the gas or liquid tank car is first subjected to mechanical abuse from impact. Most often the cars survive this in fine fashion as a result of their rugged construction and the attention paid to protection of the container. In most cases no leakage at all occurs or leakage is rather minor. However, severe mechanical damage is possible. Leakage of a corrosive could cause chemical stress to the shell of the car or exposure to other adjacent cars.

Unlike the nonpressurized liquid tank car, a compressed gas or liquefied gas tank car is under pressure and the metal shell is under stress from this pressure. The pressure might be described as a spring tightly coiled up in a keeper. Once a tear or opening starts in the tank, the pressure causes the tear to progress through the metal and the tank comes apart. When a hole appears in the container, the compressed or liquefied gas is released to the atmosphere. The escape of the gas under high pressure through the hole develops a thrust that can propel the container. As the liquefied gases escape, the liquid will vaporize very rapidly. They will produce a vapor volume 300 to 500 times greater than the volume of the liquid. This vaporization develops thrust that also can propel the container.

The cryogenic or low-temperature liquefied gas car is not subject to this phenomenon because it is under rather low pressure, usually less than 25 psi. Furthermore, its very efficient insulation slows down the rate of vaporization inside the tank.

Liquefied gas tank cars can be single unit cars with a water capacity up to 34,500 gallons (130 cubic meters). They should not be more than 85 percent full to allow for a sufficient vapor space above. A car containing 29,000 gallons (110 cubic meters) of liquid propane, for example, would release over 7.8 million gallons (29,525 cubic meters) of flammable gas if suddenly ruptured. At the lower flammable limit of propane, this would be a flammable gas-air mixture of over 350 million gallons (1.325 million cubic meters).

A similar quantity of anhydrous ammonia could release over 3 million cubic feet (85,000 cubic meters) of gas. Such a quantity of gas has a great potential for harm whether it has a subsidiary capability of flammability, toxicity, oxidation, corrosion, or even if it is inert. A sudden rupture of this type is essentially beyond fire department capability.

The danger of fire exposure to single-shell tanks causing an overpressure and rupture has had more disastrous results than mechanical puncture of a tank. The exposing fire can be from flammable or combustible liquids spilled from other tank cars, a fire in Class A combustibles, a gas burning at damaged container openings or punctures, or relief valves impinging back on the tank itself or on another tank. Fire exposure of this type on compressed or liquefied tanks causes the gas to expand and increase the internal pressure as long as the container is intact. The container has a limited pressure resistance, so the rise in pressure must be limited accordingly. Tank cars are equipped with relief valves sized to limit the pressure rise from an exposure fire to a safe level with one important "if." This is, *if* the metal container shell is not weakened.

Unfortunately, the fire that caused the pressure rise and relief valve operation is also quite capable of heating the container metal enough to weaken it to the point where it cannot withstand the pressure even with the relief valve properly functioning. Loss of strength results in the metal thinning out to a point where the increasing pressure inside causes a tear or hole to start in the thinned, weakened area. Once the tank begins to open, the vessel may come apart in one of three ways. It could fragment and blow into many small pieces; however, this is quite rare. The tear may extend to and then along the liquid-vapor level line and open up the tank longitudinally, almost like an over-roasted frankfurter. Lastly, the tear may extend around the circumference of the tank and cause it to split into two or more segments, which then rocket from the scene with the release of the pressurized contents. Recent experience has shown that although the pieces generally go in a direction in line with the tank, pieces of the tank can go in any direction.

When a single-shell car comes apart, ignition of the escaping gas, if flammable, is a virtual certainty because of the presence of the exposing fire. This results in a sizable fireball, which can extend several hundred feet (meters) into the air with tremendous radiant heat. This heating of the tank, which creates a weakening of the shell, and the increased boiling of the liquid or liquefied gas, which builds up the vapor pressure, can cause the vessel to rupture. This is known as a BLEVE (Boiling Liquid Expanding Vapor Explosion). In essence, the vessel is weakened during a pressure buildup and ruptures. The released contents vaporize and burn with explosive force. This is similar to what happens when an aerosol can is involved in a fire.

Only a few tank cars carry products that will actually explode or detonate, as a firecracker explodes. Cars of hydroxylamine could do this, and commodities like ethylene oxide can polymerize with explosive results. Under certain circumstances a burning flammable liquid car can explode.

It should not be construed that only tank cars are responsible for major hazardous material incidents. Box cars and hopper cars also carry commodities that can create numerous problems during an incident. Other types of cars, such as gondolas and flat cars, may also carry hazardous materials.

Despite efforts to eliminate causes of rail incidents, problems continue to occur. During the first two months of 1978 there have been 11 major rail incidents with 23 deaths and millions of dollars in property loss. Statistical records report a total of over 10,000 rail incidents, of which approximately 8,000 are derailments. This averages out to almost one every hour annually.

COMMODITIES

The products hauled by the railroads vary considerably. Almost any product that can be moved in bulk will be transported over the rails. Recent incidents have involved chlorine, liquefied petroleum gas, sodium hydroxide (caustic soda), vinyl chloride, butadiene, ethylene oxide, hydrocyanic acid, chlorosulfonic acid, acrylonitrile, ammonium nitrate, uranium hexafluoride, phosphorous, acetaldehyde, tetrahydrofuran, methyl ethyl ketone, and epichlorohydrin. Over the past few years, the following materials have also been involved: acid aldehyde, fuel oil, carbon disulfide, methyl alcohol, anhydrous ammonia, butyl acrylate, diethylamine, isobutane, propylene, naphtha, gasoline, magnesium, and styrene. This, of course, does not include all of the products that are moved in rail transportation, but it provides a sample of what can be expected.

There are several regulations concerning the positioning of hazardous materials in a train. These vary with the classification except for combustibles upon which there are no requirements. (For the specific regulations, see Code of Federal Regulations for DOT, Part 174, Subpart D.)

The variety of products and their many hazards clearly illustrate the need for early identification of the commodities involved. Once they are identified, information can be gathered on their suspected actions.

Warning Indications

Placards can be difficult to see from a distance or under the adverse conditions of an emergency and in most cases provide merely a warning to a problem. Formerly, the *Dangerous* placard used by the railroads had the name of the product on the placard. Now only two products can be definitely identified by placard: chlorine and liquid oxygen. Placards, however, when visible give an indication of the problem. The *Flammable Gas* placard, for example, may be identified at a distance by recognizing the two white lines on a red background. The top line is longer than the bottom one. This placard will only be a warning, not an identification of the product.

With some manufacturers, the color of a dedicated car—one that is dedicated to carrying a single product—may give an indication of the hazardous material. The color will also distinguish a single-shell pressurized car from a nonpressurized car. The single-shell pressurized car will always be painted white. A single-shell car painted black will always be a liquid carrier. Some corrosive cars can be recognized by the protective band, usually black, around the middle of the car, but this is neither a requirement nor a general rule.

The shape can give several warnings. It can give an indication of the size of the tank and thus the amount of product. The shape of the dome can indicate gases, liquids, or acids. The tank shape can help to distinguish between an insulated and noninsulated car. A noninsulated tank car will have a rounded end with large sheets of welded steel. It might be described as looking like a frankfurter. The insulated car has a flatter head (end) construction and the outer shell often has both riveted and welded construction. Figure 11-2 shows both types. A bottom-unloading outlet will also indicate a nonpressurized car. The DOT class 114A pressurized car may have a bottom arrangement, but of different design. The shape will readily distinguish a cryogenic car. The outside tank reinforcement and piping arrangements on these cars make them easy to recognize.

The military services for some time have identified enemy ships and planes by the use of silhouettes. This method can be applied to the rail industry. The exact product may not be immediately identified, but the type of

tank car and the general products carried by the car can be recognized. Silhouettes can be identified by the top or dome configuration and by the bottom-unloading outlet or lack of it. In Figure 11-3, for example, the dome configuration and lack of a bottom outlet clearly identifies the car as one containing a liquefied gas under pressure. The car shown in Figure 11-4, with its safety vent, top control for the bottom unloading outlet, large manway, bottom outlet, and top unloading/loading arrangement, is readily identified as a non-pressurized liquid carrier.

Figure 11-2. Noninsulated car with rounded end on the left; insulated car with flatter end on the right

Product Identification

The actual product identification of hazardous materials in rail transportation can be done in a number of ways. As mentioned earlier, chlorine and liquid oxygen can be identified by their placards.

The primary method of product identification for railroads is by reference to the shipping papers. These papers will be located in the caboose with the conductor. Some railroads may also carry shipping papers in the locomotive with the engineer. The engineer must be made aware of explosives when they are on the train. At an emergency, the conductor may leave the scene to notify the dis-

Figure 11-3. Noninsulated pressurized liquefied gas car

Figure 11-4. Noninsulated liquid car

patcher by phone of the incident. This will mean that the papers probably will not be available because the conductor may have taken them.

The waybill is the shipping paper used by the rail industry. This bill will be marked distinctly when carrying hazardous materials other than combustible liquids. Explosives, radioactive, and poison gas will be indicated as such and all other hazardous commodities under the marking dangerous. The shipper, route, and destination are on the waybill along with the amount, for example: one tank car. Some waybills also include emergency phone numbers. The name of the product is usually given on the waybill, which provides positive identification. The waybill, however, may not show the generic name. The waybills may or may not be in the same order as the cars in the train. In order to match the waybill to the proper rail car, the car number must be known. The car number is stenciled on the sides and ends of rail cars (Figure 11-5). The waybill carries the car number in the upper left-hand corner. Included on the waybill is the Standard Transportation Commodity Code, which is a seven-digit number. If the STCC ("stick") code number starts with the digits 49, the commodity is a hazardous material. Most railroads can use the STCC number in their computers to produce a readout of hazardous material data for the product. Recently, many are adding this hazardous material data to the information carried by the conductor in the caboose.

Several other papers may be in the ca-

Figure 11-5. Car number to be matched to waybill number for positive identification of product carried. *Courtesy of Federal Railroad Administration*

boose, depending on the particular railroad. Local railroads should be contacted to familiarize the fire department with procedures used in each area. The consist, which is a computer printout, lists the order of the cars from the front or rear of the train and includes the car number and the Standard Transportation Commodity Code number. The consist will have some products marked, usually with abbreviations, that can be difficult to read. The consist is kept by the dispatchers and may or may not be with the train.

A train list is essentially another name for a consist. A wheel report is very similar and may be found in the caboose. The switch ticket is carried by the conductor on a local train, which will be dropping and adding cars as it moves along. They may be handwritten or computer generated. They will indicate the hazardous materials that are in the train when it starts; however, any added cars of hazardous material will not be indicated.

The product name stenciled on the side of the tank is another means of identification. There are approximately 40 products required to have the name stenciled in 4-inch (10 centimeters) letters. These include all flammable gases. Table 11-2 lists most of these substances.

Another method of identification is to check with railroad dispatchers. Given the train number, which can be identified from the locomotive number, the dispatcher can generate the consist. From the consist the dispatcher can obtain the waybills for those cars necessary and, if the railroad is so equipped, the hazardous material readouts for the products indicated on the waybills. To generate the waybill, the dispatcher, of course, must be given the car number or the car position in the train, such as cars in the 15 through 23 position are derailed. From the position, the car number can be determined from the consist. With the car number or position, the STCC can be determined from the consist or waybill for product identification and to obtain further chemical data. The dispatcher, who can be of considerable assistance during a rail incident, is in charge of all train movements in the area and can stop trains, summon railroad personnel, and in some cases provide information on access to the railroad property.

Color can be used for positive identification of only one car in use today. This is the hydrocyanic acid car, which is a white, pressurized double-shell car with a horizontal red band around the tank and a red band three feet from each end circling the tank vertically.

Product identification is vital before proceeding with operations. Far too often, emergency response personnel rush into incidents without properly identifying the product and

Table 11-2.   Tank Cars Requiring Product Name in 4-inch (10 centimeters) Letters

Acrolein
Anhydrous Ammonia
Bromine
Butadiene
Chlorine
Difluoroethane[a]
Difluoromonochloromethane[a]
Dimethylamine, Anhydrous
Dimethyl Ether
Ethylene Imine
Ethylene Oxide
Formic Acid
Fused Potassium Nitrate and Sodium Nitrate
Hydrocyanic Acid
Hydrofluoric Acid
Hydrogen
Hydrogen Fluoride
Hydrogen Peroxide
Hydrogen Sulfide
Liquefied Hydrocarbon Gas
Liquefied Petroleum Gas
Methyl Acethylene Propadiene, Stabilized
Methyl Chloride
Methyl Chloride—Methylene Chloride Mixture
Monomethylamine, Anhydrous
Motor Fuel Antiknock Compound or Antiknock Compound
Nitric Acid
Phosphorous
Sulfur Trioxide
Trifluorochloroethylene
Trimethylamine, Anhydrous
Vinyl Chloride
Vinyl Fluoride, Inhibited
Vinyl Methyl Ether, Inhibited

1½-in. (4 centimeters) Lettering
Ethylene Only Methane Only

[a]May be stenciled "Dispersant Gas or Refrigerant Gas" in lieu of name

performing adequate risk analysis. As discussed in Chapter 10, there should even be an approach evaluation before initial apparatus placement; this is especially important in rail incidents because of the large volumes of hazardous materials that are involved. Emergency response personnel must endeavor not to overreact to the situation and find themselves involved to the extent that they become a part of the problem rather than an organized force to alleviate the problem. Safety demands that all precautions be taken. Remember the adage, "If you don't know, don't go, it may blow!"

PREPLANNING

Railroad preplanning is much like the planning of transportation routes described in Chapter 7. A primary consideration is to determine what railroads traverse your area, how the railroads are contacted including a complete list of telephone numbers, and what hazardous materials are frequently handled. Preplanning is initiated by surveying the route of the railroad through the area to determine access, obstructions, and difficulties in getting to incidents that could occur. Railroads often run above or below grade and this must be noted. Electric power lines, tunnels, and bridges will also be important to record. Natural bodies of water that must be protected from spills or runoff should be noted, and plans should be made on how this will be accomplished. In rural areas, because of accessibility problems, access roads and methods of getting through fields or across country must be investigated. All due consideration must be given to a safe approach. This includes access from different directions.

Of interest in built-up areas are the exposures adjacent to the railroad right-of-way. Concern must be given to commercial, industrial, densely populated residential areas, and high life exposure target hazards. A school adjacent to a railroad is a target hazard, as demonstrated by Figure 11-6. An area of high loss potential, one in which losses could disrupt the economics of the community, is important to consider.

As in any preplanning, water supply and the proximity of sources, especially static sources, along the right-of-way will be important. Special water supply operations will be needed in many railroad accidents because of the remote locations and the large amounts required for high-capacity tank cars.

Spill control operations will bear planning for adjacent bodies of water and to limit the spread of any leaking material, or possibly runoff. The planning must include provisions to handle the volume anticipated.

Local railroad dispatchers should be contacted to obtain as much information and assistance in preplanning as possible. Some idea of the commodities handled should be obtained for both trains passing through and delivering locally. Much assistance and insight to the rail mode of transportation can be gained from the dispatcher.

Figure 11-6. School adjacent to railroad is target hazard

The difficulties of handling situations in railroad yards should also be preplanned. Many of these yards are large complexes (see Figure 11-7) in which there have recently been several incidents of sizable proportion. Incidents in East St. Louis, Houston, Wenatchee, and Decatur have not only caused millions of dollars in damage, but have cost several lives. Yard installations, which are generally located adjacent to heavily built-up areas in or near population centers, present special problems. Since the yard is a virtual maze of tracks and ever-moving cars, response routes can hardly be planned. Rather, arrangements must be made with the yard master to specify one or more of several locations at the perimeter of the yard where railroad personnel will meet the emergency response team and lead them in. Definite around-the-clock personnel assignments should be made to ensure the necessary guides.

The situation can be further complicated when the rail yard is used by more than one railroad. All users should be immediately notified when an incident occurs so that train movements can be stopped and any problems that could be caused by the other railroad averted. This will help to alleviate problems of crossing blockage, hose lines disrupted, or hitting emergency response personnel. Because of the distances involved, water supplies and hose layouts may be quite complex. It may be necessary to move cars or entire trains. Whom to contact and how this is done must be included in the preplan. If the incident is small and only involves a car or two, it may be advisable to move the cars to a location accessible to the fire department. Crews doing the switching in the yard will not have the waybills, but may have a switch ticket, which would indicate the hazardous materials by name. The waybills for cars in the yard will be in the terminal office.

Figure 11-7. Railroad yards present problems in controlling incidents within them

TACTICS

The tactical operations at a railroad hazardous material incident are not appreciably different from others thus far described, other than the increase in quantities.

Full protective clothing, including self-contained positive-pressure breathing apparatus is essential. In all cases, there should be no exposed skin that could become contaminated. If there is a need for special protective clothing, sources for obtaining them and the length of time before procurement should be disseminated to the suppression forces. Since railroad problems can be of large magnitude and long duration, provisions for air supplies, compressors or cascade units, additional masks, and extra tanks will be needed.

Of vital importance is a command post. This has already been covered in Chapter 8. It must be reiterated, however, that a command post will be needed early to establish an organized approach to the incident. The location at a rail incident may be of special concern. It cannot be too close to the scene. The command post should be provided with as much information on the railroad as possible and put in contact with officials immediately. Communications are vital. Communicating to both sides of a derailment may be difficult if sufficient radios on the same frequency are not available. Since the train can be of great length, runners may not be practical. The communications with the railroad, from the train crew on up, and with technical resources must be well coordinated.

An adequate, accurate method of analyzing the risk potential must be established. This includes a comprehensive size-up, evaluation, and decision making process as described in Chapter 10. It is essential that during these procedures, the various stresses to which the cars were or are being subjected are correctly evaluated. The hazardous materials incidents with the majority of deaths and injuries recently have centered around rail incidents. In many cases, there has been a lack of or inadequate threat assessment, which has led to committing emergency response forces to actions that were highly dangerous.

The size-up must start with the approach so that personnel and equipment are not unduly exposed to risk. There must be constant awareness for the unexpected; perception must be keen for all warning signs of further deterioration of the situation. All alternatives must be carefully considered before the final decision on a course of action is made. If available through technical assistance revealed during contingency planning, computer modeling of the incident may assist the decision making.

Evacuation is a vital concern because of the great distances that large pieces of tank car can be hurled and the radiant heat exposure from the fireball of a rupture. The large quantities of hazardous material involved in a rail car increase the danger potential. Consider the problems of moving everyone within a half mile (0.8 kilometer) of a mid-town location of the railroad in your city. Then think

about doing it in 30 minutes or less. A great deal of assistance may be needed in the evacuation effort; initiate its response early! Contingency plans should include dispersion information and models from the state environmental group to aid in evacuation plans. If at all possible, evacuate for a one mile (1.6 kilometers) distance to ensure maximum safety.

The tactics for a rail incident must include product control and how this will be achieved. Product control may be as simple as stopping a packing gland leak, as described later in this chapter, or it may involve the neutralization of an entire load, as happened in Youngstown, Florida, with the tank of chlorine. It may require very special operations, such as when dealing with Class B Poisons. In many cases, product control will involve dissipation of vapors if sufficient water is available or building dikes or dams to control spills. Product control can refer to gaseous leaks, liquid spills, or solids. In some cases, control can be achieved by knowledge of specific gravity, such as with carbon disulfide, or by blanketing with foam solutions. Product control has many variables, and can involve diking, diluting, dispersing, absorbing, neutralizing, inhibiting, and directing. Many of these operations require special training and knowhow and will be carried out under the guidance of technical assistance. Once the situation has stabilized, product control is not over. Concern still has to be given to disposal and possibly decontamination of the area, equipment, and personnel. The properties of the product will influence greatly how it is going to be controlled.

Water delivery for rail incidents is critical because of the need for cooling pressurized cars. Since railroads traverse many miles of rural America, water supply in these areas may be quite remote. On a large incident, a capable water supply officer will be needed. With railroad cars, it must be remembered that they are sizable cars and carry large volumes; therefore, large volumes of water will be necessary for each burning car for adequate cooling. Proper-sized nozzles in sufficient numbers (preferably unmanned) must be provided to deliver the required quantities the distance necessary to reach the derailed cars. Solid streams may be needed for reach or penetration of the fire. Smooth-bore nozzles may be needed for exposure protection. Water delivery can be critical because of the number of cars that might be involved and that require cooling. The equipment to make the water delivery—enough large stream devices, for example—will be vital. Water must be applied where necessary for proper cooling, control, and impingement protection, and the supply must be sufficient to last through the duration.

FIRE FIGHTING

The inevitable question falls to the incident commander in decision making during a rail emergency: which alternative is the safest, most feasible, and will gain earliest control of the incident? Should a decision be made to commit forces to "fight the fire" or to withdraw? The decision can only effectively be determined after applying the decision making process described in Chapter 10.

To assist in making decisions at rail incidents, remember to review the obvious safety problems that may exist: water-reactive materials, a covered relief valve or change in relief valve activity, a nearly blue-red smokeless flame indicating the possibility of the fire being drawn into the tank, additional pressure buildup from heat or temperature exposure on a closed container, abnormal strain on the tank shell caused by mechanical damage during the accident, or chemical damage caused by the leak or spill of a corrosive material weakening the container or nearby ones.

The risk of possible or probable toxicity, explosion, rupture, or polymerization must be carefully analyzed. How these reactions would affect nonemergency response personnel also will vary on the plan of action implemented. Always remember to maintain a line of retreat and have an adequate signal established. Perhaps an adequate retreat signal can be developed using apparatus air horns.

If the action plan is one of confinement, control, or extinguishment of the fire in and around burning hazardous material rail cars, especially tank cars, the following points must be reviewed.

A frostline on the container indicates the portion of the tank with product remaining. Most importantly, this will indicate the vapor space that should not be subjected to fire exposure and must be adequately cooled. A frostline will not always be present and will not occur on insulated containers.

The position of the relief valve, fusible plug, or rupture disk on any liquid or liquefied gas tank will be important. If any of these function while the tank is in a position to discharge liquid, the pressure will not be adequately relieved and additional commodity will be added to the spill or as fuel to the fire. Under fire conditions this can cause impingement and severe thermal stress on the car itself or on adjacent tanks.

Fire at the relief valve when the tank is upright must also be carefully evaluated. An increase in the audible pitch or the volume of the fire indicates that the cooling efforts are ineffective and the situation deteriorating. The blue-red, nearly smokeless flames signal a tank almost empty of commodity and requiring special action.

Fire at a tank vent that burns with a snapping, blue-red, nearly smokeless flame indicates a vapor-air mixture within the tank that may be explosive. There is constant danger of an explosion should the flame reach the inside of the tank. The best attack procedure, therefore, is to maintain a "positive pressure" within the tank by pumping liquid into it. When a vapor-rich condition is indicated by change in the flame character to a smoky yellow-orange flame, danger of explosion has passed ... Never apply cooling water to tanks burning with a snapping, blue-red, nearly smokeless flame at the vent. Nor should such tanks be pumped out. Several of the tank vent fires reported to the NFPA featured injury to firemen when water applied to cool the tank shell caused an explosion.[1]

An intermittent relief valve operation indicates that the combination of cooling and relief valve operation is controlling the pressure buildup within the tank. A relief valve will reset when the pressure is amply decreased; however, fusible plugs and rupture disks are single usage pressure control devices. A lazy flame continuing to burn at the relief valve, even though effective cooling has been applied, would indicate that the O-ring in the relief valve has been destroyed and the valve is not resetting properly. It must be pointed out and emphasized that the operation of a relief valve does not ensure adequate pressure reduction to prohibit tank rupture. Fire fighters have been killed at several incidents where the combination of tank weakening and internal pressure buildup has caused the tank to rupture with the relief valve operating.

Flame impingement on the vapor space of a closed container is a serious and dangerous circumstance. It is recommended that under these conditions any necessary rescue be performed, evacuation completed, and withdrawal selected as the action plan if adequate cooling cannot be accomplished, as described in Chapter 10. If impingement originates after attack has been initiated, the situation will have to be reevaluated.

In all cases of fire or flame impingement, large volumes of water will be needed. Be-cause of the size of railroad cars and tanks, a minimum of 500 gallons (1900 liters) per minute must be used as a starting point. Quantities in this range and larger must be preplanned, since each tank will require this much or up to 10 percent of its volume for cooling. Preferably, water should be applied using unmanned devices. This, of course, limits the ability to fan the top of the tank with cooling water. Streams should be placed to lob the water on the top so it cascades down both sides, not aimed at the tank, which causes it to ricochet off the tank rather than create the protective water film over the tank surface.

With any decision to advance on rail tanks, the approach should be from safe or protected locations. Fire fighters must be protected from radiant and convected heat as well as the possibility of flying fragments and pieces. Personnel should not be placed at the ends of tanks or above them on ladders, embankments, or bridges. Those in hazardous locations should be protected by a protective fog cover.

When dealing with rail incidents, the availability and capability of railroad fire fighting equipment should be incorporated in the plan of action. This information should be available from the preplan data. Of major concern will be how long it will take for this equipment to arrive in the needed location.

All facets of the attack procedure, whether fighting a fire or controlling a commodity release, must be constantly monitored for the progress being made or the lack of progress and possible need for a change in direction. All sector commanders and the safety officer must be constantly alert to the signs of impending danger.

How to Stop a Train

Knowing how to stop a train at the location desired by the fire department will be important not only at hazardous material incidents, but also at structural fires, crossing accidents, and for stopping an incoming train with a leak or fire.

The normal flare or fussee is used to stop a train. A railroad torpedo can also be used or the waving of a light or other object. There are some variations among the railroads, but generally a fussee left beside the track will only bring a slowing of the train (some railroads will stop momentarily), but the engineer will proceed at a reduced speed like 20 miles per hour (32 kilometers). Similarly, a fusee placed between the rails will bring a short stop and then a continuance at a reduced speed. The only positive way to stop the train is to have someone waving the fusee, light, or flag, as shown in Figure 11-8. The person flagging the train can then give proper instructions to the engineer. The engineer is in radio contact with the conductor in the caboose, who is in charge of the train, and with the dispatcher.

An important point is that a suitable distance must be allowed for the train to stop. The weight and speed of the train develop tremendous kinetic energy that takes considerable distance to stop, even during an emer-

Figure 11-8. Stopping train with ample stopping distance by waving a fusee or flag

gency stop procedure. It is recommended that at least 1½ miles (2.4 kilometers) be alloted for stopping a train.

A train having a hotbox or car on fire will often proceed to the nearest community and pull onto a siding, where the fire department will be standing by to extinguish the fire. This procedure is not a practical one for incidents that involve hazardous materials. Once a call of this type has been received, the fire department should immediately contact the railroad dispatcher and designate the stopping point. Why bring a hazardous material problem into a populated area? Have the dispatcher contact the train by radio and have it stopped in an unpopulated area away from any exposures. There is no need to bring the train into an area and then have to evacuate the area. Good stopping points in each local area should be designated in advance and incorporated in preincident planning. Selections should be made on the basis of remoteness, water supply, nonblocking of grade crossings, and finally, the location must be accessible to the fire department at all times. In many cases, a problem can be averted by some forethought and planning that will take the emergency response team to the train, not the train to the response team.

How to Separate Cars

In some cases it will be necessary to move a car or two that are exposed to a stationary hazardous material emergency, cut a car with an overheated journal or fire inside the car from a train, or move exposed cars from a derailment site before they become involved. Separation is made in five steps:

1. Release the brakes on the car or cars.
2. Bring the cars together if necessary to remove the tension on the coupler so it can be operated.
3. Operate the uncoupling device.
4. Move the car as necessary.
5. Set the brakes.

The first step in moving cars is to release the brakes. It is recommended that railroad personnel do this type of work; however, they will not always be available. If this is the case emergency response personnel should use the following procedures.

With standing rail cars, it is necessary to check both the hand brake and the air brake before attempting to move the car. The hand brake is located on the "B" end of the car. Depending on the car, the hand brake can be released either by turning the hand wheel at the end of the car counterclockwise or by operating the release mechanism (Figure 11-9). If the chain extending from the hand wheel to the brake is loose, the brake has been released.

Figure 11-9. Handwheel to operate hand brake

Now check the air brake system. To operate the air brakes, there must be air in the system. With cars sitting there is no compressor-equipped locomotive to build up the air. Therefore, the air to operate the brake will come from the reservoir on the car. This reservoir is under the middle of the car or at one end. The air is bled into the system by operating the bleed rod (Figure 11-10), which is at the end of tank cars and the center of other cars. The air pushes the brake cylinder in and releases the brakes. Some cars are equipped with a two-position bleed rod. The first position bleeds the air from cylinder, allowing it to close and release the brakes. The second position bleeds the air from the reservoir. A car may have two bleed rods that operate in the same manner as a two-position rod. Essentially, one must remember to pull the single rod, and for the two-position type, to use the first position to release the brakes.

If the car is still connected to the train, the air can be built up to release the brakes. This could take some time, depending on the length of the train and the position of the car. It will then be necessary to close the angle cock valve at the end of the adjacent car toward the locomotive before disconnecting the air hose in order to maintain the air in remaining cars. When the air is released from the system, as in a derailment, all of the brakes set.

If there is tension on the drawbar, the friction may be great enough to prohibit the knuckle coupler from opening. The cars must then be moved together to relieve this tension. Then raise the uncoupling lever or cut lever, as it is called, which will release the couplers. The cut lever is operable from either side of the coupled cars (Figure 11-11). Disconnect the air hose if this has not already been done and move the car. The car can be pulled by a truck, a farm tractor, or moved by hand with the use of a car jack or Johnson bar. Once the car has been moved to a suitable location, the brakes need to be set again. This is especially vital on roller bearing equipped cars, which can be easily moved with the brakes released.

Once a car has been moved, especially those having roller bearing trucks (Figure 11-12), it is necessary to set the brake. This is done by setting the hand brake on the "B" end. The hand brake is set by turning the hand brake wheel clockwise until the chain is taut. If more than one car has been moved, set the brake on each car. If uncertain about the hand brake, another method of keeping the car from moving is to chock the wheels with a piece of wood.

On a car that has been standing, it will not be possible to set the air brakes. Cars that are still part of a train can have the air brakes set through the operation of the system in the locomotive, an emergency bleed valve in the caboose, opening the angle cock valve (see Figure 11-13) on the air hose to bleed the air from the car, or by breaking the air hose line.

Figure 11-10. Bleed rod used to release the air brakes on a standing car

Figure 11-12. Roller bearing trucks on car

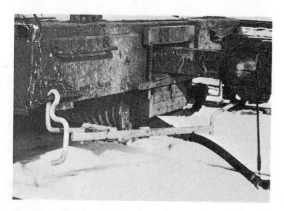

Figure 11-11. Lifting up on the cut lever will release the coupling

Figure 11-13. Opening the angle cock valve removes the air from the air system. Valve is open when handle is in-line with air hose

SPECIAL PROBLEMS

A number of special problems involving hazardous materials can occur on the railroad that will precipitate a response from the fire department. The fire service must be aware of these in order to lend the maximum assistance and maintain overall safety.

Tank cars may be subject to leaks at the top loading/unloading area or bottom unloading device for nonpressurized cars or in the dome area for pressurized cars. Tank cars opened in a derailment may have to be purged of the remaining commodity or vapors before they can be moved, to eliminate poisonous or flammable vapors and the possibility of ignition. Box cars carrying hazardous materials in large containers or packaged as ORM-D can become involved in fire. Diesel locomotives may be derailed or involved in a fire; knowledge of their safety features is vital to responding emergency personnel. Also, a new mover of hazardous materials, the tank train, should be of interest to those who may encounter it in an incident.

Tank Cars

Tank cars may be pressurized or nonpressurized, noninsulated (single shell) or insulated (double shell), with special linings such as rubber or glass, or equipped with lines for steam heaters. They can be top unloading or bottom unloading. The tank cars, of course, carry a wide range of products.

Because of the volume of material they carry tank cars cause some of the larger and most spectacular hazardous material emergencies when involved in an accident. Numerous lives can be lost and millions of dollars in property damage incurred. There are over 170,000 tank cars in use in the United States today. These cars can contain from 4,400 gallons (16.7 cubic meters) to 34,500 gallons (131 cubic meters) with a few 48,000 gallon (182 cubic

meters) cars in service. Regulations now restrict tank size to 34,500 gallons (131 cubic meters) or less. The majority of the major tank car incidents recently have involved the DOT 112A or 114A specification cars, which are single shell or noninsulated pressurized cars for carrying liquefied gases.

Tank cars are of varying construction. Most have a carbon steel fusion-welded tank, but others are of aluminum, nickel or nickel alloy or alloy (stainless) steel fusion-welded construction. There are also steel and aluminum riveted tank cars. Some commodities require the interior of the tank to be lined. The linings can be of lead, rubber, glass, elastomeric polyvinyl chloride, or elastomeric polyurethane. Shell thicknesses vary according to the products carried and the pressures the car contains. Also, a double-shell car can have a different thickness for each shell. The metal thickness may be as little as 3/16 inch (4.8 millimeters) or as much as 7/8 inch (22.3 millimeters) depending on the number of shells and test pressures. Common thicknesses are 5/8 (16 millimeters) and 11/16 inch (17.5 millimeters). The insulating material also varies and may be glass wool, cork board, styrofoam, a glass wool fiberglass combination, or polyurethane. Many cars are also equipped with heater or steam coils to raise the temperature of the product for unloading.

A construction feature that causes a problem during derailments is the lift-off type of center-pin connection between the truck (wheels) and the body of the car. Essentially the weight of the car and its contents hold the body of the car on the trucks. The body center plate on the bottom of the body bolster rests in the truck center plate on the truck bolster. A center pin, 1¾ inches (4.4 centimeters) in diameter and 18 inches (45.7 centimeters) long extends through each center plate. The body of the car must lift only about 6 inches

(15.2 centimeters) before the pin emerges from the body center plate. This allows the truck to separate from the car, except for the small pin that fastens the brake rod between the body and trucks.

This construction feature can be a factor in derailments as emphasized by the National Transportation Safety Board report from Dunreith, Indiana:

A contributing causal factor was the inadequate track maintenance which left the joint unsupported and allowed the development of the break in the rail. This initial derailment and the design of the lift-off type of center-pin connection between the truck and the body of the car, which allowed the truck to separate from the car under the impacts of a simple derailment, led to the secondary collision and general derailment.[2]

Most rail cars have a frame referred to as the sill. Tank cars may have a full or continuous sill for a completely supported tank, or a stub or draft sill for an unsupported tank. On those cars with a stub sill, the tank itself is the entire supporting mechanism for the car.

Many of the liquid cars are equipped with domes. These may or may not house loading/unloading devices. The domes are designed to hold the one or two percent of the tank contents that could be expected as volumetric increases because of temperature changes during transport. The bottom unloading valve control may be in the dome. The pressurized car domes contain loading/unloading devices only and are not designed for outage since the cars are not carried completely full and the vapor space compensates for temperature changes.

All but a few of the tank cars are equipped with a pressure-relieving device. This may be a safety vent, a relief valve, or a rupture disk. A pressure tank car is usually one having a safety device set for 75 psi or higher. Before 1959 some cars were built with relief valves set at 25 psi. Safety devices now range from 35 psi to 450 psi for relief valves; safety vents equipped with rupture disks are required based on thickness and composition to be 45 psi or 75 psi. For a few special commodities, relief valves are prohibited.

Tank cars come with a variety of openings. These depend on the use of the car and the specifications. Most are on the top, but many nonpressurized cars are equipped with bottom unloading outlets. Openings in the top include: manways, relief valves, safety vents, top unloading outlets, air connection valves, gauging devices, thermometer wells, sampling valves, and fill holes. The combination of these devices depends on the commodity; for example, liquids, liquefied by pressure, or corrosive.

A periodic testing program is established for tank cars. Each tank is tested by filling it completely, including any manway nozzle or expansion dome, with water or similar liquid and applying the specific pressure for 10 minutes for noninsulated tanks and 20 minutes for insulated tanks. Interior tank heater systems must be hydrostatically tested to 200 psi. Safety relief valves must also be tested with air or gas and must discharge within plus or minus 3 percent of the prescribed pressure for valves of 100 psi or higher. Lower operating valves must work at plus or minus 3 psi of the prescribed pressure.

After repairs requiring welding, riveting, or caulking of rivets, tanks must be retested before returning them to service. The month and year of tests on tanks, safety relief valves, heater systems, and pressure to which they were tested must be stenciled on the outside of the car.

Stenciled on each tank car is the DOT specification number which indicates the class of car and is followed by identifying letters and numbers. DOT 112 A 340 W would indicate a DOT 112 class car; the A has no significance, the 340 indicates the test pressure, and the W fusion-welded construction. The DOT classes 103, 104, and 113 will not have the A. If the material used for the tank is other than steel, it will be noted after the test pressure. The final suffix may be an F indicating a forge-welded tank, or be lacking entirely when the tank is of seamless construction.

Pressurized Car

The pressurized tank car may be either single- or double-shelled, depending on whether it is insulated. Generally, these cars are top unloading, although bottom unloading is permitted on the DOT 114A specification cars. The single-shell pressurized tank car has caused the greatest problems during derailments because of its susceptibility to rupture under flame impingement. To eliminate this problem, all new cars built to carry flammable compressed (liquefied) gases must be equipped with thermal protection and all single-shell cars currently hauling these commodities will have to be retrofitted. The thermal protection must be capable of withstanding a 100-minute pool fire test at temperatures of 800°F (424°C) or a 30-minute torch fire test. The pressurized tank cars with thermal protection will no longer be required to be painted white and as the protection exceeds the above tests, the relief valves may be reduced in size.

The dome of a pressurized car contains two liquid eduction valves, a vapor eduction valve, a gauging device, a thermometer well, and often a sampling valve. The chlorine car is an exception to this, having two angle valves into the vapor space and no gauging device, sampling valve, or thermometer well. Leaks can occur around the valves, gauging rod, or the relief valve.

Nonpressurized Car

The nonpressurized tank car may be a single-shell car or an insulated double-shell car. They are generally top loading cars, but may be unloaded either at the top or bottom, depending on the particular car. These cars will be divided into three categories: top loading/unloading, bottom unloading, and

corrosive cars. The corrosive car may be either top or bottom unloading, depending on the product.

Leaks

The subject of leaks on railroad tank cars covers where the leaks are found, the causes, and what can be done about them. The responsibility to repair leaks should fall to the railroad, shipper, receiver, or along the route to someone who has experience in working on the type of car and product involved. The nature of the leaks will be discussed here to acquaint the fire service with what to expect and give guidance in those cases where they must take repair action.

In all cases of leaks, when working around flammable gases or low-flash-point flammable liquids, nonsparking tools should be used. This ignition source can be easily overlooked and no opportunity for accidental ignition can be chanced.

Dome Leaks. A leak in the dome area of a pressurized car can be detected by a frost condition around the dome, commonly referred to as a frozen dome, by visible moisture coming from the dome or down the side of the car, by a hissing sound, or by an unusual odor. The fire department normally would follow the procedure of identifying the product, using appropriate protective clothing, summoning technical assistance, and diking any spill or dispersing any leaking vapors. If the leak has autorefrigerated and frozen shut, no water should be applied as this would merely melt the ice and cause the leak to continue. All ignition sources should be kept away from the leaking car. Repair of the leak should be left to the responsible parties who can be identified by the railroad dispatcher.

The angle valves into the liquid or vapor space of a pressurized car may leak if not closed tightly. During transit they are to have plugs installed. These may not be tight, may vibrate loose, or may not have been put in. Tightening the valve or plug can easily be done. The flange holding the angle valve to the car (Figure 11-14) may become loose and need tightening. This may require a special crowfoot wrench to reach the nuts. The final area of possible leak with these valves is the valve packing adjustment, which may have to be tightened.

The gauging rod is the problem child and the cause of most dome leaks. This is a spring-loaded, graduated hollow slip tube. It may have a clip-on cover or a screw-on type. It is lowered into the tank until liquid is emitted as detected by the freezing moisture in the air creating a cloud. Problems with this device occur when the valve at the end of the hollow slip tube (Figure 11-15) is left open and the

plug is not installed or vibrates loose in transit. The packing gland may also need adjustment, but the main cause of leaks is as mentioned. The protective cover housing may not be properly tightened on its gasket and leak. Precaution must be taken when removing the protective cover since if the tube is loose and spring loaded, it can come shooting up and cause serious injuries to the unsuspecting.

The other possible leak in the dome area is with the relief valve. Occasionally there can be a defective O-ring or rubber washer around the valve. These can be replaced with the car loaded: however, it is recommended that the job be left to experienced repair personnel.

Corrosive Car Leaks. The majority of problems with leaks in nonpressurized corrosive cars is with the top unloading cars, caused by bad gaskets around the manway or fill hole cover. Repair requires replacement of the gasket, generally with an acid-resistant material. For stronger materials the tanks may be rubber- or heresite-lined. It must be remembered that even a small leak could quickly eat a hole in the tank shell. The other source of leakage on a corrosive car is the rupture of a pressure relief device. These cars have a safety

Figure 11-14. Typical angle valve showing flange to tank, plug, and packing adjustment

Figure 11-15. Gauging rod may leak due to an open valve or missing plug. *Courtesy of Federal Railroad Administration*

vent with a rupture disk made of a resistant material like lead or stainless steel. The disk is designed to confine the contents within the car, but rupture and relieve the pressure at a predetermined setting, usually not in excess of 75 psi. The disk then must be replaced to stop all venting. Remember that the leaking material may have to be neutralized, especially on the tank shell.

Liquid Car Leaks. The top loading/unloading liquid cars may have leaks occurring in three areas. If the dome or manway is not securely closed after filling, there will be a leak. Similarly, if there is a bad gasket in this area, problems will develop. Another area would be the safety relief valve or vent, as earlier described. There could also be a leak where these devices are attached to the car; these would merely require tightening. The various piping flanges and valves for top unloading and air induction/eduction could also vibrate loose over the course of time and necessitate tightening.

The bottom unloading cars may have the valve controlled from the bottom of the car or from the top. Those operating from the top may require the valve stem protective cover to be inverted and used as a key to operate the valve. The first place to check when a bottom unloading outlet is leaking is the valve itself. Is it closed tightly? Next, using a 36-inch wrench or a smaller wrench and "cheater" tighten the outlet cap to see that it is secure. Never remove the cap from a leaker until you have opened and closed the valve to assure that it is tight. If the leak persists, the gasket in the bottom cap may need to be replaced. Attempt to catch the runoff as the cap is removed for environmental and safety considerations. The other possible source of a leak with nonpressurized cars is at the top where the packing around the operating stem may need to be replaced or the packing gland tightened.

Purging

Purging of tank cars may take place at a derailment to totally empty the car of product or residual vapors. This may require filling the car with water, inert gas, or a neutralizer. In the past, vinyl chloride has been purged from tanks by adding water and the vapors burned with a propane fed torch. A special connection or piping device may be needed to introduce the materials into the car (see Figure 11-16). Evacuation downwind may be necessary. Sufficient information should be available and preparation made before initiating purging operations. Technical assistance from railroad safety personnel will be required. Never allow anyone to attempt relieving the pressure in a tank car by shooting holes in it. This can cause a rupture of the car with explosive force.

Commodity transfer is different from purging and is used to off load a car that is normally not punctured or has had the leak stopped. Transfer saves the product, while purging removes a hazardous condition.

Punctures

Numerous recent rail incidents have resulted in the puncture of a car causing release of the commodity. In several instances, the puncture has been in the end of the tank

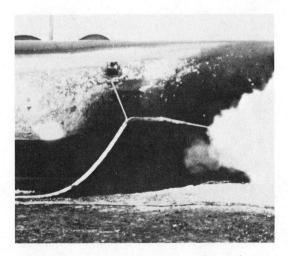

Figure 11-16. Purging may require special equipment and technical assistance

called the tank head or head sheet. The puncture of head sheets has happened both during derailments and in classification yard incidents.

In order to limit punctures, a number of regulations are currently in force. Any placarded tank car except a "Combustible" or "Empty" cannot be positioned in the train next to a loaded flat car or open top car on which the load extends above the car end. This is done to help shiftable loads from moving and damaging the end of a car in a derailment. This is a good precautionary measure, but it does not cover the sloping hopper car, which has a protruding end that would essentially function in the same way. In order to stop the occurrence of recent switching incidents that punctured head sheets, the DOT issued Emergency Order No. 5 which no longer allows DOT specification cars 112A and 114A —both single-shell cars used to transport liquefied gases under pressure—to be cut off in motion. In other words, they must be brought to rest by a locomotive. Also, no other car moving under its own power can be allowed to strike a standing 112A or 114A.

This has been followed by the regulation requiring the retrofitting of these cars. All of these tanks must be equipped with a shelf-type interlocking coupler. Two styles will be used: Type F, a bottom locking style (Figure 11-17) and Type E shelf, which is both top and bottom locking. This will eliminate the problem of the Type E knuckle coupler, which would jump up and strike the end of the car ahead of it. New tank cars are required to be equipped with interlocking couplers.

In addition, cars carrying anhydrous ammonia must be equipped with tank head puncture resistance. As stated earlier, flammable gas carriers must have thermal protection and also head shields. The head shields must cover approximately the lower half of the head, be of ½ inch (12.7 millimeters) steel, and protect the head from coupler impact up to 18 miles (28.8 kilometers) per hour (Figure 11-18). This is based on the statistics that 90

percent of the tank car punctures occur on the head of the tank and of these, 82 percent are in the lower half of the head.

Figure 11-17. Type F interlocking coupler

Figure 11-18. Head shield on lower half of head sheet on single-shell tank cars protects head against punctures

Tank Train

The Tank Train is a system developed by the General American Transportation Corporation to move bulk quantities of liquids in an interconnected unit train. Each series of interconnected cars can be loaded and unloaded by the use of a single connection. The major advantages of the Tank Train are: less handling in loading and unloading since only a single connection is used to handle up to 40 cars; greater flow rates, under proper conditions up to 3,000 gallons (11,370 liters) per minute; increased safety due to less human involvement and exposure; environmentally acceptable since vapors can be collected; reduces capital goods needed in loading and unloading facilities; and minimizes the turnaround time while obtaining the maximum efficiency from equipment and staff.

The Tank Train is designed to work in either a closed system to control the vapors of flammable or toxic products, or in an open system for less hazardous materials. Products which are adaptable to the system are as follows: alcohol, bulk liquids, bunker C, caustic soda, crude oil, ethylene glycol, fuel oil, JP4, super phosphoric acid, and sulfuric acid. The cars are each equipped with 10-inch (25.4 centimeters) flexible hose, elbow pipes, protective housings, outage control for product expansion, and pneumatically controlled tank isolation valves. The product is introduced at one end of the string of cars and the vapors are expelled through each car to a collection line at the end of the string. As each car fills, the product passes through the interconnection to the next car and fills it until the fluid-level sensing device placed in the last car shuts down the fill pump. The isolation valves on the top at each end of the cars would then be closed, with most of the product in the connecting piping having drained into the cars. The cars are also equipped for individual top loading and top or bottom unloading.

During a derailment, the Tank Train will have the following added considerations for the fire service. Some product will be in the connecting piping, which will undoubtedly be broken, and spill. The cars have two additional top outlets which, if not operating properly, could discharge liquid. The couplers of the cars will be secured so they cannot be uncoupled; thus a number of cars could be derailed or overturned during the incident. Finally, it is a unit train with a typical string being 20 or more cars of hazardous material in a row.

Box Cars

Several types of cars fall under this general heading. Fire fighting operations with all of them are similar. These cars may be hauling hazardous materials in a variety of methods including bulk, drums, wooden crates, bags, fiber drums, or cartons and cases of plastic or metal containers, to list only a few.

The first broad category is the insulated car. These cars may be either heated cars or refrigerated cars. The cars are polyurethane lined, and if this flammable insulation becomes involved in the fire, the fumes will be toxic. Also, a heated car may be poorly vented and there could be dangerous accumulations of fumes inside if the car must be entered.

Although these cars do not carry hazardous materials, they will have fuel tanks. Refrigerator cars are equipped with diesel-powered compressors and can have up to 500 gallons (2 cubic meters) of fuel beneath the car. (See Figure 11-19.) These have no history of being a problem, but during an accident could add fuel to or spark a fire. The diesel engines can be shut down if necessary by using the emergency fuel cutoff in the engine compartment. Another method is to use a CO_2 fire extinguisher and discharge it in the engine air intake.

Box cars may be carrying a variety of products. Initial operations when responding will be to perform the normal size-up, evalua-

Figure 11-19. Fuel tank on a refrigerated car contains up to 500 gallons (1,900 liters) of diesel fuel

tion, and decision making for a hazardous material incident, using all precautions previously discussed. The difficulties usually stem from floor fires, which are not registered by hotbox or dragging-equipment detectors. Floor fires can be caused by spontaneous ignition in the car lading, a contaminant that was left in the car, heat from a hot journal bearing, or from sparks generated by a dragging or overheating brakeshoe. Precautions against these are spark shields to protect the floor, roller bearing trucks, and hot-box detectors to eliminate hot journals.

The actions to take will depend essentially on the hazardous material being carried and the extent of the fire. A few older cars of all wood construction may be found; all new cars are of steel construction. The floor of the car and the inside lining may be of wood or steel construction. Newer box cars are of tight construction, may be difficult to enter, and can contain a smouldering fire, which will rapidly increase in intensity and possibly even backdraft when additional oxygen is introduced. Extreme care must be used in opening

the doors of the car. Also, since the load may not be braced, the packages or burning materials may come tumbling out. In other cases, depending on the load, it may be required to use winches in removing materials to complete the overhaul or get to ends of the car for total extinguishment. Depending on the commodity and extent of the fire, it may be best to use an indirect attack, if any, on the burning box car.

The indirect attack will get to all areas of the car without having to unload it. It also alleviates the problems of opening the car, which could require forcible entry especially during a wreck. With the doors closed, find the location of the fire if not evident. This can be done by feeling with the back of the hand, observing paint discoloration or blistering, or if necessary wetting down the side of the car and watching for the drying action of the fire inside. Once the location is known, proceed to the roof of the car and cut a hole in the roof directly over the fire area. The hole needs only to be large enough to insert a 1½-inch fog nozzle and move it around. With the nozzle on a wide fog setting, move the nozzle in a clockwise circular pattern or back and forth along the length of the car. This may be necessary in more than one location, depending upon the extent of fire in the car. Then allow the car to stand and the generated steam to do its job. When there is no longer any indication of fire coming from the holes in the roof, the car may be carefully opened for overhaul.

Flatcars

Many flatcars today carry TOFC (Trailers on Flat Cars) or COFC (Containers on Flat Cars) units. Many are loaded with hazardous materials. Fireground operations for TOFC and COFC units are essentially similar to those for a box car. They are not of as strong construction and their stability is less, certainly when derailed. If the decision is made to attack a fire in these units, the indirect method can also be used.

Hopper Cars

The hopper car, although not constructed like a box car, can carry bulk shipments of hazardous materials ranging from plastics and fertilizers to oxidizers. These units may also be polyurethane lined, which could contribute to fire problems. During a derailment, hopper cars may spill easily, and approach must be made with all due regard for the products carried. The car lining and lading could add fuel to any derailment fires.

Hazardous materials may also be carried on special rail cars, such as the flat car loaded with one-ton (900 kilograms) chlorine cylinders, containers for radioactive materials, and heavily insulated tank cars for carrying cryogenic liquefied gases.

Locomotives

Diesel locomotives fit into the rail hazardous materials scene because they may be involved in the derailment, could be a source of ignition, and carry a large combustible fuel load.

The fuel tank is located under the body of the locomotive and varies in size from 1,500 to 4,000 gallons (6 to 15 cubic meters).

On each side of the locomotive, located near the fuel fill opening, is an emergency fuel shutoff. Two types are in use, a mechanical one and an electrical one. The mechanical type (Figure 11-20) has a protective cover that is removed and a ring that is pulled. There is about a 30-second delay until the engine dies. The electrical type is a push button, which shuts down the fuel supply immediately. Older locomotives have the mechanical style, whereas newer ones use the electrical. An additional fuel shutoff is located in the cab of the locomotive. The older mechanical style painted red may be near a door. The electrical control for fuel shutdown (Figure 11-21) is well marked, also painted red, and always located on the electrical panel near the engineer's station. The purpose of the fuel shut-

offs is to stop the engine and cease the generation of high-voltage electricity, up to 2,500,000 watts. Once the engine has stopped, the high voltage is removed from the electrical compartments, engine area, and generators.

Figure 11-20. Mechanical shutoff device for the fuel supply on a diesel locomotive

Figure 11-21. Electrical controls for fuel shutdown and battery disconnection at electrical panel near engineer's station

Another source of combustible liquid on a locomotive is the 250 gallons (11 cubic meters) of lubricating oil in the crankcase. This has not been a serious problem in the past, but could add to a fire if released during a derailment.

As part of the diesel locomotive procedure, it is necessary to disconnect the battery power source of the locomotive electrical system. This is a 74-volt system and should be shut down. On newer diesels the battery switch is in the cabinet below the electrical fuel shutoff in the cab. The switch is somewhat marked but identifiable as a large knife-blade switch that must be opened. The cabinet containing the battery control may not be in the same place or well marked on older units; however, the switch is usually easy to identify.

At the engineer's station there is also an excitation switch. This would cut the power from the diesel engines to the electrical generators. In an emergency, since all power and ignition sources should be removed, the fire service will not be concerned with it. The doors to the engine compartment should not be opened until all electrical energy is shut off. If there is need to rescue personnel from the locomotive cab, there are always at least two doors to use for access.

In summary there is a great deal to know about the handling of hazardous commodities by rail. The frequency of exposure and accidents and the potential magnitude of the incident require that the emergency services be adequately prepared to handle all rail emergencies. It must not be forgotten, however, that the handling of many may be by cordoning off the area and evacuating and withdrawing all emergency forces to let the situation resolve itself. If the outcome of the incident cannot be favorably influenced by the actions of those involved, they have no reason to endanger themselves and MUST be removed.

QUESTIONS FOR REVIEW AND DISCUSSION

1. List the means of positive identification of the products carried in rail cars.
2. If your department does not have an adequate retreat signal for withdrawing from an incident, develop several alternatives that are not only practical and realistic for use, but also adequate to accomplish the objective.
3. Describe a frozen dome and how it should be properly handled.
4. What are the major hazards and how are they dealt with on a diesel locomotive?
5. List four special problems that may occur with tank cars of hazardous materials and how each should be handled.
6. What difficulties will be encountered in a box car fire involving ORM-D?

CASE PROBLEM

At 3:15 A.M. on February 27 a train was moving through a central Michigan city. It was snowing, with about 6 inches (15 centimeters) already on the ground and the wind blowing from the west at 12 miles (19 kilometers) per hour. The temperature was 0°F (–17°C).

The 98-car freight train was moving westward at 23 miles (36.8 kilometers) per hour when the 16th car was struck by a tractor trailer tank truck when the driver apparently fell asleep. The tank truck was placarded flammable and although sustaining considerable damage was not punctured.

The train cars were as follows:

Position in train	Type of car	Placard
No. 15	Hopper	Oxidizer
No. 16	Tank Car	Flammable
No. 17	Tank Car	Flammable Gas, Empty
No. 18	Tank Car	Corrosive
No. 19	Box Car	Empty
No. 20	Tank Car	Poison

Car 16 was punctured as it derailed and is burning. The probable ignition source was the truck tractor. The truck driver is in the cab. The fuel from the 16th car is flowing to the east and under the 17th car at this time. The 20th car has skidded onto the top of the 18th car, has a small leak, which is forming a pool under the 19th and the 18th cars.

The derailment is in an industrial area (Figure 11-22) with most buildings unoccupied except for a night security guard at Beeman Enterprises and Coleman's Cantina, which is open all night. The buildings are all one story, of noncombustible construction with tar and gravel flat roofs, except Cole-

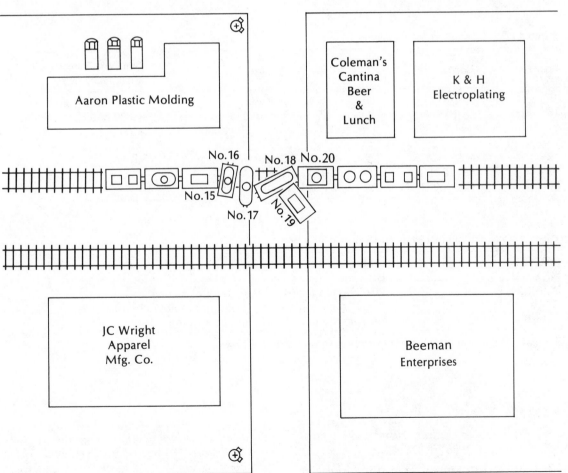

Figure 11-22. Derailment of a freight train

man's, which is a 1½-story wood-frame building with composition roof.

The local fire department consists of two stations. Station 1 has a 1,250 gpm (4,738 liters per minute) pumper and an 85-foot (25.5-meter) aerial with a 1,000 gpm (3,790 liters per minute) pump. Station 2 has a 1,000 gpm (3,790 liters per minute) pumper and a heavy-duty rescue squad. Both stations have ambulances. Each station has 7 fire fighters on duty and the duty chief responds from home in a station wagon. Mutual aid is available from two township departments. Each can supply a 1,000 gpm and 750 gpm (2,843 liters per minute) pumper and 10 fire fighters. They are 10

and 12 minutes away after alerting. The off shift can be called to operate two 750 gpm (2,842 liters per minute) reserve pumpers. This would take 30 to 45 minutes since the fire alarm operator is alone. Additional mutual aid would also be 30 to 45 minutes in arriving. The fire flow available in the area is 4,500 gpm (17,055 liters per minute).

1. What are your initial tactical considerations?

2. How would you deploy the resources available to extinguish the fire and bring this incident to a successful conclusion?

Table 11-3. Consist in the Caboose

Position	Car No.	Type	Placard	Standard Transportation Commodity Code
15	CN 9763	LH	Oxidizer	49 187 05
16	DUPX 2641	LT	Flammable	49 093 50
17	UTLX 1384	ET	Empty Flammable Gas	49 057 52
18	MCX 52475	LT	Corrosive	49 361 10
19	BN 2641	EB	None	—
20	GATX 91146	LT	Poison	—

Table 11-4. Waybill Information Denotes

UTLX 1384	Empty tank, former lading liquefied petroleum gas
MCX 52475	Bromine
DUPX 2641	20,000 gallons (76 cubic meters) of xylene
GATX 91146	20,000 gallons (76 cubic meters) of epichlorhydrin
CN 9763	Bulk ammonium nitrate, fertilizer grade

Truck bill of lading indicates the product to be methyl acrylate.

NOTES

1. "Flammable Liquids Notebook," *Firemen*, April 1956, p. 17.

2. U.S. Department of Transportation, National Transportation Safety Board, *Railroad Accident Report, Derailment and Collision, Dunreith, Indiana, January 1, 1968*, p. v.

ADDITIONAL READINGS

Accident/Incident Bulletin No. 144. U.S. Department of Transportation, Federal Railroad Administration, Office of Safety, 1975.

Best, Richard. "Futile Rescue Attempt Kills Volunteer." *Fire Command*, February 1976, pp. 22–23.

"Coordinated Attack Limits Post-Blast Damage." *Fire Command*, July 1975, pp. 14–17.

"Devastating Blast of Railroad Cargo." *Fire Command*, April 1975, pp. 32–34.

Ditzel, Paul C. "BLEVE." *Firehouse*, June 1977, pp. 29–31.

Ellis, Donald L. "Propane Cloud and a Lot of Luck." *Fire Command*, February 1974, pp. 18–21.

Federal Register, vol. 42, no. 179, Thursday, September 15, 1977, pp. 46306–46315.

GATX Tank Car Manual. General American Transportation Corporation. Chicago, 2nd ed., 1966, and 3rd ed., 1972.

Hayes, Terry. "Train Wrecks, Fire, and Explosion." In *FDIC Proceedings*, 1970, p. 163.

"Hazardous Materials Hearings Stir Action on Derailments." *Washington Scene*. Vol. 1, No. 9, April 21, 1978.

"Hazardous Materials Transportation Accidents." *Fire Command*, April 1974, pp. 11–16.

Keeping Hazardous Materials Contained in Tank Cars. Southern Pacific Transportation Company, San Francisco, 1975.

Lathrop, James K. "The Terrible Blast of a BLEVE!" *Fire Command*, May 1974, pp. 14–17.

Mooney, Eugene F. "Statement by Commonwealth of Kentucky Before a Hearing Conducted by the National Transportation Safety Board on Derailments and Carriage of Hazardous Materials." Unpublished. April 6, 1978.

Nailen, R. L. "Learning to Handle Rail Disasters." *Fire Engineering*, February 1970, pp. 40–43.

Railroad Accident Report Derailment and Collision, Dunreith, Indiana. U.S. Department of Transportation, National Transportation Safety Board, December 1968.

Railroad Emergency Response. Burlington-Northern, Inc., St. Paul, 1978.

R. M. Graziano's Tariff Publishing Hazardous Materials Regulations of the Department of Transportation. Bureau of Explosives, Association of American Railroads, Washington, D.C., 1976.

"The Role of the AAR Research Center in Solving Railroad Fire Problems." *Fire Journal* 64 (January 1970): 43.

Sharry, John A. and Wilbur L. Walls. "LP-Gas Distribution Plant Fire." *Fire Journal*, January 1974, pp. 52–57.

"Tank Car Fire Advice Offered by Safety Board." *Fire Engineering*, March 1972, p. 33.

Tank Train, General American Transportation Corporation, Chicago, Illinois, 1977.

"Two Train BLEVEs: Different Situations Require Different Strategies." *Fire Command*, September 1976, pp. 24–26.

12
Truck transportation

A rising number of highway incidents involve hazardous materials. This stems from the frequency of vehicular accidents coupled with the increasing movement of hazardous materials over streets and highways. The majority of hazardous material incident reports submitted to the Materials Transportation Bureau of the Department of Transportation originate with highway carriers. During the seven-year period 1971-77, 54,000 or 90 percent of the 60,000 reported incidents occurred in the trucking industry.

This chapter will discuss briefly the causes and commodities involved and touch on the preplanning and tactics that must be considered. Particular types of trucks and how to handle incidents with them are also covered. Hazardous materials in cargo vans, liquid carriers including corrosives and liquefied gases, compressed gas tanks, and dry bulk carriers will be addressed.

CAUSES OF INCIDENTS

The causes of the trucking incidents can be attributed to human errors in driving or product handling operations, equipment breakdown because of improper maintenance or mechanical failure, or to accidents involving other vehicles. Human factors that have been the cause of truck accidents include lack of attention, dozing or sleeping, fatigue or driving too long, use of alcohol or drugs, and illness. There are also a number of drivers who violate traffic regulations and thus cause an accident. Violations include speeding, driving too fast for weather or road conditions, careless or reckless driving, using poor judgment, failing to downshift, following too close, and a general lack of experience. Equipment failures can lead to leaks or to accidents that involve the load. Loose valves and fittings or bypassing safety devices can cause leaks. The majority of mechanical failures are with tires and brakes. Coupling devices, electrical systems, and steering mechanism also have failed causing accidents.

The majority of vehicle fires—over 80 percent—occur in tractors or semitrailers of over-the-road haulers. The causes of the fires vary, but overheated tires, fuel tanks damaged in accidents, and defective wiring are the major contributors to the fire problem. Although many of these fires are extinguished by local fire departments, it is not before extensive damage has occurred. Also a number of these fires burn themselves out. This is, in part, caused by the delay in initiating an alarm along major roadways and possible confusion concerning fire department jurisdictions, which causes delays in response.

COMMODITIES

Trucks today carry numerous hazardous materials in vans, bulk carriers, tank trucks, cryogenic carriers, containerized loads on flatbeds (Figure 12-1), and stake-sided flatbeds. The National Tank Truck Carriers, Inc., publish Commodity Data Sheets for their members, which include 92 commodities ranging from aminoethylethanol (amine) to vinyl toluene. Far more products than these are carried and they vary considerably in degree of hazard. There are only a few hazardous materials that federal regulations do not permit to be carried by the trucking industry, one being prussic (hydrocyanic) acid.

A major problem in identifying hazardous commodities arises in the placarding system as described in Chapter 6. A mixed load carrying numerous products will often bear only the *Dangerous* placard, which merely warns emergency response personnel of a hazard but does not indicate the specific type of hazard. The other major point to remember is that placards are not required for many substances unless 1,000 pounds (450 kilograms) or more are carried. A strange mixed load that may occur in the future is foodstuffs and poisons. The American Trucking Association is presently experimenting with a plastic container to adequately enclose poisons for shipment with edibles.

Generally, positive identification of the load can only be accomplished by checking the shipping papers. These would be the bill of lading, trip ticket of the driver, or load manifest. Labels on containers or placards on the truck itself will warn of possible hazards but rarely aid in product identification. The shipping papers should be in a pocket on the left-hand door of the tractor, behind the driver, or on the seat. They are required to be within the driver's reach and in sight when the driver is out of the cab, generally on the seat. In those cases where the shipping papers cannot be found, the dispatcher of the trucking firm can be called on for assistance. The tractor or trailer number will aid in identifying the shipment. If the dispatching location is unknown, contact CHEMTREC with the name of the hauler and the tractor or trailer number. CHEMTREC will initiate operations to track down the identity of the load. In some cases, the state license tag number may be helpful. With it, the hauler can be identified and the dispatcher contacted.

PREPLANNING

Preplanning for truck incidents has essentially been covered in earlier chapters. Transportation route plans should be laid out and any necessary preparations made. Local truck terminals should be visited to acquire the dispatcher's phone number, to determine commodities moving through the area, and the amount and frequency of shipments. All personnel should familiarize themselves with the pertinent features of tractors, vans, and the variety of tank trucks locally in use.

TACTICAL OPERATIONS

Because of easier accessibility, fire fighters responding from their homes or recalled may respond directly to the scene. This will require strict enforcement of regulations regarding the wearing of protective clothing. At highway incidents, protective clothing adds additional protection against being struck by traffic when of light color or clearly marked with reflective tape. Perhaps some fire department personnel may be required to direct traffic with chemical flares or explosion proof lights. Wearing proper protective clothing will add authority for this operation.

Figure 12-1. Containerized units for moving hazardous commodities. *Courtesy of Joel Woods*

Size-Up

As emergency response personnel approach they must perform adequate size-up for continuing into the incident and for apparatus placement. All precautions should be taken on the basis of the information supplied by the fire dispatcher or the warning indicated by the vehicle placarding.

Apparatus should generally not be placed too close, positioned directly behind the involved vehicle, or parked blocking other emergency response vehicles from getting to the scene, moving if necessary, or being used effectively. Apparatus must be parked to allow adequate working room and not restrict removal of equipment from the unit. Generally, apparatus should pull past the involved truck. This operation will be tem-

pered by the wind direction and slope of the land. Positioning should also consider water supply, if available, and response route where dual highways, median dividers, or cloverleafs are present.

Establish a command post based on your Standard Operating Procedure. Report its location and identify it with some marking. The command post should establish suitable communications. Either the command post or communications center will have to develop communications with outside agencies. Incidents on roadways may require technical assistance for the products involved, and cooperation of public maintenance agencies or utility companies.

Size-up also encompasses those items

discussed earlier. At a truck incident a check should be made for downed wires beneath and on all sides of the vehicle. The truck body could be energized or an ignition source readily available. The battery may need to be disconnected. Always disconnect the ground side first. It may be necessary to chock the wheels of the unit. Of course, the hazardous vapors or amount of fire may prohibit these actions. Size-up should include information on the braking system of the vehicle and if there is an air suspension system.

All the information collected in size-up must be evaluated based on the situation and a decision made as to a course of action. The course of action may include the separating of the tractor from the trailer if it is not involved. A number of alternatives are available depending on the commodity, its location, and extinguishing agents available. The plan of action chosen will have to be monitored and reassessed constantly for possible necessary changes.

Braking System

Fire service personnel should have some knowledge of truck braking systems. In addition to the air-operated service brake and emergency brake, there are other systems for the tractor and trailer. The dash is often equipped with three brake controls. The first is a yellow button to lock up the tractor air brakes. This functions as an emergency brake. The second is a red knob for controlling the air to the trailer brakes. When pulled out this cuts off the air to the trailer and sets the brakes. It must be pushed in to move the trailer. This device should be pulled out before opening the air hose glad hand connections between the tractor and the trailer when separating the two units. On some newer units these two controls work at the same time when setting the brakes; however, they must be released independently. The final dash-mounted brake control is a blue button, which controls the auxiliary safety air tank.

This is a spare emergency tank to provide air to the braking system if the normal air supply is lost. It enables the driver to move the unit in an emergency, to get it off the road, or out of an intersection.

New tractors are equipped with computerized braking systems that meet Federal Motor Vehicle Safety Standard 121. These are essentially an accident prevention device to avoid wheel lockup, skidding, and jackknifing. They are of no consequence to emergency response personnel once an accident has occurred. The main function of this antilock device is to release the brakes and keep the wheels turning to allow the vehicle to be stopped in a straight line, although it may angle somewhat off pattern to one side.

If the brakes on a tractor lock up and it is necessary to pressurize the system to move the unit, this could be done with the air brake compressor system on fire apparatus. It would require a chuck and hose, which are available on many pumpers, for air tools. The hose can be connected to the air chuck on the air starter tank of the power unit. This may be slow, depending on the intake capacity. It may be easier to merely lower the landing gear on the trailer and pull the tractor away with a tow truck. In an emergency a power unit can be impounded to move a trailer.

The brakes on a trailer except spring loaded can be released by bleeding the air reservoir under the rear axles. There is a petcock on the tank that will serve as a blowoff. With no air, there will be no brakes and the trailer can be moved.

Air Suspension System

Air ride or air suspension systems (Figure 12-2) are in use on many trucks, tractors and trailers, and most buses. These are important because if the air bag bursts due to fire exposure the unit can drop, in some cases as much as 6 to 8 inches (15 to 20 centimeters), to the tires. There will be no warning before the bag fails. When working on trucks with air bags,

Figure 12-2.   Air bag installation on the rear axles of an over-the-road truck

do not place your head or limbs under the vehicle, especially between a wheel and its housing since if the air shock bursts, they could be crushed. It is recommended that an applicator or piercing nozzle be used under these units. If not available, place the nozzle and a stiff portion of the hose under the truck, but do not put hands or arms in unsafe positions. If there is fire around the bags, they should be cooled immediately without taking unnecessary chances.

Rescue and Evacuation

A primary tactical operation will be rescue and evacuation. Suitable tools must be selected and properly used to expedite a safe, efficient extrication. The area may need to be checked for flammable vapors. Remember that all due precautions must be taken for the victim and the rescuer during a rescue operation. With hazardous materials it is not advisable to go charging in for a rescue without sufficient protective equipment, including breathing apparatus. Rescue also includes checking the drainage ditch alongside the road to see if anyone was thrown from the involved vehicles.

In incidents with explosives, reactive materials, or gases, evacuation can be extensive in populated areas. All parties involved in the evacuation will need to be coordinated by the command post. It may be necessary to designate a person in charge of evacuation. Recent incidents involving corrosives and/or toxic gases have demanded some extensive evacuations. If transportation routes pass target hazards, all evacuation possibilities should be discussed and considered.

Separation of Power Unit

During a hazardous material truck fire if the tractor of the rig can be salvaged, there will be a considerable savings to the operator or company. Whenever possible the tractor should be separated from the burning cargo van or tank. Separation is achieved by the following three steps.

First, lower the dolly wheels, legs, or landing gear on the van or tank trailer. In some cases, these devices will not support the weight of a loaded trailer. In those instances the landing device will puncture the bottom and release the load. Depending on the construction of the rear frame of the power unit, it may be necessary to lower the landing device in order to separate the units. If the tractor has a square boxed frame, the kingpin which secures the trailer to the fifth wheel latching device of the power unit will not clear. If the frame is tapered or notched (see Figure 12-3), the trailer kingpin will clear without the landing gear down.

Figure 12-3.   Tapered or notched frame, which allows the kingpin to slip off the tractor without lowering the landing gear

The landing device is usually manually operated by a crank handle folded away near the gear on the right side. It may be a two-speed type with the shaft positioned in for high gear (raising) and out for low gear (lowering), or it may have two handles with one to move the position of the device forward or backward under the trailer and the other to operate the landing device. On some units the landing gear may be air or hydraulically operated. With an air operated unit, if the air is disconnected, the landing device cannot be lowered. Under fire conditions it may be necessary for the fire department to approach the landing gear with water fog to drive the fire away and lower the device while under a protective fog cover. Remember that with a square box frame, the dolly wheels must be lowered to separate the power unit from the trailer.

Second, the electrical power cord from the tractor to the trailer and the air hoses for the brakes must be disconnected (see Figure 12-4). The electrical line is merely unplugged at the front of the trailer. The air hoses are connected by a quarter-turn device called a glad hand. Before releasing the air hoses, the trailer air brake control valve in the cab of the tractor should be shut off. The glad hands are released by lifting up on the one coming from the tractor. On newer units the air lines are color coded with red for the emergency brake, normally on the left side of the trailer, and blue for the service brake on the right side. Many trailers have the glad hand protector painted to match. If the air lines are crossed, the trailer will not move. On a trailer that has stood for a period of time and had the air bleed off with the glad hands crossed, no air would enter the trailer system, which would leave the trailer with no brakes. It would be important for emergency response personnel to check this before moving trailers, especially at a loading dock or filling rack.

When the trailer air brake control in the cab is not closed, the broken glad hand con-

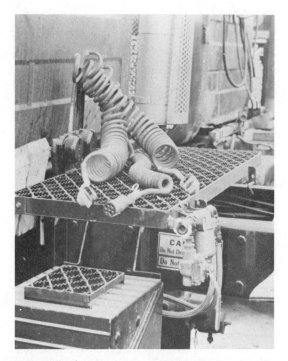

Figure 12-4. Air hose and electrical plug should be disconnected before separating the tractor and trailer

nection will cause an air loss to the power unit. Since there is an air safety valve on the tractor, when the pressure reaches approximately 50 psi, the valve will automatically close with enough air pressure to stop the tractor. Once this valve closes, the air compressor will build the system back to proper air pressure.

The third step is to release the fifth wheel which is the connection between the tractor and the trailer. To do this the kingpin must be unlatched. The kingpin is released by a handle that is pulled out or by a sliding lever moved horizontally on either side of the fifth wheel near the front (Figure 12-5). Because a variety of mechanisms are employed, several tractors should be observed to gain familiarization. Before moving the tractor, driving, or dragging, be sure the release man is clear of the area.

Figure 12-5. Lever or handle to unlatch the kingpin and release the trailer from the fifth wheel on the tractor

Release of the fifth wheel should not be confused with a sliding fifth wheel. Many tractors have a fifth wheel that moves backward and forward. This may be operated hydraulically from the cab, by removing pins between the fifth wheel and the frame, or by releasing a lever. A pin should not be taken literally when releasing the kingpin of the trailer from the fifth wheel.

With the landing gear down, the air and power cords disconnected, and the fifth wheel released, the units can be quickly and easily separated. In an emergency in which hazardous materials are involved and time is critical, it may not be possible to take all the above steps. In this case, lowering the dolly wheels and disconnecting the electrical and air lines can be eliminated. The tractor will lose air and the trailer will crash to the ground and perhaps spread the fire, especially if a tank vehicle is involved and the landing gear punctures the tank. The truck driver may be able to provide valuable assistance in separating the tractor from the trailer.

The procedure for a double-bottom unit (Figure 12-6) is quite similar when it is necessary to separate the lead trailer from the follower or pup. In addition to the above steps, the safety chains or cables must be disconnected, the air valves at the rear of the lead trailer must be closed, and the safety latch securing the pintle hook released so that the hook on the drawbar can be removed from the lead trailer. A double may have an additional air-operated safety latch on the pintle

hook. This is no problem since the air pressure is discharged and the latch released when the air valves are closed and the glad hands broken.

Product Control

The basic considerations in product control are plugging leaks and controlling spills. New devices using polyurethane currently undergoing testing by the Environmental Protection Agency may be available in 1979 to assist in plugging holes up to 6 inches (15 centimeters) in diameter. It is no longer considered acceptable to flush spills of hazardous materials; they must be absorbed or contained and picked up for proper disposal. The highway or street department should be of assistance. A bevy of federal agencies will become involved if the spill spreads or is not contained quickly. During the spring of the year with rain-swollen waterways, it may be extremely difficult to control a product after it enters a watershed. Likewise, in the western states, where the water is fast moving, it will be best to control spills and runoff before they enter a waterway. Early product control requires an accurate size-up and the summoning of those required to ensure it is completed immediately. A great deal of special equipment could be required and should be in the contingency plan.

Figure 12-6. Double-bottom tanker. *Courtesy of Charles J. Wright*

Extinguishing Agents

Extinguishing agents in proper supply bear consideration in tactical planning. This includes not only water delivery to the numerous truck transportation sites, but also to the routes they traverse. Additionally, foam, dry chemical, or dry powder agents may be necessary. In most cases, little forethought has been given to the amounts of polar solvents, those miscible with water such as the alcohols, which are crossing the highways and how to effectively control and extinguish fires involving them. Don't overlook this in your tactical considerations.

Attack Tactics

Finally, a tactical decision must be made on whether to implement an offensive attack, defensive operation, or withdraw from the area. With truck incidents, especially if an enroute accident, rescue of the driver or others involved is a primary objective. Special operations required to do this may be removing ignition sources, driving the fire away, using protective fog lines or extrication equipment, or special protective clothing for corrosives.

If the problem centers on small containers of moderately dangerous products, high-flashpoint liquids, or an easily controlled spill, an offensive attack to confine or extinguish should be chosen. Truck incidents will call for defensive operations when capabilities are not available to control the leak or fire. A defensive dike to stop the spread of flowing liquid or fire fighting operations to protect exposures may be best. If the situation is rapidly deteriorating, such as intense fire around an explosive carrier, the involvement of contaminated oxidizers, the escape of toxic vapors, or highly radioactive materials, emergency response personnel should withdraw from the area performing evacuation as necessary.

In all truck incidents several factors must be considered for fireground safety. These include keeping fire fighters and equipment safe from traffic, checking vehicle components such as bumpers and frames which could be magnesium, turning the ignition off to avoid vehicle movement, tractor fuel tanks, and auxiliary fuel supplies. Since there is always the possibility of a fuel spill at an accident, an attack line should be ready for the unexpected.

BRAKE AND TIRE FIRES

When hazardous materials are involved, extreme care must be taken with tire and brake fires to limit them from spreading to the cargo area. Brake fires are best handled by cooling with a hose line from the outside to avoid cracking of the drums. As the brakes cool, the line can be moved to the inside to complete the cooling. Check for extension of the fire to the trailer.

With tire fires, in many instances, the tube begins to burn first. This can go undetected and provide greater headway for the fire. The tire bursts into flames when a break occurs in the carcass and admits air to the innertube. Also, if one tire is flat in a set of dual wheels, a surface fire can start on the other. A tube or interior fabric fire can be extinguished by using a piercing nozzle through the wall of the tire at the top. This will allow the water to flow down the inside walls and cool the tire interior. All tire fires should be cooled down with large quantities of water because a tire can hold heat and burst into flames over an hour later if not properly handled. Check for extension of the fire to the floor of the van.

CARGO VANS

The majority of over-the-road trucks are semitrailer or straight truck vans. These carry a variety of hazardous materials in numerous types of containers. Even the common delivery truck to the grocery store carries many hazardous materials in small packages. Vans may be straight frame, trailers, doubles, or even triples. The units consist of steel frame, floor supports, and wood or steel floor and side walls of lightweight construction (Figure 12-7). The vertical framing for the walls and the roof supports are of aluminum on 12- to 20-inch (30.1 to 51-centimeter) centers. The outside walls are aluminum and the inside walls may be open or sheathed with metal or wood. The roof is generally exposed and may have plastic skylights. Some units have a tarpaulin for a roof.

Cargo vans may be of tight construction. During a cargo fire this can necessitate ventilation to alleviate a backdraft condition before opening the cargo area. Vertical ventilation is preferred because it is effective, easy to accomplish due to lighter construction, easy to repair, and the repaired area is not obvious. The thin aluminum roof (0.04 to 0.05 inch or 0.10 to 0.127 centimeter) can be quickly cut with any number of forcible entry tools. Make cuts parallel along roof supports across the van.

Gaining access to the cargo area should be done with safe procedures. Personnel should be in full protective clothing and breathing apparatus, and a charged line should be available. Keep in mind that on an unvented trailer, as swinging doors are

Figure 12-7. Lightweight vans may carry single or mixed loads of hazardous materials

opened, there could be a backdraft blowing the door forcefully around on its hinges. Thus, when doors are being opened it is best neither to stand in front of them nor behind them. A good practice is to use a pike pole to open doors from the side after unlatching the door. Another concern when opening doors, either swinging or the rolling overhead type, is the stability of the cargo at the rear of the truck. Drivers do not as a rule use restraining straps or secure the load, so there is the danger of emergency response personnel being struck by falling cargo.

Aluminum constructed vans will often melt quickly through the roof during a fire. With a heavy fire load contributing to the rapid melting of the body and loss of structural strength, there can be an early collapse at the center and/or failure of the side walls. This calls for close observation and restricting emergency response personnel from operating under the trailer or too close to dangerous side walls.

The doors on vans are at the rear and may be on one or both sides. The swinging style are usually doubled with the right door opening first. These doors are secured at the top and bottom by one to four vertical bars latched at the bottom. If the doors are sealed and must be forced open, the steel band seal should be retained. Some vans have overhead rolling type doors at the rear (Figure 12-8). These are also locked at the bottom and may have a safety latch that would require releasing before unlocking the door. Overhead doors may also need to be supported to keep them from coming down during operations.

In case of an accident, the truck may be overturned or in a position in which the doors cannot be used for entry. Having ascertained the contents and the dangers involved, the position of the unit and where the fire is, along with the structural condition of the cargo area, a decision can be made about forcing entry. A vented fire may also influence the decision. Generally, entry can be made most easily through the doors. If doors

Figure 12-8. Overhead rolling-type door on the left; swinging type on the right

are inaccessible, entry should be made through the roof, or through the side walls and/or front, and finally, through the floor. Insulated vans will be more difficult to enter.

In hazardous material truck incidents, watch for warning placards (or labels on packages), position apparatus safely, take all safety precautions, identify the contents from the driver or shipping papers and the nature of the problem, secure the area, reroute traffic, remove ignition sources, perform rescue and evacuation as necessary. With fires, separate the tractor from the trailer if necessary, watch for air-suspensioned units, ventilate the roof if required, and use precaution when opening up the cargo area. Have charged attack lines and backup lines available. Remember that vans could be carrying drums, cylinders, or small containers that can rupture with an explosive force. For a spill or leak, provide containment and stop the leak. In all cases, notify those interested parties and agencies directly or through CHEMTREC.

Extreme caution must be used in approaching and positioning units at incidents involving trucks carrying explosives. On these units, once the fire has reached the cargo area, all fire fighting operations should cease, evacuation should be completed, and personnel withdrawn.

Insulated Cargo Vans

Insulated cargo vans normally carry goods that are either refrigerated or heated. However, on the return trip, rather than returning empty, the unit may be carrying a wide variety of products, including hazardous materials. Both refrigerated and heated units will have heavy insulation on doors, walls, and ceiling. The insulation generally is a lightweight plastic base such as polyurethane and burns with toxic fumes. Fiberglass may also be used. Forcible entry can be more difficult because of the thickness of construction. The walls instead of the ceiling may be the best point of entry after the doors. This is the case with refrigerated units that carry meat carcasses. The meat hangs from steel tracks on the ceiling that would be a sizable barrier. The floors of refrigerated units are of heavy ribbed steel construction and would also be difficult to cut through.

Most refrigerated units will also be carrying additional tanks of fuel or refrigerant that could cause a small hazardous material incident regardless of the load carried. The auxiliary fuel supply is mounted below the trailer (Figure 12-9). Although other fuels may be used, diesel tanks are the most common.

Heated units may be gas or liquid fired. Wall-mounted gas-burning units are supplied from a tank beneath the trailer and use one of the liquefied petroleum gases. Liquid-burning units have small capacities of kerosene or occasionally methanol and are usually located near the rear doors where there is the most need for the heat.

Refrigerated units primarily use mechanical units mounted outside the front wall of the van and operated by diesel fuel from a tank under the truck body. A few have the refrigeration unit under the vehicle and some use gasoline or LPG for fuel. The fuel supply can range from 20 to 200 gallons (76 to 760 liters). A few long distance haulers and many local delivery units are now using a cryogenic gas refrigeration system. These systems use liquefied nitrogen at $-320°F$ ($-195.6°C$) from tanks in the front of or under the trailer or over the cab of a straight van. The nitrogen is dispersed over the load through tubing to create a freezing atmosphere. Since a fire would not burn in the inert gas, the major problem would be that of rescuing someone inside the refrigerated unit or involved in gas that leaked out during a leak or puncture. Remember that because of the vaporization rate, the nitrogen would exclude air from the area and thus serve as an asphyxiant as well as cause severe freeze burns on contact with the liquid.

LIQUID CARRIERS

Liquid vehicles will be the area of major concern in transporting hazardous materials by highway. Liquid carriers carry a wide variety of products in gallonages up to 17,400 (65,859 liters). The tanks vary considerably in construction depending on the product carried. The majority of liquid tanks are of aluminum, steel, or stainless steel construction and may be insulated or noninsulated. Tank shapes and configurations also cover a broad spectrum.

The liquid carriers are built according to Department of Transportation specifications and have a data plate attached to the right front of the tank. This plate will show the manufacturer's name and location, the tank serial number and model number, the date the tank was manufactured, the DOT specification number to which the tank was built, the nominal tank capacity in U.S. gallons, compartment sizes where existing, design and test pressures where applicable, and the maximum working pressure, material, and maximum density of the cargo in the case of a corrosive carrier.

The current DOT specification numbers are MC 306, MC 307, MC 312, and MC 331. Motor Carrier 306, diagrammed in Figure 12-

Figure 12-9. Auxiliary fuel supply mounted below refrigerated trailers

10, is authorized for flammables, combustibles, and Class B poisons with vapor pressures under 3 psi. These are considered nonpressure tanks. Only MC 306 is now manufactured for this use and it replaces MC 300, MC 301, MC 302, MC 303, and MC 305. MC 307 is authorized for similar products but has a minimum design pressure of 25 psi (172.25 KPa). Liquids with vapor pressures of 18 psi (124 KPa) or more at 100°F (57.78°C) but less than the 40 psi (275.6 KPa) or more at 70°F (21.11°C) of a compressed gas are transported in these tanks. They are considered low-pressure chemical tanks. MC 307 (Figure 12-11) replaces MC 304. MC 312 is authorized for carrying corrosive liquids. These tanks are of stainless steel or lined so that the shell is not reduced in thickness below that required during 10 years of service. These tanks have a 35 psi (241 KPa) or greater working pressure. Many of these units are top unloading. MC 312 (Figure 12-12) replaces 310 and 311. MC 331 (Figure 12-13) is for carrying gases which have been liquefied under pressure.

Some units are insulated and equipped with piping in the jacket for internal heating. These are generally called hot-product carriers and may have temperatures of several hundred degrees. Others are constructed for intermodal use and may be transferred to trains or ships as containers.

Safety Features

All liquid carriers have essentially the same safety devices. They are required by federal regulations and consist of the following:

1. Internal fire valves are required for each compartment. Bottom-unloading trucks carrying liquid hazardous materials have spring-loaded valves in each compartment which are closed during transit as a safety feature (Figure 12-14). They are operated from the fire or control box at the unloading station. Most are mechanically operated; however, some are air or hydraulically operated (Figure 12-15). The fire box door is designed to close with enough force to operate the mechanical control and close the valve. Valve controls in the horizontal position are open while those in the vertical are closed (Figure 12-16). The valve is internally mounted in the bottom of the tank and equipped with a shear section so that it will not be dislodged during an accident.

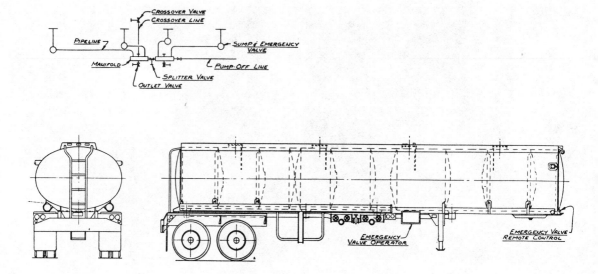

Figure 12-10.   Major features of a tank classed as a Motor Carrier 306 (DOT)

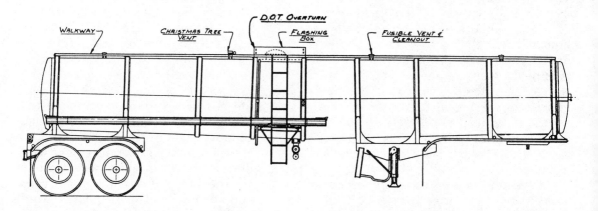

Figure 12-11.   Motor Carrier 307, low-pressure tank for chemical transportation

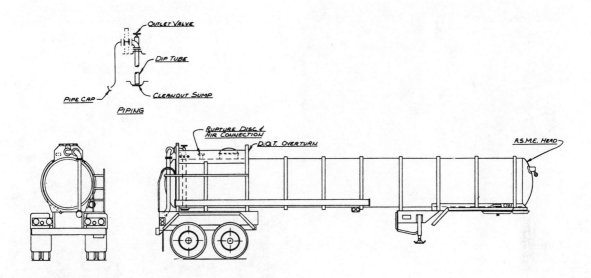

Figure 12-12.   Motor Carrier 312, for corrosive liquids, especially acids

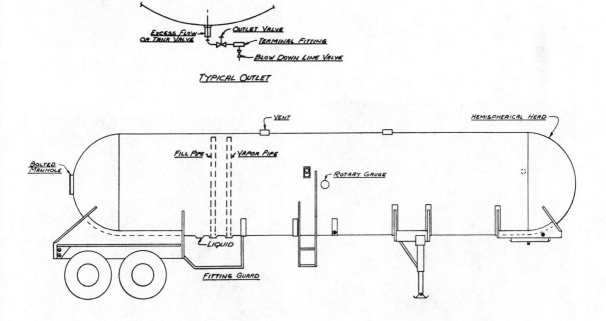

Figure 12-13.   Motor Carrier 331, for liquefied gases such as anhydrous ammonia, chlorine, and liquefied petroleum gas

Figure 12-14. Spring-loaded internal fire valve as safety feature

2. Fusible links or nuts are provided on the cable controls from the fire box to the internal fire valve. During a fire these would melt and close the internal fire valve stopping the flow of fuel (Figure 12-17).

Figure 12-16. Mechanically operated valve with lever in open position; spring-equipped door would operate the valve if closed.

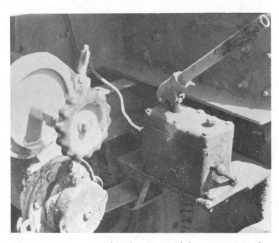

Figure 12-15. Hydraulic control for opening the valve on a corrosive carrier

Figure 12-17. Fire valves equipped with fusible links to automatically close during a fire

3. The internal fire valve also has an emergency shut off control that may be operated from remote locations. This may be mechanical, air (Figure 12-18), or hydraulically operated. Emergency controls may be located at numerous places, but generally they are on the left front of the tank behind the driver. They may, however, be on the right front, the rear, or at the midship unloading station. There may be several on the same tank. The remote location for operating the internal fire valve is well marked in Figure 12-19.

4. Breakaway protection by using shear sections (Figure 12-20) is provided for all internal and external valves and piping. The purpose is to protect the valves and keep as much product as possible within the unit and its fittings. External discharge valves are mounted in such a manner that they could be sheared off while the integrity of the piping would be maintained. With the fire valve operating properly there would be no loss of product.

Figure 12-18. Shutoff control for air-operated valve; note the breakaway section

Figure 12-19. Mechanical system for operating the internal fire valve has been operated. Control lever is in the open position

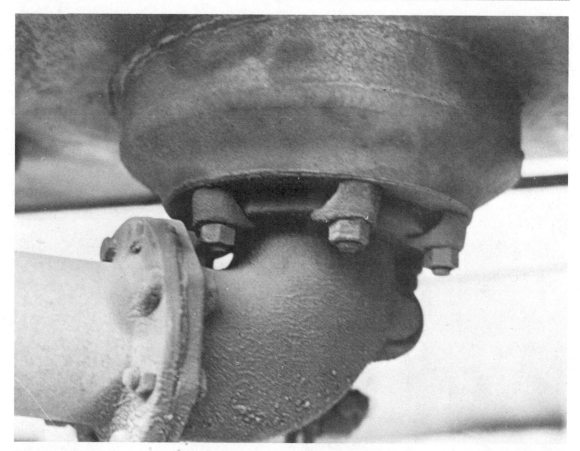

Figure 12-20. Shear point allows piping to break away and internal fire valve to remain intact and function properly

5. All valves, piping, and pumps are protected against accidents (Figure 12-21). This coupled with the use of shear sections limits the damage to a tank truck during an accident.

6. The rear of the vehicle must be protected by a suitable bumper, 6 inches (15 centimeters) from all plumbing, which could contain lading and of sufficient size to protect the tank from damage occurring during a rear end collision.

7. Safety relief devices are required on all vehicles. They enter the vapor space of each compartment and under normal venting operate at not more than 1 psig

Figure 12-21. Undercarriage protection against collision damage to the valves and piping under tank

(6.89 KPa). Various devices are used, such as relief valves, rupture disks, and univents. Their design must be such that they prevent loss of liquid through the vent in case of tank overturn. When loading and unloading is performed with the dome cover closed, the vents must limit the tank pressure to 3 psig (20.67 KPa). For emergency venting, they must open at not more than 3 psig (20.67 KPa) and close when the pressure drops below that. Additional venting may be required during emergencies and this can be provided by fusible plugs of at least 1.25 square inches (8.06 square centimeters) which operate at 250°F (121.11°C) or less (Figure 12-22). Liquid trucks carrying corrosives have similar venting requirements but generally use frangible disks.

Several important points must be remembered about relief valves. Relief valves are designed to operate whenever there is a pressure rise under any condition of vehicle rollover attitude. This means they function in any position and will discharge liquid if the tank is askew. The devices may not relieve the pressure if blocked or covered by liquid. They are designed to minimize the vapor pressure buildup by relieving vapor, not liquid. The tank shell can fail from fire exposure while the relief valve is operating. A vent fire should not be extinguished until the tank is cooled and the fire well under control. In fact, the valve should reset and extinguish the fire itself on an adequately cooled tank.

8. Overturn protection rails or guards must be provided for all fill openings, manholes, inspection openings, and relief devices. These devices must withstand a vertical load of twice the weight of the loaded tank. On many tanks they run parallel to the catwalk, while on corrosive carriers they often run across the tank at both ends or near the fill opening.

Figure 12-22. Dome cover with vents and a fusible plug on a flammable or combustible liquid carrier

Vapor Recovery Systems

In many areas, tank trucks unloading or loading gasoline are no longer permitted to exhaust the vapors from the tank being filled to the atmosphere. Thus, many tankers are being built with vapor recovery systems and set up for bottom-loading procedures. In essence, a vapor recovery system establishes a closed circuit. Vapors from an underground tank being filled are returned to the delivery vehicle and tankers being loaded exhaust their vapors into the storage tank. The transport tank is equipped with sensors to monitor the compartment's condition and shut down the system when the liquid level or compartment internal pressure exceeds predetermined levels. This protects against overflowing when filling from the bottom. Figure 12-23 shows a retrofitted tank with a vapor recovery system with domes for vapor movement and sensors to stop loading when the liquid reaches a given level or overpressure occurs. At many terminals, bottom loading cannot be initiated until the sensor connection is made. In Figure 12-24, the sensing unit is plugged into the tank and connected to the pump controls. Also note the remote emergency control for the internal fire valve on the right.

Figure 12-23. Vapor recovery system on retrofitted tank

Figure 12-24. Sensing unit plugged into tank and connected to pump controls

The problem area with vapor recovery equipped units is mainly with those that have been retrofitted (Figure 12-25). In a number of cases the rollover protection has been negated by the vapor recovery system. Some use construction that would be considered substandard by the fire service for withstanding an accident. This includes the use of plastic piping and connections made with radiator hose clamps and lightweight rubber tubing. Also of concern is the high concentration of vapors in an "empty" tank, which if subjected to severe fire conditions would build up in pressure rather quickly and could possibly rupture the tank with explosive force.

Figure 12-25. Unit retrofitted with a vapor recovery system; aluminum pipe runs from the loading/unloading station to the top of the compartments

Other Factors

Another item of interest is the number of compartments a unit has. Under the right conditions this could be determined by reading the capacities painted on the tank compartments. This cannot always be done nor can the data plate be checked. The other method of determination of the compartments would be to count the discharge outlets or valves. It may be necessary during a fire to protect an intact compartment from a burning one. This would be especially true on a tank of steel construction. Tank trucks designed to carry different products in the compartments will have a double wall or bulkhead between the compartments.

Bottom loading and unloading takes place at the unloading station. The internal fire valve control box is at this point. Several safety devices such as shear sections, fusible links or nuts, piping protection, and in some cases emergency controls for the fire valves, are in the area. Some tractors will also be equipped with power takeoff operated pumps for discharging the load.

The tanks cannot be filled to capacity, but must leave an area of 3 percent for product expansion. The placards on all four sides will be in place even when the tank is empty. The newer aluminum tanks melt easily under fire conditions and generally alleviate the potential for rupture. These tanks also are susceptible to heavy damage in collisions or turnovers.

Dangers and Precautions

Initial responding emergency response personnel must be ever-cognizant of signs indicating a further deterioration of the situation. These include any unusual color or volume of smoke; the growing intensity of the fire at the relief device, especially if water is being applied; a blue-red, nearly smokeless flame at the relief device; the audible pitch of the escaping vapors increasing; any unusual sounds of boiling or spattering; closed con-tainers subjected to flame impingement; and a bubble or blister in the shell of a closed container.

With spilled chemicals or corrosives, look for the vapor cloud or an intensely irritating smoke. Also, an unusual hissing or spattering can indicate problems. If the chemicals are flammable, all ignition sources must be kept from the area.

There are simple tactical operations that can prevent serious injuries and possible losses of lives:

1. If you don't know, don't go, it may blow. Do not expose personnel unnecessarily if the product has not been identified.
2. Use care in positioning of apparatus, personnel, and attack teams. Do not place at the ends of tanks, above the incident, downwind, or downhill.
3. Follow the decision making process for closed containers in Chapter 10.
4. Develop efficient withdrawal operations from the danger zone in your Standard Operating Procedure and train all personnel so it works effectively.

Fire Fighting and Control

Once an incident involving a tank vehicle has occurred, several types of situations may be encountered:

1. Hazardous materials not involved, but potential for danger in a tire, brake, or cab fire with a highly volatile load.
2. Hazardous materials involved, but no spill, leak, or fire as could occur with a jackknifed or overturned vehicle.
3. A spill or leak of the hazardous commodity.
4. A small fire such as a dome cover at a loading rack or a limited discharge leak fire.
5. A large spill around the tank.
6. A fully involved tank vehicle fire.
7. A loading rack fire.

Quick size-up must be made, including identification of the product. On tank trucks consider overhead or downed electrical wires, rescue of the driver, how to best protect exposures and not create an exposure problem by floating the product on top of applied water, how to stop the fuel flow and confine the situation, the extinguishing agents needed, especially if the product is miscible with water, the need for additional assistance to accomplish the objective such as advancing under protective fog cover to perform rescue or remove the tractor, and finally, the positioning of attack teams so they do not work against each other in holding the fire under the tank.

Fires not involving the load have been covered earlier. The main consideration is to stop the spread to the tank with cooling lines and by positioning attack lines to drive the fire away from the tank.

In a hazardous material carrier accident in which there is no fire or spill, take all precautions to eliminate a serious incident. Care must be taken when moving the vehicle not to add any undue stress, drop the unit, or cause a puncture. Often, unloading the tank will be required before moving it. The incident commander must exert authority and not allow actions that could endanger the safety of anyone in the area. Protective lines or foam blankets may be necessary to eliminate sparks.

Product Spill or Leak. When a tanker has a product spill or leak, several key points must be covered almost simultaneously. The driver and any victims must be rescued, all ignition sources removed, and containment of the spill initiated. Hose lines may be necessary to flush the spill away from the rescue operation, to dissipate vapors, or to provide protection to the rescuers and victims in case of ignition.

Diking may assist in exposure protection and in keeping the spill out of sewers or surface water (Figure 12-26), but on a large spill the arrival of adequate materials and equipment may not be soon enough to avoid

Figure 12-26. Diking to protect storm and sanitary sewers from product runoff. *Courtesy of Robert H. Shimer*

wide exposure problems. Do not use motorized equipment for diking around low flash point commodities. The objective of stopping the leak may be achieved with a plug, quick-setting patching compounds, or the EPA polyurethane leak control device presently under development. Large leaks, of course, could not be controlled this way, but would have to be contained or channelled to a holding area.

If the hole is near the bottom of the tank containing a liquid with a specific gravity of less than 1, control is possible by pumping water into the tank and floating the product on the water. Water would then leak from the opening. Extreme care must be exercised not to overflow the tank. The discharge of water into the tank must equal the flow rate of the leak to achieve this.

A leak at the dome lid can be wedged shut. Care must be taken not to get it too tight and release the lockbar. It may be best to secure the lockbar before stopping the leak. If the lockbar opens, the lid can open and release all contents above it. Use only non-ferrous tools and wedges. Restrict the use of road flares in the area of liquid or vapor travel. All precautions must be taken in the area to limit the effect of the incident until the product can be removed. It may be advisable to cover the spill with a foam blanket until it can be removed. Removal and cleanup may involve evaporation, absorption and removal of the earth or product used; covering the commodity; or physically pumping or vacuuming it up. As a general rule, flushing is no longer acceptable as a means of disposal unless there is an imminent threat of danger to lives or property.

Limited-Area Fires. A dome cover fire at a loading rack that uses a top-filling operation can be readily handled. The fire involves the fumes exiting the dome and once this fuel source is removed, the problem is controlled. Remember, do not move the vehicle being filled if the loading pipe is in the dome. The fill pipe should not be removed either, as both of these operations can spill or splash the fuel and cause a much larger area fire. Extinguish the fire by closing the dome covers. A straight stream or pike pole can be used to do this; however, if the fill pipe prohibits this, put out the fire with a dry chemical extinguisher, a fire blanket, or a CO_2 if available and wind is not a deterrent. Another method is to direct a straight stream across the dome opening and "blow" the fire away from the vapor source. Exercise care in doing this so that water is not directed into the tank, which could cause the fuel to overflow.

As in other incidents, consider rescue immediately. Fortunately, the loader will probably be the only person in the area. Outside loading racks may be adjacent to other buildings or essentially unexposed. An enclosed loading rack will provide added exposure problems. An immediate action at these fires is to stop the flow of fuel to the rack by closing valves or stopping pumps.

A small fire supplied by a limited discharge, as through an excess flow valve or broken hose, is little problem. Protect any exposed tank area with a cooling stream, especially on an aluminum tank. An attempt should be made to stop the flow using fog protection; if necessary, extinguish the fire with dry chemical. The spill area should be contained by diking with a charged hose line or salvage covers if other materials are not readily available.

Large Spill Fires. When a tank vehicle is engulfed in a large spill fire, the situation is serious and can deteriorate rapidly. Of prime concern is the contents of the tank and the safety of the drivers or those nearby. The decision making guidelines described in Chapter 10 will have to be applied. As fire fighting operations begin it will be important to protect exposures if the vehicle is parked adjacent to buildings, storage tanks, or other vehicles and to maintain the integrity of the tank. Be alert for any signs of water-reactive materials or an uncontrolled chemical reaction due to the heat from the fire. Initial straight streams should cool the tank and flush the burning material away from any trapped persons, exposures, and the tank. If lines are working from both sides of the tank, they should not work against each other, but through coordination work from one end to sweep the fuel and fire out and away to a safer location. Small fires in the cab or the tires should be extinguished and any points of flame contact on the tank cooled. Utilize the knowledge of the tank safety devices to control a fire being fed from the tank itself. Close the internal fire valve from one of the emergency cutoffs or from the fire box at the unloading station. These may have to be cooled before touching or closed with a tool.

During the early stages, reconnaissance should report the relief vent fire and how it is acting. If, under severe fire, the vent does not open, there should be concern, for it could be blocked or malfunctioning and lead to a pressure buildup. With the tank cooled, a properly operating device will close itself. The flushed fuel can be handled by controlled burning once the source has been stopped. In some cases, foam may be used to control and extinguish the fire. Be sure that adequate supplies are available before initiating application. Once the fire is extinguished, it may be necessary to assist in picking up the remaining product. This can be done by a

Figure 12-27. EPA recovery system with portable enclosed bag tank to hold product and vapors until properly disposed of

commercial firm in the cleanup business, utilizing a similar pump-equipped tank truck for commodity transfer, or using a vacuum truck. In an emergency a septic tank cleaner could be used, or the pump and collection bag recovery system currently undergoing development by the Environmental Protection Agency, shown in Figure 12-27. The system has a pump unit that can pick up a contained spill and vapors until properly emptied and disposed of.

Tank Vehicle Fires. A fully involved tank vehicle fire with the shell opened creates another serious problem. If the shell has been torn, punctured, or melted away, the violent rupture possiblity is removed on a single-tank vehicle. Many liquid carriers today have multiple compartments and there could be internal self-exposure to a compartment that has not been opened, which could cause a dangerous pressure buildup in a steel construction tank.

Size-up must identify the product and all signs of the situation deteriorating. If an acci-

dent has occurred, there may be a need to rescue the driver or those in other vehicles. Straight streams should be used quickly to cool enclosed compartments and flush burning product away from any involved or trapped people. Overlapping fog streams may be necessary to move in and free those involved. The difficulties of large water supplies to carry out these operations along transportation routes, freeways, or interstate highways becomes readily apparent. This becomes more critical if there are nearby exposures that must be protected. Water that runs off in the gutter is wasted. It must be used for extinguishment or keeping exposures wet down. Cooling water should be applied if necessary. With an open tank, water must be judiciously applied so that the product does not float and cause a running spill or gutter fire. The fire should be contained and kept out of sewers and open waterways by diking or pushing with hose streams. The use of foams to extinguish the fire may be the most practical operation to protect exposures such as an adjacent bridge, and expedite reopening of the roadway. Remember, however, that preparation for the safe removal of the unburned lading must be done. In other cases confinement of the fire to the tank area and controlled burning is a good solution since it alleviates many cleanup difficulties.

The following are important considerations during tank vehicle fire fighting operations:

1. When applying water to tank vehicles, the water should be lobbed onto the top and allowed to cascade down the sides of the tank, not directed straight at the tank where it will ricochet off and be wasted. The point of direct flame impingement and the vapor space—that is, the entire tank if empty—must be cooled first. Sweep the entire length of the tank from both sides, if possible, using 2½-inch or larger streams.

2. When extinguishing the fire with water, remember that metal will retain considerable heat, which could cause a reignition. Twisted and torn metal can form pockets where the fire remains and flashes back across the exposed fuel surface. All care must be used because unlike structural fire fighting where a single cooling suffices, these fires can and will easily reignite, since the remaining hot liquid will be giving off flammable vapors at a high rate.

3. Water must not be used on water-reactive substances.

4. Many components other than the commodity tank when exposed to the fire have the potential for rupture. These include the battery, air brake reservoirs, hydraulic lines, air conditioning units, and tires.

5. When extinguishing a fire with foam, use the proper percentage mixture. For example, a 6 percent foam solution used on at a 3 percent eductor setting will not create an adequate foam and in reality will be a waste of the foam. Cooling of tanks should be done prior to the foam application; otherwise the foam will be washed off or broken down by the water and its effect negated. Do not follow a foam line with a water line!

6. Maintain backup lines at all times to ensure protection for personnel should the unexpected occur, such as a burst hose line which might allow the burning fuel to overtake the fire fighters.

7. Alert other responding units to current layouts, direction they should approach from, which side of median or along the frontage road, extinguishing agent being used and needed, and their assignment as they approach.

8. Two attack groups may be required. The first group should perform immediate operations necessary to limit the scope of the incident, such as rescue, flushing fuel away, cooling tanks, and protecting exposures. The second group will lay lines for foam eductors, set up foam units, establish adequate water supplies, and mount an attack of proper magnitude for the knockdown and extinguishment. The task force concept is readily adaptable to this type of attack.

9. Use unmanned streams from hose holders or master stream devices for fires too dangerous to fight, such as when the only approach is from the end of the tank or there is evidence of a possible pressure rupture. They are valuable for adequate reach to cool a tank by delivering large droplets, for control burning, and when additional personnel is limited to provide cooling streams while available personnel are committed to handlines for approach and rescue or fuel control.

10. Maintain proper coordination of all units throughout the operation by effective command. This is the key to success.

Loading Rack Fires. Loading rack fires often involve the tank vehicle being filled. Since the racks are stationary, they should all be examined in preplanning, looking at access due to the congested area and vehicular traffic, the products involved, and the piping arrangement with the pertinent control valves in addition to items discussed earlier. Fires at loading racks are caused by human carelessness, equipment failures, and static electricity. The main ignition source is static electricity built up by allowing the fuel to splash down into the tank, by filters in the piping near the outlet, and by switch loading, which is loading a tank that previously carried a high-vapor-pressure product with a low-vapor-pressure commodity. The new bottom-loading racks with vapor recovery closed systems eliminate most of the static electricity problems.

The majority of problems arise at top filling operations where overhead piping arrangements permit filling stems to be inserted through the open domes into the individual compartments. With long fill stems that extend close to the bottom of a tank, filling rates can range from 100 to 600 gallons (380 to 2,280 liters) per minute with little splashing or vapor loss. To prevent overfilling, manual operating valves are self-closing or "dead man" type. These, however, can be wired or blocked, which would negate the safety feature. Many racks now use metered filling, which automatically halts the filling operation when a preset gallonage has been delivered.

Running Spill Fires. In a dome cover fire, discussed earlier, the problem of overfilling and creating a running spill fire is ever-present. This will occur if bottom closures have not been secured, there is an overflow due to human error, or a valve has been overridden. The possibility of a second fire should be considered because many racks are equipped with drains, occasionally from a berm-protected area. Thus, in a fire situation, it should be known where a drain discharges.

If the discharge is collected in a holding tank, then the capacity, location, and possibility of overflow from the tank should be learned.

The most essential procedure for a loading rack spill fire is to stop the flow of fuel to the rack. Good preplans will indicate where remote valves to accomplish this are located.

With the source of additional fuel controlled, the spill can be flushed from under the tank truck before its exposure would cause the situation to worsen by a release of its contents. Care must be exercised in flushing to avoid driving the spill into other exposures. Loading racks are often adjacent to sizable bulk storage facilities, sewers, or open drainage ditches or waterways. A dike may have to be built before flushing can be initiated.

With the fuel away from the vehicle any spot fires around it, and in the tires and cab, can be extinguished. Cool tires adequately, and the tank if necessary, before extinguishing any dome cover fires. Two important don'ts are:

1. If foam was used in extinguishment, don't wash it off; required cooling streams should be applied before the foam.

2. Water should not be allowed to enter open dome covers since the addition of any water will cause further overflowing and add fuel to the fire.

The burning fuel flushed away from the tank should be pushed into a safe area and control burned. If there is a severe exposure problem, it can be extinguished. Foam would be advisable since it would create a vapor barrier. All ignition sources, including heated metal, would have to be removed to avoid reignition. A cautious cleanup operation of the product should be started as soon as practicable. A combustible gas indicator can be used to determine when the area is safe and to monitor during cleanup.

At some loading racks, it will be necessary to control the product flow by closing a valve on the rack itself. Depending on the water supply and lines available, flushing and tank cooling operations should take place concurrently with the approach to the loading rack to operate the control valve. The approach will require at least two 1½-inch hoselines to provide ample protection to the officer who must mount the rack and close the valve. Extreme care must be used in positioning of the nozzles to protect this individual. Any break between the two fog patterns would cause serious burns to the officer. After controlling the fuel, the backing out of the lines must be done slowly and cautiously. The nozzles should be kept operating to cool exposed metal and maintain protection against the unexpected. The officer and nozzlemen should set the pace, not the crew on the hoseline.

Corrosives

There are some 75,000 flammable liquid carriers and over 110,000 tanks for acids and chemicals. Many of the former group contain special controls for temperature, pressure, humidity or contact with air.

The corrosive carriers are usually identified by the external rings around the tanks (Figure 12-28), however, some are insulated and these do not show. Those moving stronger corrosives are equipped for air-pressure top unloading. If the unloading pipe extends externally to the bottom of the tank, it is valved at the top of the tank and at the end, or valved at the top and capped at the end. The acid tanks tend to have a smaller capacity, indicated by a smaller diameter, due to the weight of the acids. Nonstainless steel tanks are rubber or resinous lined with a single unobstructed compartment for ease of cleaning and lining. Many have a 35 psi working pressure. Stainless steel may also be used for piping, pumps, and braided unloading hose. A spill dam or splash guard will be around the loading position with a drain to keep the material from running down the side of the tank and corroding the tank shell.

The safety devices are essentially the same as described for other liquid carriers.

Operations Involving Corrosive Carriers

The main operations with corrosive carriers will be to contain the spill, generally limit vapor production, and minimize exposure to contact or vapors. In most cases, technical assistance will be necessary for proper neutralization.

It must be kept in mind that all acids are not corrosive. Corrosives are most of the inorganic acids and the alkalis. The problems vary, depending on the specific material, but the following are some general hazards that can be expected. Although corrosives do not burn, they give off hazardous vapors when exposed to fires, cause tissue burns on contact, give off corrosive vapors, act as oxidizers and can ignite combustibles, react with water, and can react in closed containers and cause a rupture.

The best procedure with these materials is to dike the product and then neutralize it. These commodities must be kept out of the sewers, since they will react with the water and cause explosions that would damage the sewer system. Small spills can be handled without a great deal of difficulty; however, a large spill may require truck loads of sodium or potassium bicarbonate (generally for weak acids under 20 percent concentration), lime, or soda ash. Diatomaceous earth may be mixed in to act as an absorbent. For small spills, absorbent clay will be required in about two volumes to one.

The fuming inorganic acids produce very irritating dense vapor clouds. Neoprene or vinyl acid suits and breathing apparatus are required since prolonged exposure will destroy the skin. In this category are sulfur trioxide, hydrofluoric, and chlorosulfonic acids.

Essentially, two operations are possible for first-arriving apparatus. The first is to blanket the diked material with high-expansion foam and obtain assistance and materials for neutralization. For some acids the vapors from a diked spill can be contained with the use of a white mineral oil. The other possible operation is to flush the material. This is advisable when high expansion foam and neutralizers are not available and only then where no additional damage will occur. It may be best to obtain advice from the manufacturer before making the decision. The procedure is to flush the product to a control area and flood it until safely diluted. This can only be done with fog streams upwind, using breathing apparatus and all protective clothing, from a maximum distance since the product will splatter and create a large vapor cloud. There must be little or no exposures downwind.

Liquefied Gas Carriers

The liquefied gas carriers are grouped into two categories based on the manner in which the gas is liquefied. They are liquefied by increasing the pressure or by reducing the temperature. When liquefied by reducing the temperature, the gases are referred to as cryogenics.

Pressure Liquefied Gases. The most common product used in this form is liquefied petroleum gas. Considerable amounts of anhydrous ammonia and chlorine are also carried over the highways.

The tanks are built to DOT specifications for pressure vessels. They are a single compartment and may have up to four baffle plates to reduce surges during transit. They are manufactured of quenched tempered steel and have a burst pressure four times that of the relief valve setting. The tank shells are approximately ¼ inch thick.

The safety devices on these units are similar to those described earlier and include: internal fire valves, fusible links or nuts, emergency controls for fire valves, collision and breakaway protection for plumbing and valves, rear end protection, and safety relief valves.

These tanks are carried approximately 85 percent full to maintain a vapor space and allow for temperature change expansion. In order to properly fill the tank with varying temperature changes, four additional devices are found on these units. A thermometer is provided for the product temperature, a pressure gauge for internal pressure, and two devices to measure the capacity, a fixed-level gauge and a rotary gauging device.

The tanks can be unloaded by pressure differential, but normally use a PTO (power take-off) pump mounted on the trailer. These gases are transported and stored in the liquid state and most often used in the gaseous state since they increase 300 to 500 times in volume as they vaporize. Anhydrous ammonia expands over 800 times from a liquid to a gas.

Fire fighting operations for these tanks are generally known. If the fire occurs in piping or hoses during a loading or unloading operation, stop the flow of fuel by closing the internal fire valve and the valve to the storage tank. While this is being done, protect any area of flame impingement on the tank, especially the vapor space, and attempt to drive the fire away from the tank and any exposures.

If the vehicle is involved in an accident or adjacent to a storage area fire where it is exposed due to fuel running beneath the tank, the same procedures apply. Keep the tank cool and direct streams under the tank to push the fire away. If this or any exposure fire causes the relief valve to operate, begin cooling the vapor space immediately. When sufficiently cooled, the valve will reset. If cooling is not adequate, the audible pitch and intensity of the relief valve fire will indicate that changes are necessary.

If a tank is punctured, there can usually be no control of the fuel source. The major problem here is that of flame impingement back on the tank itself. The answer is large cooling streams on the area exposed to flame impingement. Water running over the metal surface of the tank will keep the tank shell at 212°F (100°C) or below, which is a safe temperature.

A similar problem occurs when a unit

overturns. This also could bring a self-imping-ing fire on the tank. The major difference is that in this case the relief valve will not assist in relieving the buildup of vapor pressure as it will in the preceding case. The relief valve will be discharging liquid—fuel for the fire—which does not substantially reduce the vapor pressure. Thus, the chance for rupture in-creases if there is not adequate cooling and flushing. The puncture fire, if in the vapor space, would not only relieve pressure, but would not be as hot a fire as one from a punc-ture in the liquid area.

Do not attempt to relieve the pressure buildup in a tank of this type by piercing the shell. This can cause a violent rupture. Be alert for blisters or bubbles starting in the shell. Watch for severe stresses caused by impact or crushing which could cause failure of the tank under thermal or mechanical shock. Don't move or drag an involved tank.

Tanks that are liquefied under pressure fail either by hydrostatic destruction or by an impinging fire. An overfilled tank or a faulty or obstructed relief valve can lead to hydro-static destruction since the valve cannot pro-vide the venting required. As the fire heats the tank the vessel will become entirely full of product. As the heat continues, the tank will burst about one-third of the way from one end. The major portion of the tank will stay in place and the ends will rocket.

An impinging fire weakening the tank shell while there is an internal pressure build-up will eventually create a blister and burst the shell if not properly cooled. This causes instantaneous release and ignition of the va-por with an explosive force and fireball. For these reasons the level of the liquid is impor-tant and can be determined by the frost line so that the area above can be cooled.

When vehicles carrying commodities that are liquefied by pressure sustain a leak with no ignition, the operations required to handle the situation will depend on the prod-uct and its characteristics. If flammable, all ig-nition sources must be removed. Remember

that for some of these products a leak in the vapor space will not be visible. Use gas detec-tion equipment if a leak is suspected. Water can be used to help disperse the vapors of liq-uefied petroleum gas, but should not be used on a leaking chlorine tank because it will form hydrochloric acid and further eat the tank and most other materials it contacts. The vapors of chlorine, vinyl chloride, and heavy concen-trations of anhydrous ammonia are all irrita-ting and will cause injuries. In incidents with these products, evacuation may be required. An anhydrous ammonia leak can be con-trolled with water due to the absorption rate as long as the runoff does not get into a water-way or can be controlled. Anhydrous and wa-ter form ammonium hydroxide, which will kill fish. The above examples indicate the variety of difficulties that can be encountered with a leak. The best procedure is to identify the product and obtain technical assistance immediately.

Cryogenics. The second type of liquid carrier is that hauling material liquefied by re-ducing the temperature. These commodities are transported at very low temperatures and are referred to as cryogenics. The tank used is a well-insulated double-walled container very similar to a thermos bottle. The outside shell can range from 3/16 (.48 cm) to 1/2 inch (1.27 cm) in thickness. Reinforcing rings are used for additional strength on those using the thinner steel. The insulated area between the shell and tank is kept in vacuum and pro-tected from overpressure by frangible disks. The liquid container on the inside is also pro-tected by relief devices. All tanks will have a relief valve set at 25 psi (172.35 KPa) or less and a frangible disk set somewhat higher. Many trucks will have a second relief valve set be-tween these two to avoid bursting the disk, which would dump the entire load. Normal pressure during transit will run between 2 and 8 psig (13.788 KPa and 55.15 KPa).

These inner tanks are constructed of alu-minum, copper, chromium nickel, stainless steel, or nickel alloys so that they do not become brittle and crack under the temperature extremes. The tanks can be as large as 13,200 gallons (49,962 l). Table 12-1 shows many of the gases that are transported in cryogenic form and indicates the hazards of each.

During an emergency situation the cryogenics have three basic hazards that emergency response personnel must contend with:

First, the extreme cold of the materials can cause freeze "burns" or frostbite to body tissues. Delicate tissue such as the eyes can be damaged by the cold gases before there would be damage to the skin tissue of the hands or face. Contact with the liquid—the cold gaseous product—or metal cooled by these liquefied atmospheric gases should be avoided. Flesh will adhere to uninsulated cold metal such as exposed pipes at the unloading compartment and tear as it is withdrawn to cause painful injuries.

Second, there is a tremendous liquid-to-vapor ratio with these commodities. As the liquid boils off, it will increase in volume 600 to 800 times. Many of the vapors will not be seen, which extends the hazard beyond the visible cloud. The visible cloud is the condensed moisture in the air, which creates a fog cloud due to the extreme cold of the liquid or the gas.

Third, there are the hazards of the particular gas itself as listed in Table 12-1. Remember that due to their vaporization ratios these gases can displace air to the point where there is not sufficient oxygen to support life. Fluorine is also a very reactive material.

Suggested procedures for emergency response personnel include the basic operations of approaching from high ground and with the wind at your back, evacuating the immediate area, and securing assistance to divert traffic and technical assistance on how to handle the product and the container. Hose lines should be placed for exposure protection from any fire, if there would be ignition, and to divert vapors from the exposures. Of course, any fires or ignition sources in exposures should be extinguished to remove all possibility of the vapors lighting off or reigniting. Water is not needed to cool cryogenic tanks because of sufficient insulation, but may be needed to protect the outer

Table 12-1.   Chemicals Transported in Cryogenic Form

Product	Temperature		Hazard
	°F	°C	
Argon	−302	−185.6	Asphyxiant
Air	−318	−194.4	
Carbon Dioxide			Asphyxiant
Carbon Monoxide			Flammable Toxicant
Ethylene	−155	−103.9	Flammable
Fluorine	−306	−187.8	Oxidizer
Helium	−452	−268.9	Asphyxiant
Hydrogen	−432	−257.8	Flammable
Krypton	−244	−153.3	Asphyxiant
LNG	−258	−161.1	Flammable
Methane	−259	−161.7	Flammable
Neon	−411	−246.1	Asphyxiant
Nitrogen	−320	−195.6	Asphyxiant
Oxygen	−297	−182.8	Oxidizer
Zenon	−162	−107.8	Asphyxiant

shell from losing its strength and subsequently causing a loss of the protective insulation. If a substantial area of the insulation was lost due to an unusual accident ripping off the casing and insulation and exposing 10 percent or more of the inner container, the recommended procedure is to evacuate and withdraw. This action is suggested because the pressure relief devices are designed without allowances for disruption of the integrity of the insulation. Under these unexpected circumstances there is a remote possibility of tank rupture even though the relief valve and frangible disk have functioned properly. Due to the low temperatures of the cryogenic gases, water should not be applied to the bare inner container surface since it would actually add heat and cause the material to boil and vaporize faster. Hose streams directed on the bare liquid container will only worsen the situation.

During operations around a cryogenic spill, leak, or fire, restrict the number of operating personnel to a minimum and remove all others. The best control method is to shut off the flow of commodities or stop the leak. Water may have to be used to keep the frost down during this operation, but care must also be used with water not to freeze the relief device solid. A leak may be stopped by freezing it shut. Watch the container jacket for frost spots, which could cause the carbon steel to become brittle and fracture. Avoid any impact or shock to these areas. When the tank pressure gauge nears the normal safety device setting of 17 psig (117,198 Kpa), an attempt should be made to move the tank if possible to an isolated area and relieve the excess pressure through the manual blowoff. Where there is a leak or the fire is controlled, request another tank truck in order to transfer the contents from the involved cryogenic tank.

For personal safety do not get any cryogenic liquid on your clothing. If there is an accident such as a splash, remove the clothing immediately. Remember the frostbite danger to hands and feet from the liquid and the *vapors*. Do not damage your eyes by freezing the liquid in them. Also, refrain from touching the metal container or piping as the skin will freeze to the cold metal.

Spills of hydrogen should be confined and allowed to dissipate by evaporation. Use hose streams discriminately to dissipate vapors. Hydrogen has a flammable range from 4 to 74 percent and will auto-ignite in the proper air mixture at 1000°F (537°C).

Since hydrogen burns with a nearly invisible flame and has a low heat radiation, it is possible to approach a fire without feeling a significant amount of heat. However, contact with the flames will cause severe burns immediately.

Although liquid oxygen does not burn, it vigorously supports the combustion of the material around it. Use large gallonage fog streams to cool the burning materials below their ignition temperature. Remember that since liquid oxygen is an oxidizing agent, fires it is supplying cannot be effectively extinguished by blanketing with foam, CO_2, or dry chemical agents. Driving apparatus, walking, or dropping equipment on liquid oxygen impregnated roads of asphalt or oil-soaked concrete or gravel can cause explosions. If water has been applied on a concrete pavement where liquid oxygen has seeped between the cracks, the pavement can blow up because of the expanding gas and pressure buildup under the forming ice.

A cryogenic fluorine tanker provides additional hazards due to the reactiveness of the substance. Although fluorine is nonflammable, it is classed as a flammable since it reacts with most combustibles and causes their ignition. It is a strong oxidizer and in high concentrations is toxic; low concentrations cause lung edema and corrosive thermal burns. Because water directed on fluorine will cause violent reactions, extinguish secondary fires after the fluorine has been consumed. Use unmanned streams to cool containers or extinguish fires exposing fluorine containers.

Use protective clothing (neoprene) known to be impermeable to fluorine and hydrofluoric acid, which it forms in moist air, and positive pressure self-contained breathing apparatus. Although the NFPA 704 identification is 4, 0, 3, W, the DOT classification is flammable gas.

COMPRESSED GAS TANKS

Although compressed gas tank trucks are in service they are not as common as the liquefied gas carriers (Figure 12-29). In a compressed gas unit, the product is under pressure but still in the gaseous state. This is the same as the air tank used on breathing apparatus. In essence, by doubling the pressure, the amount of gas in a container can be doubled. However, by liquefying, several times the volume can be placed in a container, which is the reason for wider use of liquefied carriers. The major problem with the compressed gas carrier is that under fire conditions and flame impingement, there is nothing inside the tank to absorb the heat. Because the tank is all vapor space, there will be a rapid pressure buildup inside and, with only the shell to absorb the heat, a very short time before it will begin to weaken. Fortunately, most of the compressed gases will not burn, except for hydrogen and some refrigerant gases. A few will support combustion if fuel is available.

The emergency response team will be concerned with assisting in dispersing the gas, keeping people out of the vapor area to avoid asphyxiation, and, in those cases where there is a fire using large quantities of water, keeping the entire tank cool and intact. If water is not available to cool the tank or drive the fire from under it, personnel must be withdrawn and the area evacuated in anticipation of a tank overpressure and rupture.

Figure 12-28. Close-up of top of a corrosive carrier showing the rupture disk, relief valve, and rollover protection

Figure 12-29. Compressed gas unit carrying gaseous hydrogen

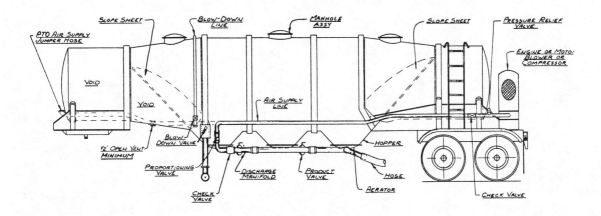

Figure 12-30. Diagram of major features of a dry bulk barrier

DRY BULK CARRIERS

Dry bulk carriers, (Figure 12-30) haul a variety of products. In the hazardous materials realm there are fertilizers including ammonium nitrate and plastic pellets. Some units can also move liquids and may carry phosphoric acid or fertilizer solutions.

The tanks are constructed of aluminum, steel, or stainless steel. Capacities range up to 1,500 cubic feet (42.45 cubic meters) or 11,200 gallons (42,392 liters). These units can be gravity unloaded through 10-inch (25.4 centimeter) plenums, but normally are air unloaded. Some units extend the air lines to the top of the tank and use a top aeration assembly to create a venturi inside each tank hopper. Others utilize the air through bottom assemblies.

The units have their own unloading air system. This may be a PTO-operated blower mounted on the tractor or an engine-powered blower or air compressor at the rear to the trailer. Those with the auxiliary engine will, of course, have an additional fuel supply at the rear of the trailer. The units are equipped with safety relief valves, pressure gauges, and check valves as safety features. The tanks are constructed for working pressures of 15 or 25 psi (103.4 or 172.35 KPa).

The major problem encountered with these units will be the toxic products of combustion given off as the fertilizers or plastics burn. Downwind evacuation may be necessary. Conventional means can be used in extinguishment with self-contained positive pressure breathing apparatus.

MOVING TANK VEHICLES

After the fire or leak is controlled, it is necessary to move the truck. Often the tanker has rolled over and must be uprighted. The first thing to remember is that the towing service personnel are not hazardous material experts. In many cases the fire department will have to oversee operation for safety.

Although some steel tanks can be up-righted while loaded, it is not recommended except for the pressurized liquid carriers, which must be uprighted to unload. Liquid carriers generally should be unloaded before being uprighted, which requires obtaining another tank to off load into. Figure 12-31 shows load being transferred from an over-turned tanker by means of a vacuum pump. Since many tractors are equipped with PTO pumps, the transfer can be accomplished without difficulty. The area should be check-ed for flammable vapors before the tractor is brought into the area. Another concern is whether the commodity to be transferred is compatible with the construction and/or the prior lading of the unit. Some products can only be placed in stainless steel or lined tanks. An oxidizer could not be placed in a tank that formerly carried a flammable liquid. Other noncompatible chemicals could cause severe unwanted reactions. All safety precautions should be taken during the transfer, including bonding the tanks and charged hose lines in position.

To upright a large tanker, two heavy tan-dem wreckers should be used as a minimum. Since it is best to upright the unit as one piece with the tractor and trailer connected, they should not be separated nor should an over-turned unit be broken in two. An aluminum tank is easily cut or damaged with the use of a wire rope or sling. Extra care must be used in working with aluminum tanks. Under no cir-cumstance should they be uprighted while loaded. Figure 12-32 illustrates an incident in which the product was not transferred, which created a dangerous spill and ignition hazard during the uprighting operation. In some cases a foam blanket may be advisable to avoid sparks. Never drag the tank, which could cause additional damage or sparks.

Figure 12-31. Load being transferred from overturned tanker. *Courtesy of Robert H. Shimer*

The roadway then must be properly prepared before it can be opened to the public. This includes removing debris, picking up spilled product, removing slick spots, and in some cases, decontamination of the area.

All fire departments are confronted with the problem of possible fires in trucks carrying hazardous commodities. Special consideration has to be given to loading, unloading, and when the trucks are involved in accidents.

Every person in the fire department should thoroughly realize the individual responsibility in connection with a fire of this nature in which a single error in judgment may result in disaster. Such errors can be basic, such as the misuse of a hose stream, or highly technical and almost undetectable, such as an incipient crack in a container.

Only through a thorough knowledge of trucks and their loads, preplanning for the situation, and hours of strenuous training can we be well prepared for a hazardous materials truck incident and then we must continue to anticipate the unexpected.

Figure 12-32. Dangerous spill and ignition hazard created by failure to transfer load before proceeding with uprighting operation. *Courtesy of Leonard King*

QUESTIONS FOR REVIEW AND DISCUSSION

1. What types of truck incidents have occurred recently in your locale?
   Discuss pertinent points of each incident, such as where it occurred and why.

2. List eight tactical considerations in truck hazardous materials incidents and how they apply in your area.

3. What are the major safety devices on tank trucks?

4. What danger signals should be watched for at truck fires?

5. List several important considerations in tank truck fire fighting.

6. Make a study of corrosives transported in your area and determine what would be necessary to neutralize a spill.

7. When and why is water applied to a cryogenic tank? When is it not applied?

8. Describe the precautions to take at a cryogenic incident and why these are important to emergency response personnel.

9. Visit a loading rack or truck terminal in your area. Describe the features you observe that will be important in controlling a spill or fire at that location or around the vehicles you observed.

10. List several unanticipated events that have occurred at truck fires you were at, or have knowledge of. What other unexpected events can happen?

CASE PROBLEM

About 9:00 A.M. on a Wednesday morning in early April, a tractor-trailer tank truck carrying 6,000 gallons (22,710 liters) of acetone cyanhydrin slid on the rain-covered pavement, jackknifed, and rolled over. The truck was eastbound on the Heatherton Freeway, which is elevated on a built-up earthen embankment. With the rain and heavy traffic, an oncoming car struck the rear of the tanker. This resulted in a personal injury accident. Drainage for the area is by means of storm sewers from the elevated highway, which drain into a small creek paralleling the freeway.

The freeway at this point passes a small community located to the south, adjacent to the creek. The temperature is 70°F (21.11°C) and the wind is from the northwest at 25-30 miles (40-48 kilometers) per hour due to the thunderstorms.

The makeup of the community in the vicinity of the wreck was mainly residential, single family and smaller apartment buildings. A commercial area and shopping center are approximately 0.5 mile (0.8 kilometer) away. There are about 3,500 people in a 1 mile (1.6 kilometer) radius.

The local fire department has two pumpers, a tanker, a light attack unit, and an ambulance. They have 20 gallons (76 liters) of AFFF. There is no water along the freeway. The street beyond the creek, which is within the city limits, has a 6-inch (15.24 cm) main with a flow capability of 1,000 gallons (3,755 liters) per minute. There is wide hydrant spacing with one 700 feet (215 meters) east and the other 800 feet (245 meters) west. There is a fence running along the creek on the freeway side. The nearest access to the freeway is 0.75 mile (1.2 kilometers) from the accident site.

1. What operations should be initiated, including the priorities to minimize the impact of the situation?

2. Will emergency response personnel need to take special precautions?

3. Are outside resources or assistance required? Whom do you recommend? Will they have to be cautioned as to proper protection?

4. Would you evacuate? How?

ADDITIONAL READINGS

"Acid Tanks Are Special." Modern Bulk Transporter, February 1975.

Analysis and Summary of Accident Investigations 1973-1976, Washington, D.C.: Department of Transportation—FHA Bureau of Motor Carrier Safety, 1977.

Bahme, Charles W. Fire Officers Guide to Dangerous Chemicals, Boston: National Fire Protection Association, 1972.

Baker, Charles J. "Fighting Truck Cargo Fires," Fire Chief, April 1969.

"Chemical Transportation." Modern Bulk Transporter, February 1975.

Coleman, Ronny J. Management of Fire Service Operations. North Scituate, Mass.: Duxbury Press, 1978.

Gerton, Myron P. Hazardous Materials—A Guide for Fire Fighters, Denver: Division of Highway Safety, 1974.

R. M. Graziano's Tariff Publishing Hazardous Materials

Regulations of the Department of Transportation, Washington, D.C.: Bureau of Explosives, Association of American Railroads, 1976.

Handling Liquefied Hydrogen Tank Truck Incidents, Union Carbide Corp., Linde Division, 1976.

Lesson Plans, Denver: Colorado Committee on Hazardous Materials Safety, 1975.

Recommended Practice on Nomenclature and Terminology of Tank Trailers and Containers, Washington, D.C.: DOT - FHWA, April 1972.

"Safe Handling of Liquid Oxygen, Nitrogen, and Argon." Union Carbide Gas Products, Fire Engineering, 1972.

Special Fires I College Park: University of Maryland Fire and Rescue Institute, 1973.

Tank Vehicle Fire Fighting, Boston, Mass.: National Fire Protection Association.

13
Pipeline transportation

The term *pipeline* can be interpreted in many different ways. For the purposes of this text, a *pipeline* is defined as the pipe, fittings, pumping apparatus, measuring techniques, storage facilities, and maintenance requirements to transport a commodity from its point of receipt to its point of delivery. The present state of technology, under this definition, permits the transportation of both liquids and gases. However, experimentation is now going on that will permit the pumping of very thick liquids and some solids, which will make pipeline transportation available to a larger body of users.

Pipelines transport numerous liquid materials with widely varying hazards to emergency response personnel. The liquids range from heavy crude oils to liquefied petroleum gases, to refined petroleum products, to anhydrous ammonia. In addition, both natural and manufactured gas are transported by pipeline.

HAZARDS OF A RUPTURE

The heavy crude oils and other heavy hydrocarbons present less of a hazard after rupture because of their slowness in vaporizing. On the other hand, many of the refined petroleum products, such as gasoline or liquefied petroleum gas and anhydrous ammonia, will vaporize upon release. Gases, of course, present an immediate hazard upon release.

Vapor clouds can thus travel into adjacent areas, find an ignition source, and ignite. Depending upon the population density of the area, evacuation may be impossible before the vapor cloud is ignited. Dispersion of the vapor cloud will be dependent upon the following:

1. Wind direction
2. Pipeline pressure
3. Flash point of product being pumped
4. Size of leak
5. Vapor density of product being pumped

In its 1976 Annual Report, the National Transportation Safety Board (NTSB) reports that:

> Accidents involving highly volatile liquids have thus far been confined to

areas which were not densely populated, but these accidents have a high probability of occurring in heavily populated areas traversed by these liquid pipelines. A rupture in a densely populated area will most likely result in an accident of catastrophic proportions since the monitoring equipment of most liquid pipelines is not sufficient to promptly identify the location of the rupture and to isolate rapidly the failed section to prevent additional liquids from being pumped to the rupture location.[1]

This means that even after a leak is detected, there will be a time delay in determining which valves to close to isolate the area of the break. Then, the product in the line between the break and the valve has to be bled out. When dealing with pipelines up to 48 inches (1.2 meters) in diameter and many miles (kilometers) long, there is a large amount of fuel for the fire. In addition, there is the problem of structural fires, which are set by igniting the spreading vapors.

SAFETY RECORD

Damage to underground facilities by excavation equipment is the largest single cause of pipeline accidents and leaks. The NTSB made over 30 recommendations for correcting this cause of pipeline rupture.

One of the major recommendations deals with a one-call notification system by which a contractor can notify all of the utilities of the plan to perform excavating work. Then, all of the affected utilities can mark the location of pipelines, gas and water mains, sewers, electric lines, and telephone cables. At that time, special precautions can be discussed to emphasize the need for safety. At the present time there are 85 of these one-call systems operating in 36 states.

Another recommendation is to require the capability for rapid shutdown of the failed

section of pipe. This would reduce the hazard to both life and property. To quote NTSB again:

> Sixteen Board recommendations pointed out the need to revise requirements for action by a pipeline operator in an emergency. Operator actions were unprecisely defined. The Department of Transportation now has revised its regulations to require the operator to improve public and employee education programs, advise the public how it should act in emergencies, developed advance plans for the operator's employees to follow in an emergency, and make police and fire units aware of the operator's emergency plans.[2]

PIPELINE GENERAL INFORMATION

In 1973, the trade association of the general contractors estimated that between 1973 and 2000, the United States will match all construction that has taken place throughout history.[3] During 1972, 41 percent of all accidents on gas transmission pipelines were caused by excavation damage, while 71 percent of the distribution system accidents involved outside force damage.

Pipelines can be divided into two general classes by the end use of the product: transmission lines, which move the product from the producers to the retailers, and distribution mains, which move the product from the retailers to the users. In the United States there are more than 600,000 miles of gas distribution mains and 335,000 miles of gas transmission lines.

Population densities are greater for the distribution mains because the products are needed for public use. On the other hand, many of the transmission lines go through unpopulated areas.

The National Transportation Safety Board reports that one of the major obstacles to overcome is the difficulty in determining the actual location of the pipeline. They state:

> If pipeline operators are to fulfill their postnotification responsibility adequately, equipment should be available to detect accurately and economically the location of underground facilities. The several commercial units currently available have limitations; they detect steel pipes most effectively. A device is needed to detect effectively other material, such as plastics, cast iron, and ductile iron. Some operators have buried steel wire with plastic piping so that the location of the plastic line can be sensed by currently available equipment.
>
> Public Technology, Inc. of Washington, D.C., a private, nonprofit corporation established to apply available technology to the solution of city, county, and State problems, initiated, in cooperation with the National Aeronautics and Space Administration (NASA), Office of Technology Utilization, a review of available devices, including military,

that could help locate underground pipes and conduits. Public Technology concluded after the review that a downward-looking radar system seemed to be the most promising approach available.[4]

An example of a pipeline locator is shown in Figure 13-1.

Pipelines for the transmission of gas can be as large as 50 inches (127 centimeters) while those for liquid petroleum products can be 40 inches (1 meter) in diameter. With operating pressures as high as 1,200 psi, there is a great danger should a leak occur. Pipelines are located from 30 to 36 inches (76 to 92 centimeters) underground.

Figure 13-1. Electronic equipment for locating pipelines

LIQUID PIPELINES

There are three different types of pipelines which are used to move the product from the oil well to the end user:

1. Gathering system
2. Trunk line system
3. Distribution system

The three systems are illustrated in Figure 13-2

The gathering system transports the crude oil from the well head to a central distribution system. The pipes are from 4 to 12

inches (10 to 30 centimeters) in diameter and may be large enough to require several pumping stations.

The trunk line system is used to bring the crude oil to the refinery or to a storage area in preparation for shipment by one of the other modes of transportation.

Finally, distribution systems carry the product to the market place. The major products shipped as a liquid are gasoline of various grades, jet fuels, kerosene, and heating oil. In addition, liquefied petroleum gas is shipped

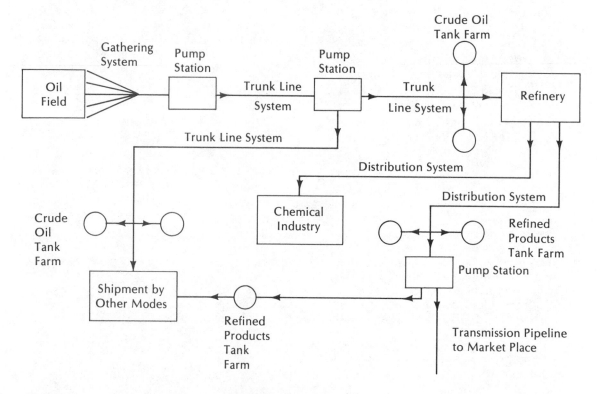

Figure 13-2. Liquid petroleum products transported by pipeline

for use as a fuel as well as in manufacturing processes. These pipelines usually start out as large-capacity systems and branch into smaller ones as deliveries are made at various locations.

Petroleum Product Pipeline System Operation

Crude oil starts its journey to the marketplace in the oil wells drilled into the earth's crust. Some are on land; others are located in the ocean, miles (kilometers) offshore. As the product is brought to the surface, the natural gas is taken off and processed separately, while the crude oil goes to heat treaters.

With the use of heat and chemicals at the heat treaters, water is removed from the crude oil. Now, the product is ready to enter the pipeline system.

As most fire service personnel are aware, a liquid resists moving through a conduit. Water being forced through a hose line suffers a pressure drop as it moves along. This pressure loss is termed friction loss, and represents all of the pressure drops which are dependent upon the size of the carrier, amount of product being pushed through, the material the carrier is made of, and the roughness of the inside of the carrier. The petroleum products going through the pipeline suffer the same pressure drops. This requires that prime moving equipment be installed in the line to keep the product flowing. Pumps are used as the prime movers. The spacing of the pump stations on the pipeline is dependent on the pressure needed and the hydraulic losses that must be overcome. A discussion of pump station spacing is contained in a following section of this chapter.

Figure 13-3. Petroleum product tank farm storage. *Reproduced with permission of Chevron U.S.A.*

Crude oil storage tank farms are installed at a number of places along the line. These are used for testing the product in transit before it reaches the refinery and serve as storage facilities. Some are used for sorting, measuring, rerouting, or for temporary holding. Figure 13-3 illustrates a typical petroleum product tank farm.

Leading from the pipeline to the tank farm is a manifold that permits directing the product to the proper storage vessel (Figure 13-4). A manifold (Figure 13-5) allows for:

1. Pumping product through main line with tanks inactive

2. Routing production from field into any tank

3. Routing main line product into any tank

4. Transferring product from one tank to another

5. Pumping product from any tank into main line

6. Isolating all pumps and tanks while the pump upstream pumps right through

7. Injecting product into main line stream as it passes through

By the nature of a pipeline system, the only way to get a product out of the line at the downstream end is to put an equal volume in at the upstream end. Therefore, there is always a balance of the product in the system.

In order to keep the pipeline operating as

close as possible to its maximum capacity, the pipeline company checks with the shippers in advance for their estimates of the amount of product to be moved in the coming month. The pipeline company will schedule pumping capability so the product will arrive at its destination on schedule.

As the product passes intermediate take-off points, the stream is divided in an operation called *stripping*. The remaining product continues to its destination.

A pipeline dispatcher runs a large portion of the system by remote control. Monitoring of the line pressure, rates of flow, and gravity of the product are all done from the control room. From this point there also can be remote control of the valves to add or take off product to and from the line. It is at this point that emergency action in the case of a line break takes place.

Figure 13-4. Petroleum product manifold

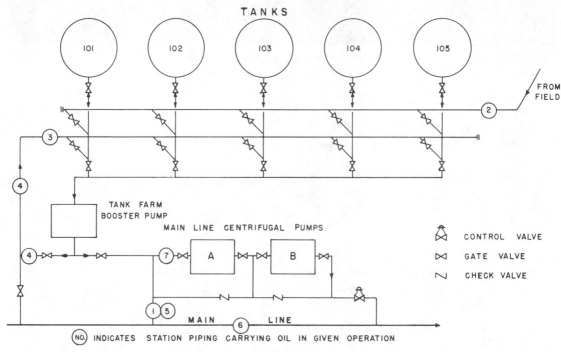

Figure 13-5. Schematic diagram of product manifold. *Courtesy of the University of Texas, Petroleum Extension Service*

Product Delivery Cycle

At any time the pipeline could contain many different products simultaneously. The pipeline companies usually work on a 10-day cycle, meaning that during this period they will ship at least one batch of every petroleum product. Figure 13-6 shows a typical product shipping cycle. The exact injection date and hour of the particular product is noted and its delivery date is projected. As the product gets close to its arrival point, a gravity sensor in the line signals the arrival of the shipment. The interface between products ranges from a few barrels to a couple of hundred. Sometimes a physical separator such as a rubber sphere is used, but usually there isn't any separator.

Verification of the arrival of the shipment is made by examining a sample of the incoming batch. Generally, the color and appearance of the product serve to verify the change from one shipment to another.

As an example of the activity on a pipeline, the Colonial pipeline from Houston, Texas, to Linden, New Jersey, will be used. This line is 1,531 miles (2,450 kilometers) long and, counting lateral lines, runs almost 3,000 miles (4,800 kilometers). At a time when deliveries are being made in Linden, and the entire line is in service, a total of 300,000 horsepower is engaged in moving upward of 1,000,000 tons (900,000 metric tons) of product at about 5 miles (8 kilometers) per hour.

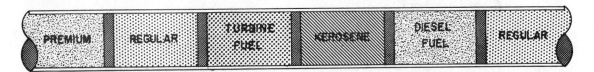

1. Premium-regular commingling can be split both ways with no loss or degradation.
2. Turbine fuel commingling usually must be cut out and blended to regular.
3. Turbine fuel-kerosine commingling can be cut directly to turbine fuel within certain limits.
4. Kerosine-diesel fuel commingling can be cut directly to diesel fuel within certain limits.
5. The kerosine buffer batch between diesel fuel and turbine fuel is cut both ways.

Figure 13-6. Typical products cycle. *Courtesy of the University of Texas, Petroleum Extension Service*

GAS PIPELINES

Gas pipelines function in much the same way as the liquid lines. The product originates in the gas fields of the southwest United States and is brought to market by the pipeline.

Whereas crude oil is first sent to a refinery where it is divided into many different products, natural gas is just cleared of impurities and then placed in the line. A series of intermediate pumping stations is necessary to overcome the hydraulic losses as the product flows through the line. (A later section of this chapter describes spacing of the intermediate pump stations.)

After arriving at the local gas company, the pipeline is tapped so that the gas can be prepared for the end user. A gas pipeline tap-off point is shown in Figure 13-7. From here, large compressors (Figure 13-8) push the gas through distribution mains until it finally arrives at the home or factory where it is burned for fuel.

Figure 13-7. Gas pipeline tapoff point

Figure 13-8. Gas compressors for moving product in the distribution mains

Natural gas is composed principally of methane. It is nontoxic, but because it displaces oxygen, it can cause death by asphyxiation. It is lighter than air and rises when it escapes. As it comes from the wells it is usually odorless, but chemicals are added, both by the pipeline company and the local utility, to give it a strong odor (Figure 13-9).

Gas can be distributed at high, intermediate, or low pressure. Before going into the customer's lines, the pressure is reduced to ¼ psig by going through a regulator. From there it goes through a meter and then on to the gas appliance for use.

Leak Detection

Gas mains are monitored on a continuing basis by remote sensors placed in the line. In addition, a continuous readout (Figure 13-10) is made locally of the pressure in the mains. Small variations are noted from the graph. Large leaks will be immediately noted by the drop in pressure at the control point. Isolation of the broken main, however, may take time and bleed down of the pipe will continue to provide fuel.

Emergency response personnel can determine the presence of a combustible gas through the use of a combustible gas meter (Figure 13-11). This meter samples the air and determines whether there is a fuel-to-air mixture in the flammable range. Most meters

Figure 13-9. Odorant-adding facility at a local gas company

Figure 13-10. Gas main pressure printout

Figure 13-11. Combustible gas meter

read from 0 to 100 percent of the lower explosive limit (lel). The area between 80 and 100 percent is shown in red, indicating that this is the danger point. If the needle goes to 100 percent and then in a few seconds goes back to zero, the mixture is above its upper explosive limit (uel). This is also dangerous because as the gas disperses it will have to pass through the explosive range and could ignite.

Gas Distribution System

Gas from the local utility is provided to customers through distribution mains. These mains are tapped off to provide the service. A service line is connected to the main without shutting down through the use of special fittings. If plastic pipe is to be connected, a cutter is used to create the connection (Figure 13-12a). After the hole is made, the cutter is removed, the top sealed with the cap, and the gas flows into the new line (Figure 13-12b). The service line is connected to the fitting by means of compression. Therefore, in case of an accidental striking by construction equipment, the service line can be easily pulled free from its connection to the distribution main. A broken plastic service line is shown in Figure 13-13 after it was pinched off with a pliers.

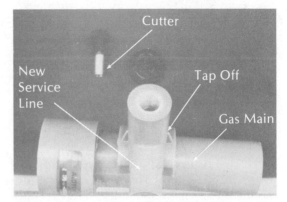

Figure 13-12a. Fitting for connecting service line to distribution main

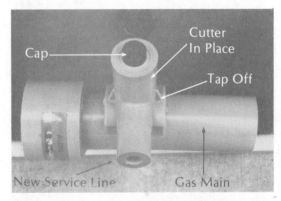

Figure 13-12b. Sealed fitting connecting service line to distribution main

Figure 13-13. Broken plastic gas service line pinched off with pliers

Service systems coming from the distribution mains can be divided into the following systems:

1. Low-pressure, single-family residential
2. Low-pressure, multiple occupancy
3. High-pressure, single and/or multiple occupancy
4. High-pressure, industrial

Low-Pressure, Single-Family Residential Gas Service System. This system operates at about ¼ psi. The meter could be located inside in a utility area or, in the newer installations, outside the house (Figure 13-14a and b). This service system has the following characteristics:

1. No means of identifying the type of service line from the outside of a building.
2. Normally there is no shutoff valve between the main and the building wall, except if the line is large [over 2 inches (5 centimeters) in diameter] or the supply goes to a place of public assembly.
3. The shutoff valve is usually located at the inlet to the meter.

Low-Pressure, Multiple Occupancy Gas Service System. This system operates at about ¼ psi (0.11 kilogram). The meters can be located in the basement, in a separate utility room, or outside the building (Figure 13-15). Shutting off the gas at one meter regulates only the flow to that apartment. A whole system shutdown may be provided.

High-Pressure, Single or Multiple Occupancy Gas Service System. In this type of service, gas enters the house or apartment at intermediate or high pressure. It goes through a regulator, which reduces and regulates the pressure entering the piping to approximately ¼ psi (0.11 kilogram) and then into the meter before entering the rest of the house piping. A shutoff valve is located in the riser ahead of the regulator (Figures 13-16a and 13-16b). A vent line goes from the regulator

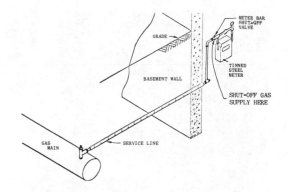

Figure 13-14a. Low-pressure, single family residential gas meter installed inside. *Courtesy of the Michigan Gas Association*

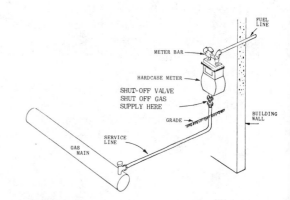

Figure 13-14b. Low-pressure, single-family residential gas meter installed outside. *Courtesy of the Michigan Gas Association*

to the outside building wall near where the underground service line enters the building. An underground shutoff valve in a box or an above-grade shutoff valve is usually installed on all service lines operating at pressures exceeding 10 psig.

High-Pressure Industrial Gas Service System. Some large commercial or industrial users may have a separate room or even a separate building for the location of the meter(s). A regulator is installed to reduce the pressure and a shutoff valve is located ahead of the reg-

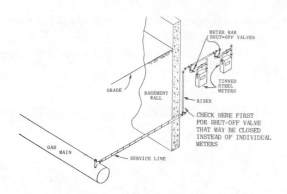

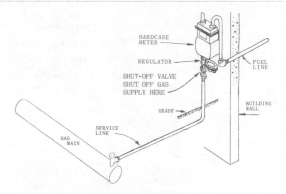

Figure 13-15. Low-pressure, multiple-occupancy gas service system. *Courtesy of the Michigan Gas Association*

Figure 13-16b. High-pressure gas service system installed outside. *Courtesy of the Michigan Gas Association*

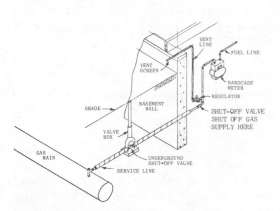

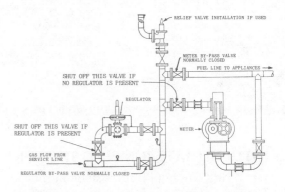

Figure 13-17. High-pressure industrial gas service system. *Courtesy of the Michigan Gas Association*

Figure 13-16a. High-pressure gas service system installed inside. *Courtesy of the Michigan Gas Association*

ulator. An underground shutoff valve is usually installed between the main and where the service line enters the building (Figure 13-17).

Shutting Down a Gas Line

An interior gas line can be shut down by turning the quarter-turn valve perpendicular to the pipe (Figure 13-18). Once the valve is shut it must remain closed until all appliances are checked by gas company personnel.

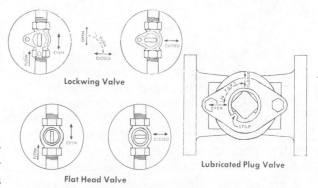

Figure 13-18. Shutting down a gas service line. *Courtesy of the Michigan Gas Association*

A service line can be shut down by special wrenches that fit the particular valves. Figure 13-19 illustrates four common types of valve shutoff wrenches. Some underground valves are located far below the surface and an ex-

tension is needed on the wrench to reach the valve. Most of these underground valves require only a ¼ turn clockwise to shut off the gas supply.

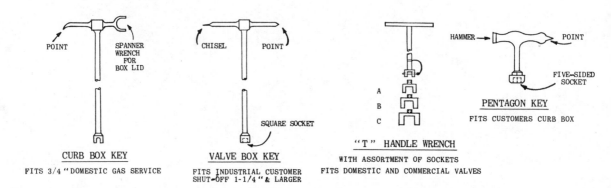

Figure 13-19. Service line shutoff wrenches. *Courtesy of the Michigan Gas Association*

INTERMEDIATE PUMPING STATIONS

As with any liquid, oil resists being moved. The factors that have to be overcome include:

1. Inertia
2. Friction caused by pipe walls (the larger the line the more friction)
3. Gravity
4. Viscosity

As the oil leaves the pump it is under high pressure. As it moves along the pipe, this pressure drops. When it is projected to reach at 25 to 50 psi, another intermediate pump station is inserted in the line. The distance between the stations is based on the calculations for the factors stated above.

In order to understand how the location of the intermediate pump station is determined the configuration of an actual pipeline will be used as an example. There is a 427-mile

(563 kilometers), crude oil pipeline running from Cushing, Oklahoma, to Wood River, Illinois. It is 22 inches (56 centimeters) in diameter and carries 190,000 barrels a day.

Based on the viscosity of the liquid and the type of pipe used, the pressure drop was determined to be 8.5 psi per mile.[5] Since the line is 427 miles (563 kilometers) long, the total pressure is

$$\text{pressure drop} = 427 \text{ miles} \times 8.5 \frac{\text{psi}}{\text{mile}}$$
$$= 3{,}629 \text{ psi}$$

Now, the effect of gravity and inertia must be taken into account. Cushing, Oklahoma, has an elevation of 875 feet above sea level, while Wood River, Illinois, is 440 feet above sea level. This means that there is a net lowering of elevation which makes it easier to pump the oil because the effect of gravity will help it move downhill. The net difference

875 feet – 440 feet = 435 feet

reduces the required pressure by 159 psi. This value is determined by mathematical calculations beyond the scope of this text. This means that the pressure necessary for the system is

3,629 psi – 159 psi = 3,470 psi

On the basis of the type of product to be pumped, the size, grade, and wall thickness of the pipe is determined. Once this is accomplished there is a maximum working pressure, in this case 725 psi, arrived at.

With the maximum working pressure of 725 psi to be spread out over 3,470 psi there would be

$$\frac{3,470 \text{ psi}}{725 \text{ psi}} = 4.8 \text{ pump stations}$$

The locations of the five pumping stations for this line are shown in Figure 13-20. They are not evenly spaced because they must be closer together when the product goes uphill to overcome gravity and inertia. When going downhill they can be spread further apart. The dotted lines represent the amount of pressure developed at the station and the pressure remaining as it moves along the line until it is only between 25 and 50 psi when it arrives at the next station. The actual distances between pumps are:

Cushing to Chelsea	85.35 miles (136.56 kilometers)
Chelsea to Diamond	71.92 miles (115.07 kilometers)
Diamond to Buffalo	82.81 miles (132.5 kilometers)
Buffalo to Bland	92.39 miles (147.82 kilometers)
Bland to Wood River	100.84 miles (161.34 kilometers)

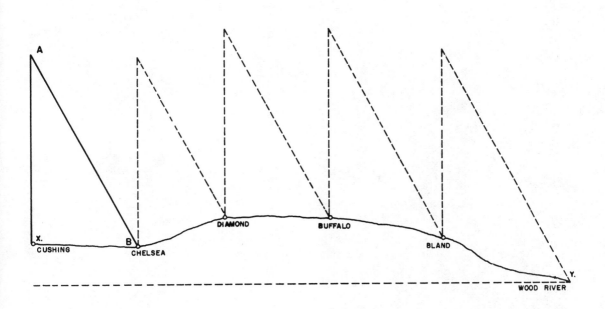

Figure 13-20. Pipeline intermediate pump stations. *Courtesy of the University of Texas, Petroleum Extension Service*

PIPELINE MARKINGS

As explained in the first few paragraphs of this chapter, the major cause of pipeline incidents is their being struck by an outside force. It is, therefore, very important to accurately know the location of the pipe. This is accomplished by placing markers over the buried line at road crossings and railroad crossings. In addition, the route the line travels is also indicated with markers.

It must also be remembered that some pipelines pass through urban areas where the placement of markers at every road would be impractical. In these cases no marker or identification is necessary.

The American Petroleum Institute (API) has published a voluntary standard entitled *Recommended Practice for Marking Liquid Petroleum Pipeline Facilities.*[6] The standard marking as recommended by API is shown in Figure 13-21.

However, API in the Foreword to their standard states:

The Standard Pipeline Warning Sign and the pipeline marking criteria herein are recommended for new construction and for use during normal marker replacement programs and relocations. Immediate replacement of existing markers for the sole purpose of standardization would be unduly costly and inefficient and is not recommended.[7]

As a result, there are still several different variations for marking the locations of liquid petroleum lines (Figure 13-22). In addition, markers along the pipeline right-of-way do not follow the standards (Figure 13-23). If there is more than one pipeline in the right-of-way, then a marker over each pipe is necessary (Figure 13-24).

In some cases the pipeline is patrolled from the air. In this case it is important for the marker to also be visible from the air (Figure 13-25). The number on the marker is the sequence number so the pilot can ensure that the entire route is surveyed.

Figure 13-21. Recommended standard for marking liquid petroleum lines at public road crossings

Figure 13-22. Old style liquid petroleum pipeline marker

The recommended standard suggests the following criteria be used for determining the locations for pipeline markers:

> The use of conventional Pipeline Warning Signs is not recommended in heavily developed urban areas such as downtown business centers where placement is impracticable, or would not serve the purpose for which markers are intended, or when the local government maintains current substructure records. In such areas the carriers may elect to indicate the presence of a pipeline by means of stenciled markings, cast monuments, plaques, signs, or other devices installed in curbs, sidewalks, streets, building facades or wherever else practicable.

Figure 13-23. Liquid petroleum pipeline right-of-way marker

a. Pipeline Markers should be located at each public road crossing, at each railroad crossing and in sufficient number along the remainder of each buried line so that its location is accurately known. The latter may be accomplished by placing markers at fence lines, property lines, right-of-way boundaries and in open areas wherever the party exerting control over the surface use of the right-of-way will permit such installation.

b. Pipeline Warning Signs should be installed at locations where the pipeline is above ground and is accessible to the public. Aerial crossings on any type of structure either publicly or privately owned should be considered for this purpose as being accessible to the public. Pipeline Warning Signs should be placed on each side of a traversed waterway or any impoundment which is an active source of water supply.

c. Navigable Waterway Pipeline Warning Signs should be placed on each side of all crossings of navigable waterways. Such installation should also include a Standard Pipeline

Figure 13-24. Multiple pipeline marking in a single right-of-way

Figure 13-25. Air survey liquid petroleum pipeline marker

Warning Sign, for each buried pipeline crossing, which should be installed above the normal high water line. Special Pipeline Warning Signs should be used at non-navigable waterways where there are activities or operations that could damage the pipeline.[8]

Figure 13-26 illustrates a typical warning sign for a navigable waterway pipeline.

Markings for gas pipelines have not yet been standardized. There are, therefore, many variations which require that emergency response personnel become familiar with those used in a particular area (Figures 13-27 and 13-28). In fact, it is not even necessary to mount the marker on a pole. Figure 13-29 shows a marker buried in the ground. Painted circles on the asphalt serve as markers in a parking lot.

Figure 13-27. Gas pipeline marker

Figure 13-28. Gas pipeline marker

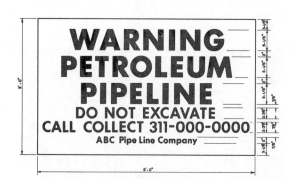

Figure 13-26. Navigable waterway pipeline warning sign. *Courtesy of the American Petroleum Institute*

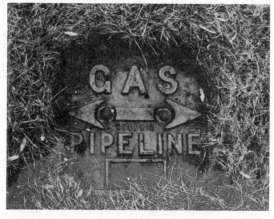

Figure 13-29. Gas pipeline marker on lawn

PIPELINE CROSSING RAILROADS, HIGHWAYS, AND WATERWAYS

When a pipeline is placed under a roadway or railroad bed it will be subjected to considerably more stress than for most other placement. Not only is there the additional loading from the road material, but the weight of the vehicles or trains has to be considered. In addition, the passing vehicles set up vibrations, which over a period of time, can cause pipe weakening.

It is necessary, therefore, to overcome these problems in the design stage of the pipeline. The methods used are:

1. Direct installation underground with the pipe designed to withstand the load

2. Installation of a casing to absorb the load and installing the pipeline inside the casing.

3. Installation of a bridge

The direct method has no effect on the activities of emergency response personnel and will not be covered any further in this text. However, cased pipe and bridge crossings must be discussed.

A railroad crossing and a highway crossing for a cased pipeline are shown in Figures 13-30 and 13-31. The carrier pipe must be kept from touching the casing to ensure that mechanical stress is not transferred and to keep the two pipes electrically insulated from each other. This is accomplished by uniformly spacing insulators and supports to hold the carrier pipe inside the casing.[9]

To prevent water from entering the casing and causing damage by freezing or corrosion, the ends of the casing pipe are sealed with a flexible material. Again this is done to isolate the carrier from the casing.

However, once all of this is accomplished the casing pipe has a large amount of void space. Should the carrier pipe rupture where the casing is located, pressure could build up fast, making the casing a bomb. To avoid this possibility, vents are installed in the casing to relieve any pressure buildup (Figure 13-32). For protection from moisture entering the space, the vent is fitted with a suitable weather cap.

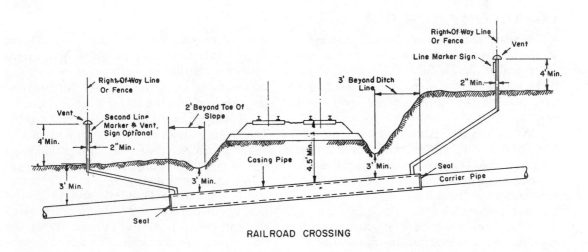

RAILROAD CROSSING

Figure 13-30. Cased pipeline crossing railroad tracks. *Courtesy of the American Petroleum Institute*

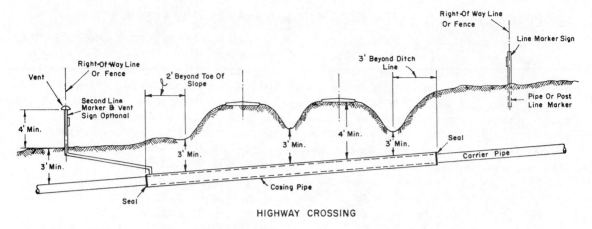

HIGHWAY CROSSING

Figure 13-31. Cased pipeline crossing a highway. *Courtesy of the American Petroleum Institute*

Bridge crossings are used when economics makes it unfeasible to bury the line. Here, the danger comes from being struck and causing a break in the pipeline if the supports give way.

Figure 13-32. Pipeline casing curved vent pipe

CORROSION PROTECTION

External corrosion of a buried pipeline is an electrochemical reaction. This can happen when two dissimilar metals are used, when there are differences in oxygen content in the soil, or when the moisture levels in the soil vary.

When new pipe is installed between sections of old pipe, an electrical flow is set up (Figure 13-33). This flow results from the difference in age and exposure of the metal surfaces as well as the soil serving as an electrolyte due to the chemicals it contains.

When soil that is deficient in oxygen is encountered, an electrical flow can also be set up. This results from the fact that the oxygen deficient areas produce a more concentrated electrolyte than the well-aerated soil (Figure 13-34). This difference in electrolytes establishes the current flow. The current flow can be measured by connecting a meter between the pipeline and the soil (Figure 13-35).

Finally, a moist soil will compact tightly. This in turn reduces oxygen content and, as stated above, changes the concentration of the electrolyte (Figure 13-36). This again produces an electrical current.

In Figures 13-34, 13-35, and 13-36, the flow of current is shown from the anode (+) to the cathode (–). As the current leaves the anode it causes corrosion. The job of the engineer, then, is to provide just the right amount of current to the pipeline to make it

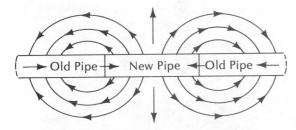

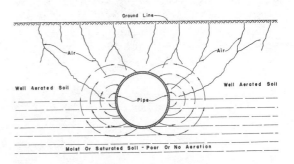

Figure 13-33. Corrosion caused by dissimilar surfaces of old and new pipe

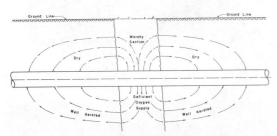

Fig. 5–25. Oxygen deficiencies cause corrosion.

Figure 13-34. Corrosion caused by oxygen deficiencies. *Courtesy of the American Petroleum Institute*

Figure 13-35. Measuring current between pipeline and soil

the cathode and prevent the current from leaving. The system that accomplishes this is called *cathodic protection.*

Figure 13-36. Corrosion caused by differences in moisture and aeration conditions. *Courtesy of the American Petroleum Institute*

Coatings of the pipe can help to insulate the line from the currents. However, the coatings deteriorate in use and additional protection is necessary.

Since the current flows from the anode and since the anode is destroyed, the objective is to make the pipe the cathode: to make currents flow through the soil toward the pipe rather than away from it.

The American Petroleum Industry describes the equipment used for cathodic protection as:

Power is taken from the power line in the form of alternating current. The transformer reduces the voltage and the rectifier converts it to direct current (Figure 13-37). The direct current flows to and through the carbon anodes and into the surrounding earth. Where there is a break in the insulating coating on the pipe the current will reach the line at that point. It will then flow along the line to the rectifier and complete the circuit. Where power lines are not in reach of the pipeline, similar results may be achieved by using wind-powered generator, motor-generator sets or by using magnesium anodes instead of carbon. Magnesium anodes generate their own electric current, gradually being consumed in the process.[10]

Measurements of the cathodic protection being supplied are made at test points along the pipeline. These test points vary in design (Figures 13-38 and 13-39), but all provide access to the current being supplied.

Figure 13-38. Cathodic protection test measurement point with taps

Figure 13-37. Cathodic protection rectifier

Figure 13-39. Cathodic protection test measurement point with bare wires

LIGHTNING PROTECTION

Many portions of a buried pipeline extend above the ground. Items such as vents, valves, tapoff pipes, as well as metering devices are all targets for a lightning strike. If lightning should hit one of these devices, it could travel many miles and cause damage along the way. To prevent this from happening, lightning arrestors (Figure 13-40) are installed. This simple device provides an acceptable path for the lightning charge to follow and thus dissipating it harmlessly.

Figure 13-40. Lightning protection for the pipeline

HANDLING A PIPELINE INCIDENT

As with any hazardous materials incident, preplanning for a pipeline break will enable emergency response personnel to react better should an incident occur. To aid in this planning the following items must be determined:

1. Location of transmission and major distribution mains. Plot on maps showing street and utility locations.

2. Routes of travel and alternate routes in case one becomes blocked by the incident. Plan for a higher initial level of response than normal, including mutual aid if necessary. If evacuation is necessary, large amounts of personnel will be needed quickly to help. Plan for this in advance.

3. Since the pipeline may be in a rural area or in an urban area, water supplies will become critical. For the rural area, static sources and tanker operations must be planned. Disrupted water utilities may make this service necessary even in the urban areas. Even where hydrants are available they may be a great distance from the line and a pumper relay will have to be set up.

4. The tactics for approaching the scene are the same as for any other emergency or fire. Emergency response personnel must be concerned with traffic, road conditions, accessibility, ground slope, water availability, and exposures. One problem to be particularly alert for is high-tension lines in the same right-of-way as the pipeline. In case of a rupture the electrical lines will produce an additional hazard.

5. The location of special extinguishing agents is also important, as well as how much is available locally and how long will it take to get additional supplies.

6. In addition to structural buildings located near the pipeline, exposures to be noted in preplanning include streams and sewers, which can spread the problem far beyond its original confines.

7. Finally, what clean-up is necessary for the product? Will the responsible environmental agencies have to be notified? What equipment is available for protecting the product from spreading? Where will it come from and how long will it take? In Figures 13-41a and 13-41b, polyurethane foam is being sprayed to close off sewer and prevent product from spreading.

So, in addition to the normal problems of fighting a structural fire, emergency response personnel must be ready to handle many more difficulties.

Figure 13-41a.,
Figure 13-41b.   Using polyurethene foam to close off sewer and prevent product from entering

348   Hazardous Materials

EXAMPLES OF PIPELINE INCIDENTS

Pipeline accidents continue to plague emergency response personnel. During 1976 the National Transportation Safety Board investigated the following incidents:

- Natural gas explosion and fire, Brewton, Alabama, January 6.
- Natural gas gathering line explosion and fire, Mooreland, Oklahoma, January 7.
- *Natural gas explosion and fire, Fremont, Nebraska, January 10.
- Gas explosion, Queens, New York City, January 30.
- Natural gas explosion and fire, Phoenix, Arizona, February 8.
- Liquid petroleum gas explosion and fire, Whitharral, Texas, February 25.
- Natural gas explosion, Cotati, Texas, March 11.
- Propane gas explosion and fire, Lehigh Acres, Florida, March 23.
- Natural gas explosion and fire, Phoenix City, Alabama, March 27.
- Natural gas explosion and fire, Niles, Michigan, June 9.
- Natural gas pressure release, Abbeyville, Louisiana, June 10.
- *Gasoline pipeline fire, Los Angeles, California, June 16.
- Natural gas explosion and fire, Pennsboro Township, Pennsylvania, June 19.
- *Natural gas explosion and fire, Allentown, Pennsylvania, August 8.
- *Gas transmission line fire, Cartwright, Louisiana, August 9.
- Natural gas explosion and fire, Bangor, Maine, August 13.
- Gas gathering line fire, Cabell County, West Virginia, August 26.
- Natural gas explosion and fire, Kenosha, Wisconsin, August 29.
- Natural gas explosion and fire, Blue Island, Illinois, September 10.
- *Natural gas explosion and fire, Robstown, Texas, December 7.

* Major Investigation

Los Angeles, California, Gasoline Pipeline Rupture,[11] June 16, 1976

At 10:32 A.M. on June 16, 1976, an 8-inch pipeline owned by the Standard Oil Company of California was struck and ruptured by excavation equipment which was working on a road widening project. Gasoline sprayed from the rupture and drenched nearby buildings. Within 90 seconds after being struck, the leaking gasoline ignited. The resulting fire killed 9 persons and injured 14 others. In addition, 7 commercial buildings were completely destroyed and 9 others were damaged. Fifteen vehicles traveling on the street or parked were also destroyed. The damaged area is shown in Figure 13-42.

Review the conditions existing at this incident with the list developed in the paragraph on handling a pipeline incident. What would you and your emergency response organization have done?

Fairbanks, Alaska, Pump Station Fire,[12] July 8, 1977

Because of a leak at Pump Station 8 on the Alaska pipeline, the flow of oil had been stopped on July 4. Personnel worked around the clock to get the problem corrected and on July 8 the flow began again. By 3:39 P.M. on July 8, the oil was 30 miles (48 kilometers) south of Pump Station 8.

One of the employees on duty noticed that there was a drop in the pressure being supplied to the oil. As the crew was checking for the reason for the diminished pressure, an explosion occurred.

One employee was killed and five were injured. The explosion lifted the shell of the

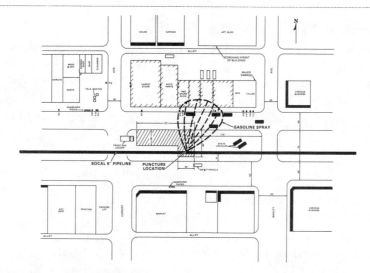

Figure 13-42.   Los Angeles, California, pipeline rupture, June 16, 1975. *Courtesy of the National Transportation Safety Board. Report PAR-76-8*

building 300 feet (90 meters) into the air, started many forest fires, and created a large oil spill. The flowing oil began burning and created a running spill fire about 25 yards (22.5 meters) wide.

In addition to local control of the valves, they are also controlled from the main operations center at Valdez. The Valdez operator was in radio contact with the intermediate pump station when the explosion occurred, but it still took four minutes to shut the line down. Then, the oil still in the pipe on the upstream side of the valve continued to flow out.

The fire was extinguished when all of the fuel was burned up. The structural fires were extinguished with water brought in by tankers.

Because of the remoteness of the pump station several problem areas became apparent:

1. Access was very difficult.
2. The size of the fire overwhelmed the local emergency response personnel very quickly and mutual aid companies were far away.
3. There was little extinguishing agent on hand.

4. There was little water available for fighting structural fires.
5. Mutual aid personnel were unfamiliar with the facility.
6. Evacuation of injured personnel was difficult.

How would you have handled this incident?

Aurora, Colorado, LPG Pipeline Leak,[13] September 28, 1977

A Phillips Petroleum LPG pipeline was discovered leaking near a new subdivision under construction. The leak was about 1½ miles from a shut-off valve but it took five hours for all of the product to escape. While the gas continued to leak, emergency response personnel evacuated 100 individuals from nearby homes. Fire service personnel used unmanned monitor nozzles at the point of the leak to disperse the vapors.

What would have happened had the leak ignited; or if there was no water available, or if the valve could not be shut down quickly?

QUESTIONS FOR REVIEW AND DISCUSSION

1. What products are transported by pipeline?

2. What are the differences between distribution and transmission pipelines?

3. Prepare a map of the transmission pipelines in your area.

4. Prepare a map of the major distribution mains in your area.

5. What hazards must be considered when a pipeline leaks or ruptures?

6. Explain the procedures used by the gas utilities in your area for detecting leaks in a gas main.

7. Describe the system and equipment necessary to bring natural gas from a distribution main to a consumer.

8. Describe the procedures that emergency response personnel must use to shut off a residential gas line.

9. Explain the need for intermediate pumping stations on a pipeline route.

10. Describe the requirements for marking a pipeline route.

11. Explain the need for a casing when a pipeline goes under a road or a railroad bed. Why are vents installed in the casing?

12. Describe the techniques used to prevent pipelines from corroding. What measurements are performed to ensure that the protection is continuing?

13. Outline the steps that emergency response personnel must perform to handle a pipeline incident.

CASE PROBLEM—PIPELINE

A 36-inch petroleum products pipeline is located on the outskirts of East Overshoe, Pennsylvania. At 10:30 A.M. on a Thursday in November, a roadgrader, leveling ground for a new interstate highway, strikes the pipeline. The pipeline ruptures and kerosene begins to flow from the break. Ignition occurs from the motor of another piece of construction equipment. The wind is blowing at 12 knots from the break toward the town. The temperature is 34°F (1.1°C). The town is located downhill from the break, approximately 1 mile away. It has a population of 25,000, most of whom work in a large city about 50 miles away.

The town's fire department has 2 pumpers, each 1,250 gpm, with 1,000 gallons of water. Ambulance service is provided by the local fire department. A 100-foot tractor-drawn aerial ladder is also located in the fire station. The department is an all-volunteer department, with 22 individuals available on weekdays.

There is no hydrant system near the break. The closest water supply is the hydrants in town, about 5,000 feet away. There is a single, 60-bed hospital in town.

Mutual aid is available from the town on the other side of the break, 12 miles away. This town also has an all-volunteer department. Equipment available consists of two 1,000 gpm pumpers with 500 gallons of water and a tanker with 1,500 gallons of water. Total manpower available is 25 people.

Mutual aid is also available from two other departments. One is 6 miles away and has 3 pumpers and 12 career people on duty. Each pumper carries 500 gallons of water. The next is 11 miles away and has 32 volunteers manning a pumper (500 gallons of water), an ambulance, and a heavy-duty rescue squad truck.

The police department consists of 6 officers on duty in the town. There is a county sheriff's department with 26 individuals and a state police contingent of 14 individuals.

1. Give a detailed explanation of how you would handle this incident.

NOTES

1. National Transportation Safety Board, *Annual Report to Congress, 1976* (Washington, D.C.: Government Printing Office, 1977).

2. Ibid.

3. National Transportation Safety Board, *Special Study, Prevention of Damage to Pipeline*, NTSB-PAR-73-1. (Washington, D.C.: Government Printing Office, 1973).

4. Ibid.

5. Petroleum Extension Service, *Introduction to the Oil Pipeline Industry* (Austin: University of Texas, 1973), p. 32.

6. American Petroleum Institute, *Recommended Practice for Marking Liquid Petroleum Pipeline Facilities*, API RP 1109 (Washington, D.C.: American Petroleum Institute, October 1977).

7. Ibid.

8. Ibid.

9. American Petroleum Institute, *Recommended Practice for Liquid Petroleum Pipelines Crossing Railroads and Highways*, API RP 1102 (Washington, D.C.: American Petroleum Institute, September 1968).

10. Ibid.

11. National Transportation Safety Board, *Standard Oil Company of California, Pipeline Rupture, Los Angeles, California, June 16, 1976*, NTSB-PAR-76-8 (Washington, D.C.: Government Printing Office, December 9, 1976).

12. Scott Yates, "The Hall Was Full of Fire and the Oil Was Flowing Toward Us," *Fairbanks Daily News-Miner*, July 9, 1977, p. 1.

13. "Gas Leak Forces Evacuation," *Rocky Mountain News*, September 29, 1977, p. 5.

14
Airline transportation

Aircraft incidents can create two types of hazardous materials problems for emergency response personnel. First, hazardous materials carried as cargo can present a danger because of mishandling during normal flight operations or from the shock incurred in a nonstandard landing. Second, the plane itself can present several hazardous materials difficulties because of the large quantities of fuel, the piped oxygen, and the large amount of flammable metals such as magnesium and titanium.

While this text is mainly designed for handling the hazardous materials incident, some coverage of procedures for handling a downed aircraft incident are included. This is done to ensure that emergency response personnel have some basic idea of what to do at the crash site. However, the emergency response personnel should familiarize themselves with other textbooks that provide more detailed coverage. Such books as *Aircraft Fire Protection and Rescue Procedures*[1] and *Manual on Aircraft Rescue and Fire Fighting Techniques for Fire Departments Using Structural Fire Apparatus and Equipment*[2] are excellent sources of information of use and interest to fire fighters.

During 1975 there were 4,431 accidents to United States civil aviation vehicles.[3] This means that on the average, there were over 12 incidents per day. No matter how isolated or rural the area, the potential for an aircraft accident is there. If you add to this the fact that small private plane airports are springing up, the potential becomes even more apparent.

When talking about aircraft, most emergency response personnel think in terms of large commercial planes. They must keep in mind, however, that many hazardous goods are shipped by small plane.

For example, consider the crash of a single-engine plane in Anne Arundel County, Maryland, on January 22, 1978.[4] This aircraft crashed 300 yards (270 meters) from the runway. Its cargo was 30 containers of molybdenum 99, a radioactive material used in medical treatments. The packages were lead containers surrounded by styrofoam, inside cardboard cartons. Fortunately, under impact the boxes remained intact. However, there is no assurance that this would be true in similar incidents.

Remember, all planes, no matter what their size, can carry hazardous materials. Are you prepared? If not, follow the suggestions in the remainder of this chapter to get your emergency service personnel ready. It can happen in your area.

LEGAL RESPONSIBILITY

The Code of Federal Regulations 14, Part 430, details the rules pertaining to the notification and reporting of aircraft accidents. In paragraph 430.10 and 430.11 the responsibilities of emergency service personnel are spelled out.[5] They are included here for your information.

§430.10 Preservation of aircraft wreckage, mail, cargo, and records.

(a)   The operator of an aircraft is responsible for preserving to the extent possible any aircraft wreckage, cargo, and mail aboard the aircraft, and all records, including those of flight recorders, pertaining to the operation and maintenance of the aircraft and to airmen involved in an accident or incident for which notification must be given until the Board takes custody thereof or a release is granted pursuant to §430.11.

(b)   Prior to the time the Board or its authorized representative takes custody of aircraft wreckage, mail, or cargo, such wreckage, mail and cargo may be disturbed or moved only to the extent necessary:

(1)   To remove persons injured or trapped;

(2)   To protect the wreckage from further damage, or

(3)   To protect the public from injury.

(c)   Where it is necessary to disturb or move aircraft wreckage, mail or cargo, sketches, descriptive notes, and photographs shall be made, if possible, of the accident locale including original position and condition of the wreckage and any significant impact marks.

§430.11 Access to and release of aircraft wreckage, records, mail and cargo.

(a)   *Access to aircraft wreckage, records, mail and cargo.* Only the Boards accident investigation personnel and the persons authorized by the Investigator-in-Charge or the Director, Bureau of Aviation Safety, to participate in any particular investigation, examination or testing shall be permitted access to aircraft wreckage, records, mail or cargo which is in the Board's custody.

(b)   *Release of aircraft wreckage, records, mail and cargo.* Aircraft wreckage, records, mail and cargo in the Board's custody shall be released by an authorized representative of the Board when it is determined that the Board has no further need of such wreckage, mail, cargo or records.

Although the above-cited regulations discuss the responsibilities of the operator, many times this individual is injured. Then it falls to the emergency response personnel to assume the specified duties. In addition, the regulations discuss the procedures that must be observed before the National Transportation Safety Board investigators arrive at the crash site, probably hours later. Again, the responsibility for enforcing these requirements rests with the emergency response personnel. Finally, the requirements for noting the positions of parts of the wreckage must be performed by emergency response personnel if they have to move the wreckage to make a rescue.

PREPLANNING

As with all of the other modes of transportation, preplanning for an incident involving aircraft is essential. Crashes can occur on the airfield, just off of the field, or in a remote area. So, the first step in preplanning is to survey your response area. You should first select the most likely places for an incident, which include the airfield itself and the approaches to the runways. While surveying the area, prepare a map showing all roads, access points, and possible water sources. This map will be of great help when trying to locate a downed aircraft in sparsely settled areas.

Next, meet with the local airport officials to determine what assistance they can provide. Find out if they have an automatic locator that sounds a warning when a special device on the aircraft is activated. If so, with the use of radio direction-finding techniques, the aircraft can be located. Obtain a list of the types of aircraft using the airport. For these aircraft, you should record information such as types of engines, height, width, length, fuel capacities, passenger capacities, crew size, exit locations, and exit door operation.

Meet with emergency response personnel of surrounding jurisdictions to determine if any specialized equipment is available. Establish the minimum number of apparatus, ambulances, and support vehicles you will need and arrange for them to be available when requested. If formal mutual aid agreements are necessary, negotiate them before the incident.

Locate extra sources of specialized extinguishing agents and determine in advance how they can be ordered, delivered, and paid for should they be needed at an incident.

Arrange for common communications capabilities for all emergency response personnel who would be on such a call. The ability to coordinate the rescue of survivors, extinguish the fire, and assist with site security require that all personnel on the scene be able to talk on a common frequency. The time to add the extra frequency or establish a common frequency with all agencies is now, at the preplanning stage, and not at the height of a disaster.

Plan for emergency medical services personnel and equipment at the scene as well as at the various medical facilities. If there are many survivors at the site, then the local emergency medical services system can easily become overtaxed. Personnel on the scene will have to triage the injured to ensure that those with the most severe injuries are transported first and that no time is spent initially on those who have already died. Hospitals must be prepared for the large influx of critical-care patients and should have a disaster plan to put in operation.

A temporary morgue may be necessary. Sites for such use must be surveyed in advance, proper officials contacted so that arrangements for the facility can be made quickly, and any special equipment arranged for. This is a critical need and preplanning will make the job of establishing the morgue much easier.

Heavy equipment such as bulldozers, cranes, or graders may be needed. They can be used to prepare roads, lift heavy portions of the aircraft, as well as bring in equipment to aid to the operations. The availability of this equipment around-the-clock must be determined before the incident occurs.

Of course, there are many more areas that must be preplanned. This material is covered in detail in Chapter 7 of the text. This section is just meant to point out the special problem areas that must be considered in order to handle an aircraft accident in the best possible way.

Finally, test out your preplan with frequent drills. On a short-term basis, check small portions of the plan, while at least yearly, run a full-scale simulated emergency. This activity cannot be stressed enough. It is the only way to ensure that what is placed on paper will translate into a functioning team at the scene of an incident.

NOTIFICATIONS

When an aircraft accident occurs, there are many notifications that must be made, depending on whether a civilian or military plane is involved. Emergency response personnel should prepare a listing of all of the required agencies and their 24-hour telephone numbers, to avoid delay. To aid in describing the accident scene, some basic definitions are included in the glossary.

Civilian Aircraft Notifications

The National Transportation Safety Board (NTSB) is charged with the responsibility to "make rules and regulations governing notification and report of accidents involving civil aircraft".[6] As a result of this requirement, NTSB has adopted the following regulations:[7]

1. Immediate notification of NTSB when there is:
   a. Aircraft accident
   b. Flight control system malfunction
   c. Inability of any required flight crewmember to perform his normal flight duties as a result of injury or illness
   d. Motor failures
   e. In-flight fire
   f. Aircraft collision in flight
   g. Aircraft overdue and believed to have been in an accident

2. Information needed and to be supplied to NTSB includes:
   a. Location of accident
   b. Date and time of accident
   c. Aircraft make, model, registration number, and nationality
   d. Operator and crew
   e. Number of persons involved
   f. Type and extent of injuries
   g. Weather conditions

Manner of Notification

The most expeditious method of notification to the National Transportation Safety Board by the operator will be determined by the circumstances existing at that time. The National Transportation Safety Board has advised that any of the following would be considered examples of the type of notification that would be acceptable:

(a) Direct telephone notification.

(b) Telegraphic notification.

(c) Notification to the Federal Aviation Administration who would in turn notify the NTSB by direct communication; i.e., dispatch or telephone.

In addition to the requirements of the aircraft operator to notify NTSB, emergency response personnel should prepare a list of other agencies that should receive prompt information about the incident. Your list should include:
National Transportation Safety Board
Federal Bureau of Investigation
Airline or Owner
U.S. Postal Service (if mail carried)
Energy Research and Development Administration (if radioactive material carried)
If the crash occurs away from the airport, then notification should include:
Nearest Airport Tower
Federal Aviation Administration
Airport Fire Station
As explained in the preplan section, a great deal of outside assistance will be needed at a major disaster. The notification book for an airplane accident should therefore include the following:
Owners of heavy construction equipment
State police
Local law enforcement personnel
Hospitals
Clergy
Ambulance Services
Red Cross
Local civil defense personnel

Military Aircraft Notifications

The military has prepared a booklet for use by emergency response personnel that provides information on the data needed.[8]

Call collect to the nearest military installation and tell the operator (or dispatcher if radio controlled) that you wish to report a military aircraft crash.

When the call is answered, state that you are reporting an aircraft crash, and complete a form similar to that shown in Figure 14-1a.

MILITARY AIRCRAFT NOTIFICATION FORM

Name _____   Call Back Number _____

Time of Crash _____   Date _____   Fire   yes ☐   no ☐

Location of Crash_____
(Closest address if available)

_____
(Roads, cross-streets, miles to closest intersection)

_____
(Compass bearings from major landmarks)

AIRCRAFT

Type of aircraft   _____
Number on tail section _____
Description of damage   _____
Nearest helicopter   _____
 landing site   _____

PROPERTY

Property Damage _____
                _____
                _____
                _____

PERSONNEL

Number of Crew members _____
Condition of Crew
                                uninjured
                                injured
1. _____   dead
                                uninjured
                                injured
2. _____   dead
                                uninjured
                                injured
3. _____   dead
                                uninjured
                                injured
4. _____   dead
                                uninjured
                                injured
5. _____   dead
                                uninjured
                                injured
6. _____   dead
                                uninjured
                                injured
7. _____   dead
                                uninjured
                                injured
8. _____   dead
                                uninjured
                                injured
9. _____   dead
                                uninjured
                                injured
10. _____   dead

CIVILIANS

Number of Civilian Injuries _____
Condition of Civilians
                                uninjured
                                injured
1. _____   dead
                                uninjured
                                injured
2. _____   dead
                                uninjured
                                injured
3. _____   dead
                                uninjured
                                injured
4. _____   dead
                                uninjured
                                injured
5. _____   dead
                                uninjured
                                injured
6. _____   dead
                                uninjured
                                injured
7. _____   dead

Special Medical Help Needed _____

_____

Figure 14-1a.   Aircraft crash form

AIRCRAFT CONFIGURATION

Emergency response personnel must become familiar with the configuration of the various airplanes that fly in their area. However, because of the many types that are available, it is not possible to cover them all in this text. Some of the common ones are included. Further information can be obtained directly from the manufacturer or the airlines.

Emergency service personnel must be familiar with the following:

1. Normal and emergency exits
2. Seating configuration
3. Cargo configuration
4. Fuel tank locations and sizes
5. Oxygen system storage location
6. Hydraulic fluid reservoir
7. Military armament
8. Ejection seat operation

Figures 14-1 through 14-13 cover the most common aircraft manufactured by the Boeing Company and in use in most commercial airports. Douglas Aircraft flammable material locations and emergency rescue access areas are shown in Figures 14-14 through 14-16. The Japanese-built YS-11 is detailed in Figure 14-17. The flammable materials locations and emergency rescue access areas for the Fairchild F-27/FH-227 are shown in Figure 14-18, while the same information for the Lockheed Electra is contained in Figure 14-19.

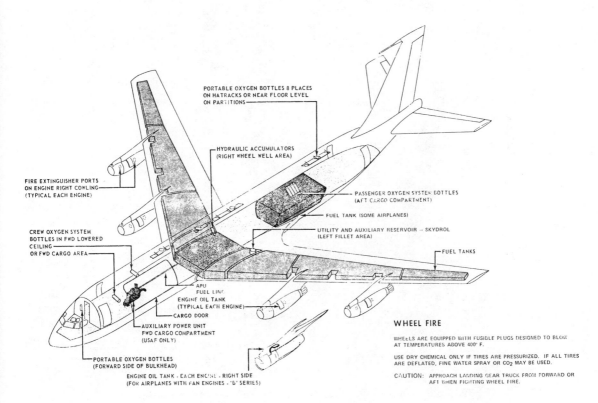

Figure 14-1b. Boeing 707-300 and 400 flammable material locations. *Courtesy of Boeing Co.*

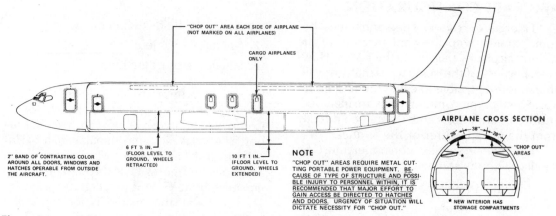

Figure 14-2. Boeing 707-300 and 400 emergency rescue access. *Courtesy of Boeing Co.*

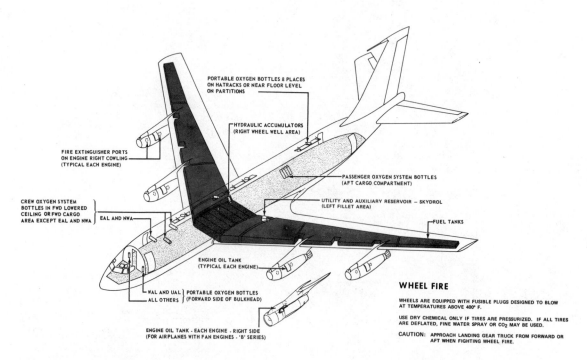

Figure 14-3. Boeing 720-720B flammable material locations. *Courtesy of Boeing Co.*

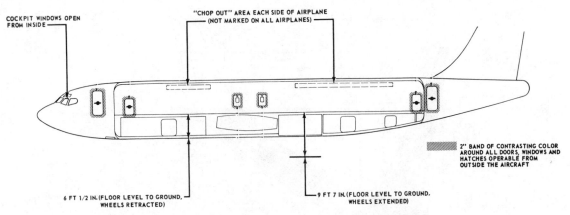

COCKPIT WINDOWS OPEN
FROM INSIDE

"CHOP OUT" AREA EACH SIDE OF AIRPLANE
(NOT MARKED ON ALL AIRPLANES)

2" BAND OF CONTRASTING COLOR
AROUND ALL DOORS, WINDOWS AND
HATCHES OPERABLE FROM
OUTSIDE THE AIRCRAFT

6 FT 1/2 IN. (FLOOR LEVEL TO GROUND,
WHEELS RETRACTED)

9 FT 7 IN. (FLOOR LEVEL TO GROUND,
WHEELS EXTENDED)

Figure 14-4. Boeing 720-720B emergency rescue access. *Courtesy of Boeing Co.*

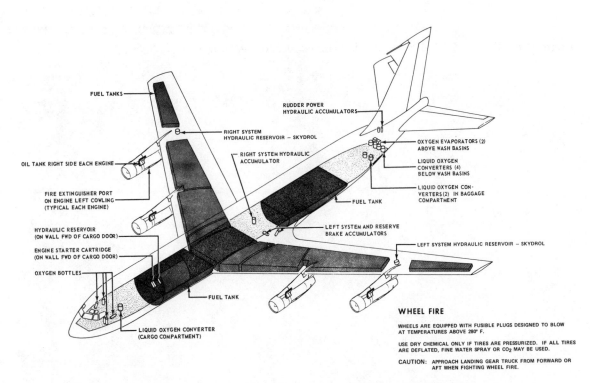

FUEL TANKS

RUDDER POWER
HYDRAULIC ACCUMULATORS

RIGHT SYSTEM
HYDRAULIC RESERVOIR – SKYDROL

OIL TANK RIGHT SIDE EACH ENGINE

RIGHT SYSTEM HYDRAULIC
ACCUMULATOR

OXYGEN EVAPORATORS (2)
ABOVE WASH BASINS

LIQUID OXYGEN
CONVERTERS (4)
BELOW WASH BASINS

FIRE EXTINGUISHER PORT
ON ENGINE LEFT COWLING
(TYPICAL EACH ENGINE)

LIQUID OXYGEN CON-
VERTERS (2) IN BAGGAGE
COMPARTMENT

FUEL TANK

HYDRAULIC RESERVOIR
(ON WALL FWD OF CARGO DOOR)

LEFT SYSTEM AND RESERVE
BRAKE ACCUMULATORS

ENGINE STARTER CARTRIDGE
(ON WALL FWD OF CARGO DOOR)

LEFT SYSTEM HYDRAULIC RESERVOIR – SKYDROL

OXYGEN BOTTLES

FUEL TANK

LIQUID OXYGEN CONVERTER
(CARGO COMPARTMENT)

WHEEL FIRE

WHEELS ARE EQUIPPED WITH FUSIBLE PLUGS DESIGNED TO BLOW
AT TEMPERATURES ABOVE 280° F.

USE DRY CHEMICAL ONLY IF TIRES ARE PRESSURIZED. IF ALL TIRES
ARE DEFLATED, FINE WATER SPRAY OR CO2 MAY BE USED.

CAUTION: APPROACH LANDING GEAR TRUCK FROM FORWARD OR
AFT WHEN FIGHTING WHEEL FIRE.

Figure 14-5. Boeing VC-135B flammable materials locations. *Courtesy of Boeing Co.*

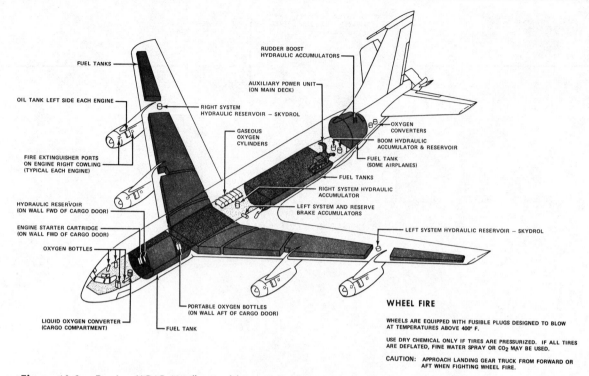

FUEL TANKS

OIL TANK LEFT SIDE EACH ENGINE

RIGHT SYSTEM
HYDRAULIC RESERVOIR – SKYDROL

FIRE EXTINGUISHER PORTS
ON ENGINE RIGHT COWLING
(TYPICAL EACH ENGINE)

GASEOUS
OXYGEN
CYLINDERS

HYDRAULIC RESERVOIR
(ON WALL FWD OF CARGO DOOR)

ENGINE STARTER CARTRIDGE
(ON WALL FWD OF CARGO DOOR)

OXYGEN BOTTLES

LIQUID OXYGEN CONVERTER
(CARGO COMPARTMENT)

PORTABLE OXYGEN BOTTLES
(ON WALL AFT OF CARGO DOOR)

FUEL TANK

RUDDER BOOST
HYDRAULIC ACCUMULATORS

AUXILIARY POWER UNIT
(ON MAIN DECK)

OXYGEN
CONVERTERS

BOOM HYDRAULIC
ACCUMULATOR & RESERVOIR

FUEL TANK
(SOME AIRPLANES)

FUEL TANKS

RIGHT SYSTEM HYDRAULIC
ACCUMULATOR

LEFT SYSTEM AND RESERVE
BRAKE ACCUMULATORS

LEFT SYSTEM HYDRAULIC RESERVOIR – SKYDROL

WHEEL FIRE

WHEELS ARE EQUIPPED WITH FUSIBLE PLUGS DESIGNED TO BLOW
AT TEMPERATURES ABOVE 400° F.

USE DRY CHEMICAL ONLY IF TIRES ARE PRESSURIZED. IF ALL TIRES
ARE DEFLATED, FINE WATER SPRAY OR CO₂ MAY BE USED.

CAUTION: APPROACH LANDING GEAR TRUCK FROM FORWARD OR
AFT WHEN FIGHTING WHEEL FIRE.

Figure 14-6. Boeing KC/C-135 flammable material locations. *Courtesy of Boeing Co.*

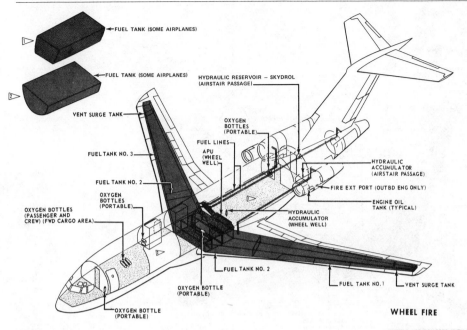

Figure 14-7. Boeing 727 100-200 flammable material locations. *Courtesy of Boeing Co.*

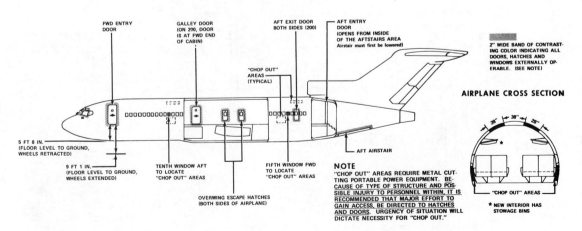

Figure 14-8. 727 100-200 emergency rescue access. *Courtesy of Boeing Co.*

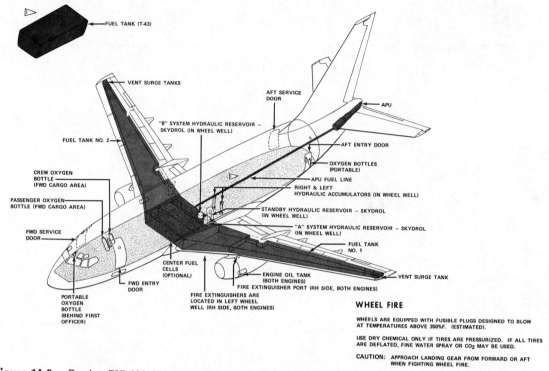

FUEL TANK (T-43)

VENT SURGE TANKS

AFT SERVICE DOOR

APU

"B" SYSTEM HYDRAULIC RESERVOIR – SKYDROL (IN WHEEL WELL)

FUEL TANK NO. 2

AFT ENTRY DOOR

OXYGEN BOTTLES (PORTABLE)

CREW OXYGEN BOTTLE (FWD CARGO AREA)

APU FUEL LINE

RIGHT & LEFT HYDRAULIC ACCUMULATORS (IN WHEEL WELL)

PASSENGER OXYGEN BOTTLE (FWD CARGO AREA)

STANDBY HYDRAULIC RESERVOIR – SKYDROL (IN WHEEL WELL)

FWD SERVICE DOOR

"A" SYSTEM HYDRAULIC RESERVOIR – SKYDROL (IN WHEEL WELL)

FUEL TANK NO. 1

CENTER FUEL CELLS (OPTIONAL)

VENT SURGE TANK

FWD ENTRY DOOR

ENGINE OIL TANK (BOTH ENGINES)

FIRE EXTINGUISHER PORT (RH SIDE, BOTH ENGINES)

PORTABLE OXYGEN BOTTLE (BEHIND FIRST OFFICER)

FIRE EXTINGUISHERS ARE LOCATED IN LEFT WHEEL WELL (RH SIDE, BOTH ENGINES)

WHEEL FIRE

WHEELS ARE EQUIPPED WITH FUSIBLE PLUGS DESIGNED TO BLOW AT TEMPERATURES ABOVE 350%F. (ESTIMATED).

USE DRY CHEMICAL ONLY IF TIRES ARE PRESSURIZED. IF ALL TIRES ARE DEFLATED, FINE WATER SPRAY OR CO_2 MAY BE USED.

CAUTION: APPROACH LANDING GEAR FROM FORWARD OR AFT WHEN FIGHTING WHEEL FIRE.

Figure 14-9. Boeing 737 100-200-200C flammable materials locations. *Courtesy of Boeing Co.*

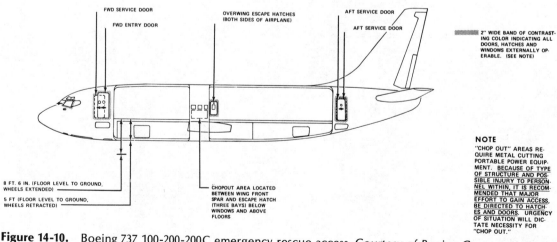

FWD SERVICE DOOR

FWD ENTRY DOOR

OVERWING ESCAPE HATCHES (BOTH SIDES OF AIRPLANE)

AFT SERVICE DOOR

AFT SERVICE DOOR

2" WIDE BAND OF CONTRASTING COLOR INDICATING ALL DOORS, HATCHES AND WINDOWS EXTERNALLY OPERABLE. (SEE NOTE)

8 FT. 6 IN. (FLOOR LEVEL TO GROUND, WHEELS EXTENDED)

5 FT (FLOOR LEVEL TO GROUND, WHEELS RETRACTED)

CHOPOUT AREA LOCATED BETWEEN WING FRONT SPAR AND ESCAPE HATCH (THREE BAYS) BELOW WINDOWS AND ABOVE FLOORS

NOTE

"CHOP OUT" AREAS REQUIRE METAL CUTTING PORTABLE POWER EQUIPMENT. BECAUSE OF TYPE OF STRUCTURE AND POSSIBLE INJURY TO PERSONNEL WITHIN, IT IS RECOMMENDED THAT MAJOR EFFORT TO GAIN ACCESS BE DIRECTED TO HATCHES AND DOORS. URGENCY OF SITUATION WILL DICTATE NECESSITY FOR "CHOP OUT."

Figure 14-10. Boeing 737 100-200-200C emergency rescue access. *Courtesy of Boeing Co.*

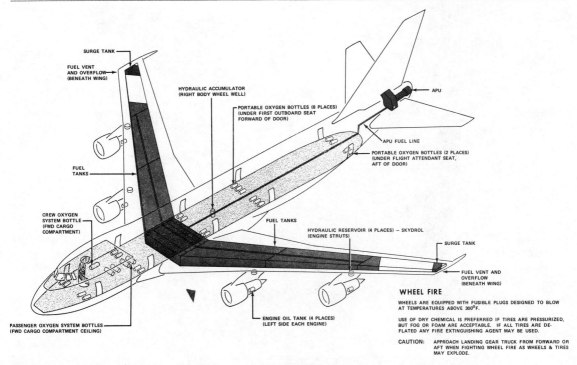

Figure 14-11. Boeing 747 flammable material locations. *Courtesy of Boeing Co.*

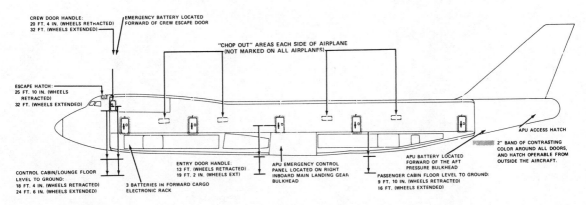

Figure 14-12. Boeing 747 emergency rescue access. *Courtesy of Boeing Co.*

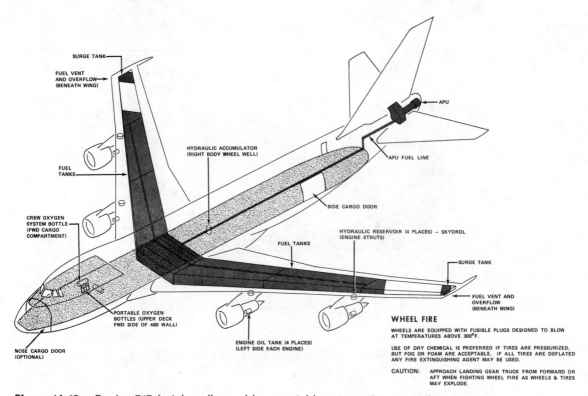

Figure 14-13. Boeing 747 freighter flammable material locations. *Courtesy of Boeing Co.*

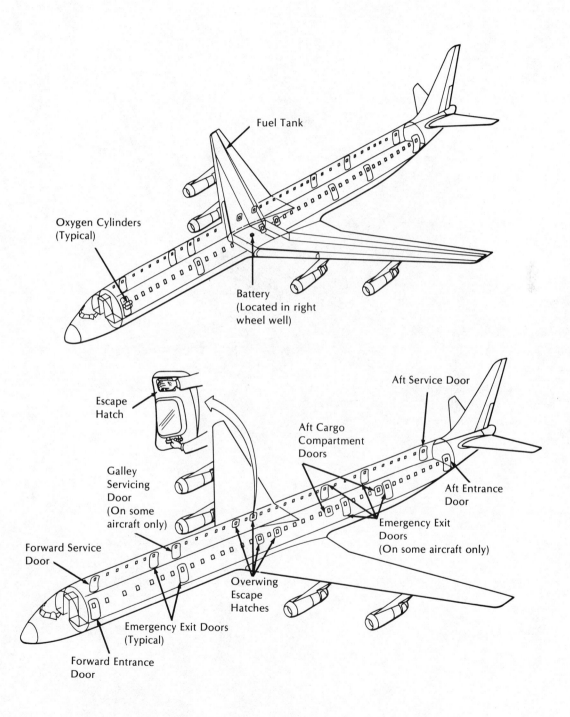

Figure 14-14. Douglas DC-8 flammable material locations and emergency rescue access

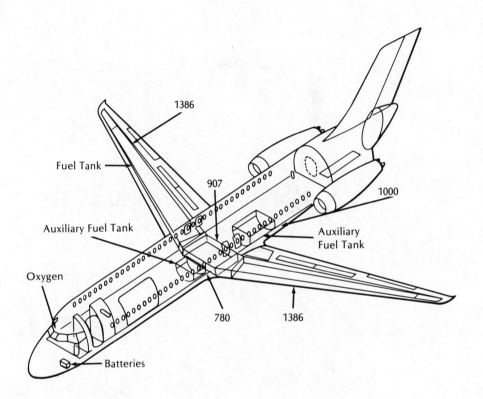

Figure 14-15. Douglas DC-9 flammable material locations and emergency rescue access

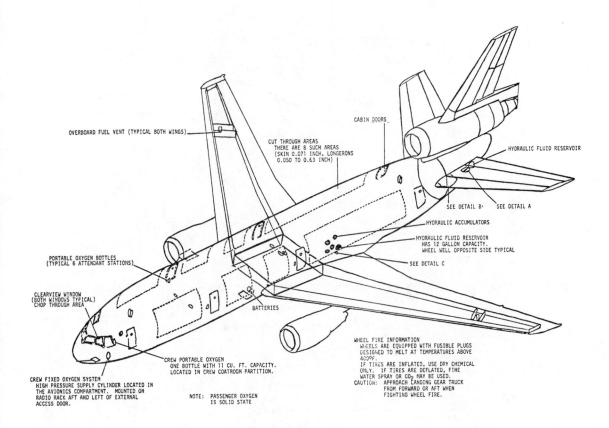

OVERBOARD FUEL VENT (TYPICAL BOTH WINGS)

CABIN DOORS

CUT THROUGH AREAS
THERE ARE 8 SUCH AREAS
(SKIN 0.071 INCH, LONGERONS
0.050 TO 0.63 INCH)

HYDRAULIC FLUID RESERVOIR

SEE DETAIL B'   SEE DETAIL A

HYDRAULIC ACCUMULATORS

HYDRAULIC FLUID RESERVOIR
HAS 12 GALLON CAPACITY.
WHEEL WELL OPPOSITE SIDE TYPICAL

SEE DETAIL C

PORTABLE OXYGEN BOTTLES
(TYPICAL 6 ATTENDANT STATIONS)

CLEARVIEW WINDOW
(BOTH WINDOWS TYPICAL)
CHOP THROUGH AREA

BATTERIES

WHEEL FIRE INFORMATION
WHEELS ARE EQUIPPED WITH FUSIBLE PLUGS
DESIGNED TO MELT AT TEMPERATURES ABOVE
400°F.
IF TIRES ARE INFLATED, USE DRY CHEMICAL
ONLY. IF TIRES ARE DEFLATED, FINE
WATER SPRAY OR CO_2 MAY BE USED.
CAUTION: APPROACH LANDING GEAR TRUCK
         FROM FORWARD OR AFT WHEN
         FIGHTING WHEEL FIRE.

CREW PORTABLE OXYGEN
ONE BOTTLE WITH 11 CU. FT. CAPACITY.
LOCATED IN CREW COATROOM PARTITION.

CREW FIXED OXYGEN SYSTEM
HIGH PRESSURE SUPPLY CYLINDER LOCATED IN
THE AVIONICS COMPARTMENT. MOUNTED ON
RADIO RACK AFT AND LEFT OF EXTERNAL
ACCESS DOOR.

NOTE: PASSENGER OXYGEN
      IS SOLID STATE

Figure 14-16. Douglas DC-10 flammable material locations and emergency rescue access

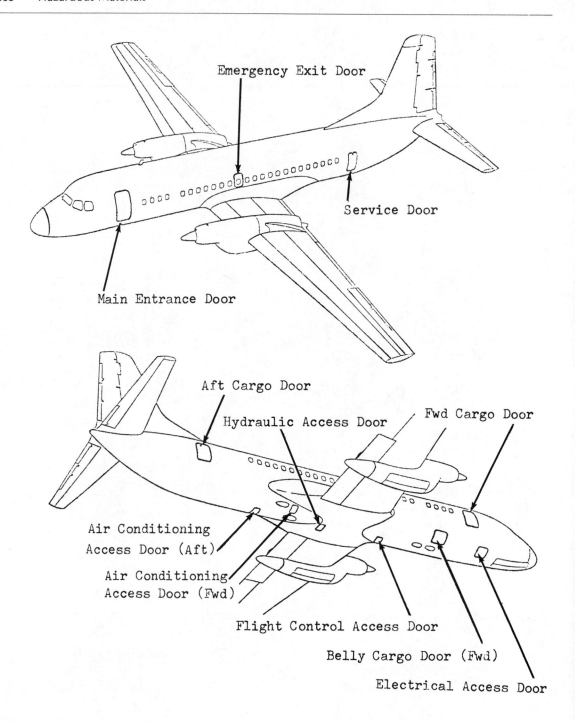

Figure 14-17. YS-11 emergency rescue access. *Courtesy of Piedmont Aviation, Inc.*

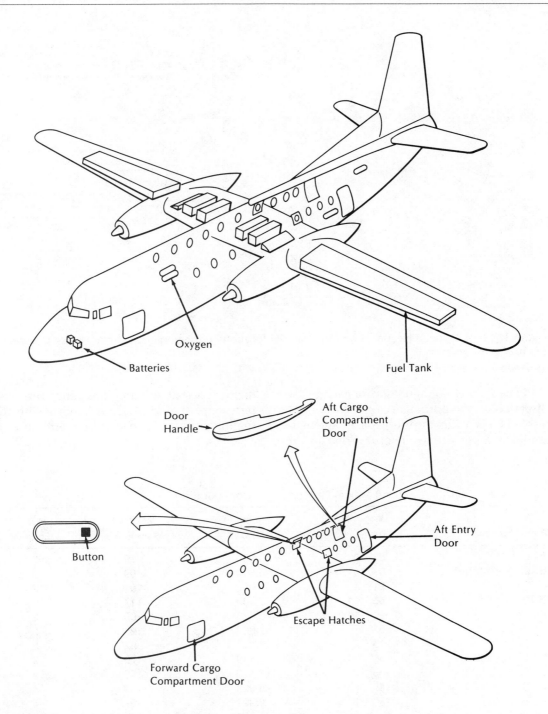

Figure 14-18. Fairchild F-27/FH-227 flammable material locations and emergency rescue access

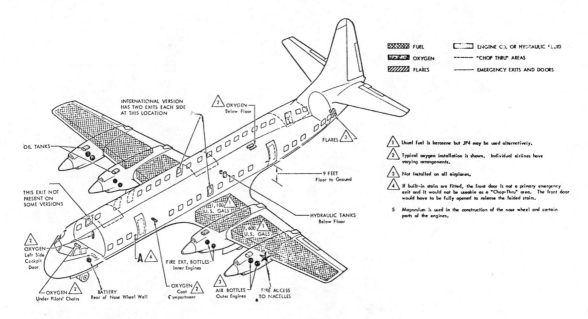

Figure 14-19. Lockheed Electra flammable material locations and emergency rescue access. *Courtesy of Lockheed-California Co.*

The seating configuration of each of the aircraft is very important in a rescue situation. Table 14-1 provides the fuel capacity and maximum number of passengers carried in the major types of aircraft. Emergency response personnel should become familiar with the seating patterns for each of the aircraft.

Table 14-1. Fuel Capacity and Maximum Passenger Configuration

Type Aircraft	Length of Fuselage[a]	Maximum Fuel (U.S. Gals.)	Maximum Passengers[b]	No. of Engines
BAC 1-11	83′ 10″	5,900	79	2
BAC 1-11 Stretch	97′ 4″	5,700	99	2
B 737	91′ 1″	8,000	101	2
B 737 Stretch	96′ 11″	7,800	117	2
DC 9	92′ 11″	3,724	109	2
DC 9 Stretch	101′ 7″	7,000	125	2
B 727	116′ 2″	11,415	82	3
B 727 Stretch	136′ 2″	11,689	170	3
B 707 (100/200)	138′ 10″	17,406	179	4
B 707 (300/400)	145′ 6″	28,923	189	4
B 707	130′ 6″	9,321	167	4
DC 8	135′ 9″	28,996	176	4
DC 8 (62)	153′ 0″	29,800	189	4
DC 8 (61/63)	183′ 0″	30,100	251	4

Table 14-1.   (continued)

Type Aircraft	Length of Fuselage[a]	Maximum Fuel (U.S. Gals.)	Maximum Passengers[b]	No. of Engines
DC 10	181′ 0″	26,400	238	3
L 1011	177′ 11″	26,593	230	3
B 747	228′ 6″	50,000	490	4
F 27/FH 227	72′ 6″	1,333	44	2
YS 11	86′ 3″	1,466	60	2
L 188 (Electra)	105′ 0″	5,400	98	4

[a]Nose to tip of Tail Cone.

[b]Each Airlines seating configuration may change maximum number of passengers carried.

TRANSPORTING HAZARDOUS MATERIALS

The carrying of hazardous materials on an aircraft has been of great concern to the airlines, pilots, and cargo handlers for quite some time. On August 27, 1974, the member carriers of the American Transportation Association adopted a position on transporting hazardous materials.[9] The airlines make the following recommendations:

The airlines recommend the following immediate and subsequent steps to be taken regarding air transportation of potentially hazardous materials:

(a) As an immediate step, on passenger-carrying and all-cargo aircraft, the federally authorized transportation of radioactive materials must be limited by the Department of Transportation to those materials used for medical diagnostic or treatment purposes and to those industrial radioisotopes which have a half life requiring their movement by air. With appropriate guidance from nuclear experts, more stringent limitations should be established promptly by the DOT as to the maximum transport indexes (measurement of radiation) permitted per package and as to the number of transport indexes permitted on an aircraft. The separation distances of such pack-

ages from passengers, crew and animals, should be increased. Central monitoring facilities for radioactive materials should be established at airports and operated or licensed by the federal government.

(b) A thorough and prompt review by the DOT, to be completed within 60 days, of the regulations governing the air transportation of all potentially hazardous materials (with special emphasis on radioactive materials) should be undertaken, with the assistance of experts in the field, and carriers, shippers (including manufacturers, packagers, and forwarders), and other interested parties, to determine, in the light of today's knowledge, environment and needs:

(1) whether safety requires that certain materials presently acceptable for air transportation should be temporarily or permanently prohibited from transportation on passenger or all-cargo flights;

(2) whether the packaging, marking, labeling, handling and loading regulations require additional changes to further ensure safety; and

(3) whether shippers' educational programs are adequate.

"Following such determinations, appropriate regulatory action should be immediately undertaken."

(c) Effective federal agency monitoring and enforcement must start with the shipper (including manufacturers, packagers and forwarders) at his facilities. Funds must be appropriated to assure adequate inspection of shippers and carriers by the responsible federal agencies.

(d) Considering the basic responsibility of shippers (including manufacturers, packagers, and forwarders) to adhere to hazardous materials regulations and the potentially serious impact which violations may have upon flight safety, severe civil and criminal penalties must be imposed against violators of federal regulations governing the transportation of potentially hazardous materials.

The regulations covering transportation of hazardous materials by air are contained in Title 14, Part 103, of the Code of Federal Regulations. Excerpts of these regulations which are of interest to emergency response personnel are quoted here:

§103.3 Certification requirements.

(a) No shipper may offer, and no person operating an aircraft may knowingly accept, any dangerous article for shipment in an aircraft unless there is accompanying the shipment a clear and visible statement that the shipment complies with the requirements of this part. In the case of shipments in passenger-carrying aircraft, the shipper shall also state that the shipment complies with the requirements in this part for carrying dangerous articles in passenger-carrying aircraft. The shipper or his authorized agent shall sign the statement or stamp it with a facsimile of his signature. The person oper-

ating an aircraft may rely on the shipper's statement as prima facie evidence that the shipment complies with the requirements of this part.

(b) The shipper shall execute the required certificates in duplicate. One signed copy accompanies the shipment and the originating air carrier retains the other signed copy.

§103.11 Packing and marking requirements.

Except as otherwise provided in this Part, each shipper who packs or marks a dangerous article for shipment under this Part shall pack or mark that article in accordance with 49 CFR Parts 172, 173 and 178 applicable to rail express. (This means that all marking must conform to the requirements for shipment by rail.)

§103.13 Labeling requirements.

Except as otherwise provided in this part, the shipper shall label each dangerous article, that is acceptable under this part for transportation in air commerce, with the appropriate label required by 49 CFR Parts 172 through 178, even though that article is exempt from those labeling requirements because of quantity and packing limitations. (This means that all labels must be applied in accordance with the same requirements used for highway or rail transportation.)

§103.28 Reporting certain dangerous article incidents.

(a) Each carrier who transports dangerous articles shall report to the nearest ACDO, FSDO, GADO or other FAA facility by telephone at the earliest practicable moment after each incident that occurs during the course of transportation (including loading, unloading or temporary storage) in which as a direct result of any dangerous article—

(1) A person is killed;

(2) A person receives injuries requiring his hospitalization;

(3) Estimated carrier or other property damage, or both, exceeds $50,000; or

(4) Fire, breakage, or spillage or suspected radioactive contamination occurs involving shipment of radioactive materials.

(5) Fire, breakage, spillage, or suspected contamination occurs involving shipment of etiologic agents. In place of the report required by paragraph (a) of this section, a report on an incident in-

volving etiologic agents may be made by telephone directly to the Director, Center for Disease Control, U.S. Public Health, Atlanta, Ga., Area Code 404-633-5313.

(6) A situation exists of such a nature that in the judgment of the carrier, it should be reported to the Department even though it does not meet the criteria of subparagraphs (1), (2), or (3) of this paragraph, e.g., a continuing danger to life exists at the scene of the incident.[10]

HANDLING HAZARDOUS MATERIALS INCIDENT

Handling a hazardous material incident which results from the cargo in an airplane is very similar to handling an incident in any other mode of transportation. Emergency response personnel should take the following steps when an aircraft has made a nonnormal landing:

1. Check the shipping papers to determine if any hazardous materials are contained in the cargo. When examining the cargo, check for any hazardous materials label.

2. If radioactive material is carried as cargo, follow the procedures in Chapter 4. Keep all personnel away from the scene until

the amount of radioactivity can be determined.

3. Follow the notification procedure outlined in this chapter.

4. If the aircraft is from the military, care must be exercised because armament might be present. If some type of armament is being carried evacuate a 2,000 foot (600 meters) radius, including emergency response personnel. All personnel must exercise extreme caution and not touch or disturb any armament exposed to fire as it might be unstable.

HANDLING A DOWNED AIRCRAFT INCIDENT

When handling a downed aircraft incident, emergency response personnel have two basic problems: rescue of victims and control of hazardous materials within the aircraft (fuel, oxygen, hydraulic fluids). Some possible problems and recommended procedures to follow are given below. However, since the purpose of this text is not aircraft rescue, the reader is directed to the refer-

ences given earlier (see Notes 1 and 2) for more detailed coverage of this subject area.

1. Once the plane is on the ground, rescue becomes the primary concern of the emergency response personnel. If there is no fire upon arrival on the scene, then any spilled flammable materials must be blanketed to reduce vapors. Ignition sources, including the craft electrical system must

be eliminated. Shut off the fuel supply if possible. Entry to the crew and/or passenger area can be made from the exterior (refer to the section Aircraft Configuration in this chapter).

2. If the emergency occurs in the air, the plane's crew will begin to operate their emergency systems. This includes oxygen and electrical shutdown. Once on the ground, the crew will deploy the emergency slides and begin the evacuation. Emergency response personnel will have to assist from the outside.

3. If an engine fire occurs in flight, the pilot will shut down the engine by turning off the fuel. The extinguishing system will be used in an attempt to put out the fire. If the engine is still burning after landing, use carbon dioxide or dry chemical for extinguishment. Water can be used to keep surrounding areas cool. If it appears that the carbon dioxide or dry chemical will not do the job, the decision must be made on whether to let the fire burn itself out or to extinguish with water. However, keep in mind that water will cause severe damage to a jet engine. This must be weighed against the damage caused by the fire.

4. If the landing gear fails on touchdown, the aircraft could veer in any direction. Emergency response personnel should stay out of the way until touchdown and then follow craft down the runway. The approach to a landing gear failure should be from the front or back and never the sides (in-line with the axle). It is possible that too rapid cooling or the heat from the fire could cause an explosive failure of the wheel. Fires of this type should be extinguished with short bursts (5 to 10 seconds) of water. On most jet aircraft, there is a fusible plug that melts and deflates the tire before there is a dangerous buildup of pressure.

5. If the aircraft is involved in fire, a coordinated attack is necessary to protect the escape routes while rescue is going on. Two examples of the type of procedures which are necessary are shown in Figures 14-20 and 14-21. A combination of handlines and crash units can be used to accomplish this. The characteristics of aviation fuel are detailed in Table 14-2.

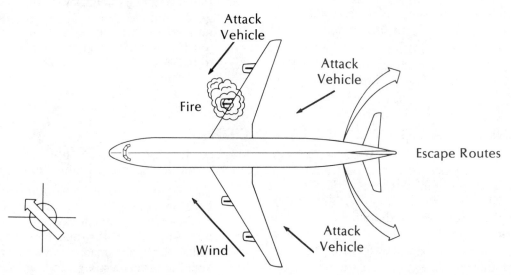

Figure 14-20.   Protecting the area escape routes

6. Remember, if at all possible, carefully observe the scene as you arrive. Those individuals who are definite fatals should not be moved unless absolutely necessary. In addition, parts of the aircraft, switches, or controls should not be touched or moved unless absolutely necessary for rescue. Emergency response personnel should make a mental note of the position of the equipment before moving or changing it.

7. If the aircraft is from the military, be careful of the ejection seat. It is possible for this to fire, causing injury to emergency response personnel in the cockpit area.

8. In many aircraft accidents, identification of victims becomes very difficult. To aid in this process the FBI has organized a Disaster Squad. As of June 30, 1976, this squad has furnished identification assistance at the scene of 114 major disasters, including over 80 plane crashes.[11] A request for the assistance of the FBI Disaster Squad may originate from the ranking law enforcement official at the scene or from a representative of the National Transportation Safety Board or from an official of any public transportation agency or facility involved. The request may be transmitted through a representative of the nearest FBI field office or resident agency. Requests from other sources submitted through official channels are also considered, as are invitations from foreign governments or U.S. citizens. This is a cost-free, humanitarian, and cooperative service offered by the FBI. (Title 28, Code of Federal Regulations, section 0.85, authorizes the Director of the FBI, subject to the general supervision of the Attorney General, to provide identification assistance in disasters.)

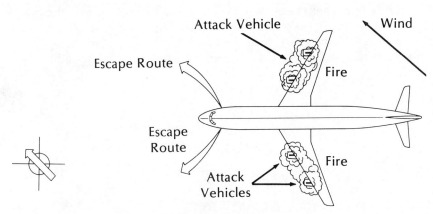

Figure 14-21. Protecting the front escape routes

Table 14-2. Physical Properties of Aviation Fuel

Characteristics	AVGAS	Jet A, JP-5 JP-6	Jet A-1	Jet B, JP-4
Common Name	gasoline	kerosene	kerosene	blend of kerosene and gasoline
Freeze Point	–76° F	–40° F	–58° F	–60° F
Flash Point	–50° F	+95° F to +145° F	+95° F to +145° F	–10° F to +30° F
Flammability limit	1.4 to 7.6%	0.6 to 4.9%	0.6 to 4.9%	0.8% to 5.6%
Autoignition Temp	+825° F to +960° F	+440° F to +475° F	+440° F to +475° F	+470° F to +480° F

RECENT AIRCRAFT INCIDENTS

By reviewing past incidents, emergency response personnel can determine whether or not they would be prepared to handle a similar situation.

Aircraft Accident, Chicago, Illinois, August 6, 1976[12]

According to the National Transportation Safety Board, a North American TB-25N crashed last August 6 near Midway Airport, Chicago, Illinois, because smoke in the cockpit from an engine fire prevented the flight crew from maintaining control in an attempted emergency landing.

Smoke and fire from a massive failure in the left engine of the converted World War II bomber traveled through the bomb bay to the cockpit before the crash, the Board found; an inspection system which was used was not effective in detecting the impending engine failure. Both pilots died when the twin engine B-25 crashed in a residential section. One person on the ground was killed and another seriously injured. Two houses, two garages, three cars, and a boat were destroyed in the crash and postimpact fire.

The Safety Board determined that the probable cause of the accident was the deterioration of the cockpit environment, due to smoke, to the extent that the crew could not function effectively in controlling the aircraft under emergency conditions.

Jumbo Jets Crash, Tenerife, Canary Islands, March 27, 1977[13]

On March 27, 1977, a Pan Am 747 and a KLM 747 collided on a foggy runway causing the worst disaster in aviation history. There were 577 fatalities from the accident.

In addition to the fatalities, however, it was reported that, "about 950 pounds of depleted uranium are added to the tail of every 747 to counterbalance the upper rudder and outer elevators, which help the planes go up or down." Authorities said there was no radioactive problem from the uranium. It was used solely because a small quantity is so heavy.

QUESTIONS FOR REVIEW AND DISCUSSION

1. Describe the hazards that emergency response personnel can experience from an aircraft incident.

2. Outline the preplanning steps necessary to be accomplished before an aircraft incident.

3. What notifications must be made when a civilian aircraft crashes? Prepare a notification list for your area.

4. Describe the parts of an airplane that emergency response personnel must be familiar with and explain their importance.

5. Prepare a drawing for each type of commercial aircraft that passes your area.

6. Survey a local, small-plane airport in your area and determine the types of hazardous materials carried.

7. Outline the steps that emergency response personnel should follow when handling an aircraft hazardous materials incident.

CASE PROBLEM—AIRCRAFT

On its approach to New Town (California) International Airport, a commuter aircraft carrying 12 passengers and 3 crew members crashes into a row of residential houses. The crash occurs at 6:30 P.M. on a Monday night in January. The temperature is 44°F (6.7°C), with fog and drizzle.

The crash sets 5 single-family residences on fire. Each house is occupied by 1 to 4 members of the family. Most are young children.

Debris from the wreck is scattered over a wide area. Fuel has spilled and is running down the street and into the storm sewer. The spill has ignited. Of the 12 passengers and 3 crew members, several of the passengers are still alive. The fate of the others is unknown at this time.

The fire department has 28 persons on duty at the time of the crash. It has 4 pumpers, 2 ambulances, 1 ladder truck, and 1 heavy squad truck. In addition, the airport fire department, 2 miles away, has 2 crash trucks available.

Mutual aid is available within a 10-mile radius of 40 persons with 6 pumpers, 3 ambulances, and 2 ladder trucks.

1. Describe in detail how you would handle this incident.

NOTES

1. E. Hudiburg and C. E. McCoy, *Aircraft Fire Protection and Rescue Procedures*, IFSTA No. 206, 1st. ed. (Stillwater, Okla.: International Fire Service Training Association, 1970).

2. *Manual on Aircraft Rescue and Fire-Fighting Techniques for Fire Departments Using Structural Fire Apparatus and Equipment*, NFPA 406M (Boston, Mass.: National Fire Protection Association, 1975).

3. *Listing of Accidents/Incidents by Aircraft Make and Model, U.S. Civil Aviation, 1975*, National Transportation Safety Board, NTSB-AMM-77-1 (Washington, D.C.), 1977.

4. "Plane with Radioactive Matter Crashes," *Washington Post*, January 23, 1978, p. C3.

5. *Code of Federal Regulations, Part 430, Rules Pertaining to the Notification and Reporting of Aircraft Accidents, Incidents and Overdue Aircraft, and Preservation of Aircraft Wreckage, Mail, and Records* (Washington, D.C.: Government Printing Office, 1976).

6. "Title VII—Aircraft Accident Investigation, Accidents Involving Civil Aircraft," Sec. 701 (72 Stat. 781 as amended by 76 Stat. 921, 49 U.S.C. 1441) (a) (Washington, D.C.: Government Printing Office, 1976).

7. "Civil Aircraft Accident Investigation Guidelines," National Transportation Safety Board (Washington, D.C.: Government Printing Office).

8. *What to Do and How to Report Military Aircraft Accidents*, Department of Defense (Washington, D.C.: Government Printing Office).

9. *Airline Position on the Air Transportation of Potentially Hazardous Materials* (Washington, D.C.: American Transportation Association, 1974).

10. *Transportation of Dangerous Articles and Magnetized Materials*, Code of Federal Regulations, Title 14, Part 103 (Washington, D.C.: Government Printing Office, 1974).

11. "FBI 'Disaster Squad' Aids Uniformed Services," *Eastern Division News*, International Association of Fire Chiefs, January 1978, p. 2.

12. "Accident Report: Safety Recommenda-
tons and Responses," National Transportation Safety Board, *Federal Register*, June
9, 1977, p. 29579.

13. "Jumbo Jets in Tenerife Crash Carried
Uranium as Ballast," *Washington Post*,
p. A3.

15
Marine transportation

In addition to shipment by the other modes of transportation discussed in the preceding chapters, a number of hazardous materials are carried over the nation's waterways and into our ports. In emergency incidents on marine vessels the situation changes somewhat because of the size of the quantity involved, the accessibility into various sections of a ship, and the difficulty of boarding a vessel which may not be docked. One advantage is that in some cases a ship or barge loaded with hazardous materials can be moved to a safer location for fire control or controlled burning.

The vessels that carry hazardous materials are ships or barges. The ships may be freighters, tankers, supertankers, LNG carriers, or container vessels. The barges can be covered hoppers, tank barges, or open-hopper barges fitted with tanks. Often on tank vessels the deck piping will be a sign of possible hazardous materials.

In this chapter we will discuss identification of hazardous materials on a ship, a pre-planning outline for ships, and details on the construction of vessels used for hazardous materials transport. This is followed by a description of shipboard fire fighting equipment and shipboard fire fighting procedures in general, as well as for specific hazardous material incidents.

IDENTIFICATION

Of initial concern will be the identification of the product or products on board. If because of language or other barriers cargo information cannot be obtained from the crew, three documents will be of assistance. The first of these is the bill of lading, which would be in the shipping office of the carrier. This could take some time to locate and may require the assistance of the Coast Guard National Response Center in Washington, D.C., for a vessel under way, but is helpful for goods at the terminal. The second is the dangerous cargo manifest, which will be the most useful document. It is carried on the bridge or pilot house and by the master and mate. The document lists the material, the quantity, and its location. The master should also have a hatch plan for a freighter, which is a schematic of the load. If additional information is required, the ship's manifest details where the material originated and where it is going so these parties could be contacted. The third document is the Cargo Information Card. These cards must be carried on the bridge of a ship or pilot house of a tow; on unmanned barges they are mounted near a warning sign, posted to be read by a person standing on the deck of the barge. The warning sign, which faces outward, states, "WARNING—DANGEROUS CARGO, NO VISITORS, NO SMOKING, NO OPEN LIGHTS." On a manned barge, the Cargo Information Card and other shipping documents are with the operator.

PREPLANNING

As with other hazardous material locations, the best place to start preparation for an incident on a vessel is to preplan where it may happen. The items that should be considered on land have already been covered, but there are several special features about ships. These will not be discussed in detail, but will be listed so they can be considered by those who must face a hazardous material incident aboard a vessel.

General Information

Preplanning general information should include the name of the ship; the type of ship and its class; the general use; the line owning the vessel; the flag it sails under; the language spoken aboard; the number of crew members; the number of passengers, if any; the general specifications, such as length, beam freeboard, and empty and loaded draft; the type of power for propulsion and the number of boilers; the types of fuel and the capacities of the storage tanks; the method used to identify the compartments; names and emergency phone numbers for the owner or agent; and the manning practices in use.

Construction Information

As part of the construction information, it is important to include a diagram of the general layout of the ship. Figure 15-1 illustrates the type of diagram that should be made during preplanning to familiarize personnel with the ship. Additional data should be gathered on appropriate forms. The diagram should also include the deck arrangement, including names and usage of each; the frame numbering system used; a description of the cargo spaces including the capacity of each hold and their location, additional tanks, and the access routes to the holds, such as manholes and possibility of removing the hatch covers; the machinery spaces, including the engine room, machinery casing, fire pump, and main switchboard; the crew and passenger accommodations, noting the deck, type of accommodations, and the number of rooms.

Location Lists

Preplanning should include lists of the location of items that will be vital during an incident. These include the location of informational data such as the hazardous cargo manifest, the cargo manifest, and the blueprints of the ship; the escape trunks, their location and where they go; the various instruments and control systems, including the clineometer, smoke detection system, communication system, the carbon dioxide system, the steam smothering system, emergency remote controls for CO_2 systems, ventilation, and fuel; and the space assignments listed alphabetically by name, indicating deck, frame number, side, and primary access.

Systems Information

Systems information goes into detailed descriptions and locations of the major systems on the vessel and how to operate them. This covers the communication systems such as the public address system, sound-powered

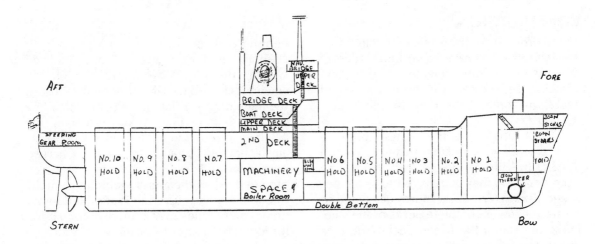

Figure 15-1. Sample diagram of ship layout

telephone system, and dial telephone system; the electrical systems; the fire protection systems, which includes alarm and detection devices, extinguishing systems detailing CO_2, foam, steam smothering, and sprinklers, and the fire main, which includes the fire stations and shore connections, fire doors, fire walls, and remote watertight doors; the hatch system with the weight of the doors and how they are removed; and the ventilation systems with the emergency shutoffs and controls.

Tactical Information

Tactical information preplanning deals with details important to the tactical fire fighting operations that have not been covered elsewhere. It includes such vital data as the location of keys for all locks, location of the pyrotechnic locker, special locations for hazardous cargo, location of the fire control plan, location of the international ship-to-shore coupling, and location and means to supplement the shipboard extinguishing systems. Data should be collected on any internal exposure problems such as open shafts or chutes, auxiliary fuel supplies, enclosed spaces or storage areas, and any unusual operations

concerning the ventilation system, especially when used in coordination with a fixed extinguishing system.

The information collected should be done for as many vessels as possible and at least for each class visiting the port or passing through the jurisdiction. Some of the data will vary; for example, a liquid natural gas (LNG) carrier and a barge would have some different considerations.

Tentative actions should be developed for the land-based equipment available and the different conditions under which a fire might be experienced: while under way, moored, tied to a dock, or during loading or unloading operations. Keep in mind the difficulties encountered with the hazardous materials on board. Changes in the tide can shift a vessel, modify fireboat operations, and require changes in land-base fire fighting equipment.

Upon final approval, the preplan material should be properly distributed to those companies and staff officers who will be involved in a ship fire. Distribution should include fireboat personnel and special boarding parties if they are available.

CONSTRUCTION

In order to confront an incident in a ship, fire fighters must have some knowledge of the way it is laid out. Freighters, tankers, and containerized vessels will have a series of decks, in some cases only in the super-structure.

The main deck is the highest complete deck extending from stem to stern. Second deck is the first complete deck below the main deck and may be referred to as the be-tween deck. Third deck, fourth deck, and so on are the lower complete decks of the ship. These are called the tween decks. The area below the lowest deck is called the lower hold. The forecastle deck is a partial deck above the main deck on the forward part of the ship. The half deck is a partial deck above the lowest complete deck. The platform deck is a partial deck below the lowest complete deck. The poop deck is a partial deck at the stern above the main deck. The superstruc-ture deck is any partial deck above the main deck which does not extend to the sides of the ship. The upper deck is a partial deck amid-ship above the main deck which extends to the sides of the ship.

The superstructure areas may be divided into three areas. The forecastle (fo'c'sle) is at the bow, the bridge is amidship, and the poop area at the stern.

Freighters

The configuration in the cargo spaces of a freighter may be holds that extend from the main deck to the lower hold. These may or may not be divided into upper and lower be-tween decks. There may also be an undivided shelter deck directly below the main deck. The tween decks are separated by watertight bulkheads.

Access to the cargo hold is through the hatch opening in the main deck. With the hatch covers open (Figure 15-2), outside fire

fighting and ventilation operations can pro-ceed. For an interior attack or use of a fixed system, the covers would remain closed. Lower decks also have hatches below those in the main deck for the loading and unloading of cargo (Figure 15-3). When removing the hatch covers to fight a fire a winch will gener-ally be needed to handle the weight. Charged hose lines and protective clothing must be provided. All in the area should be prepared to combat a flare-up as air is admitted to the hold. Lower holds may also have deep tanks for low flash point oils or fuel.

The other access to the holds is provided by one or two vertical ladders extending from the main deck to the lower hold. A small hatch cover or only part of a hatch may have to be removed to reach the ladder. Descend-ing the ladder under fire conditions is danger-ous. Never step off backwards and never let go of the ladder until a firm footing is secured since floor plates or hatch covers may be removed.

Emergency access to a hold can be through a ventilator, which usually has lad-

Figure 15-2. With the hatch covers open, outside fire fighting and ventilation can proceed

Figure 15-3. Winches may be required to move the hatch covers to reach lower decks

ders. During heavy fire conditions, however, these would act as chimneys and probably could not be used. In some cases, these ventilators can be used to deliver water to inaccessible holds if blueprints available indicate that the water will go where needed. In some cases, it will be necessary to block off the ventilators to keep air out of the hold or toxic fumes from escaping.

The fuel storage is normally in the lower portions of the ship and may be in bunker tanks, the double bottom, or deep tanks. Any escape of fuel could create a potential hazard. A fire from a leak would dangerously expose hazardous materials in the cargo area.

Containerized Ships

Containerized ships are not significantly different from a freighter. Figure 15-4 shows the layout of a containerized ship. A con-

tainer of hazardous materials would be placarded on the outside and loaded in a hold along with other containers and may thus be anywhere in the hold. The containers are stacked in the holds and after the main deck hatch covers are in place; additional containers are often placed on top. The holds are generally separated by bulkheads. Access to a fire in a containerized ship is very difficult. The containers may have to be removed one by one until the seat of the fire is found. This will require special equipment and could cause the fire to grow in intensity as additional air is introduced.

Barges

Barges for hazardous materials may be of the closed hopper type for bulk dry products or tank barges. The hopper barge is a double-walled steel floating tub with a metal cover for keeping the commodity dry and out of the weather. Most barges have no propelling machinery or living spaces.

There are approximately 3,000 barges carrying bulk liquids, which vary from petroleum products to chemicals in special purchase barges. The barges are rectangular and can be several hundred feet (meters) long. They are compartmentized by oiltight longitudinal and transverse bulkheads. Each tank has piping, venting, and an access hatch. Engines and pumps for loading and unloading may be on board or there may be no mechanical equipment.

River barges designed for large integrated tows of 20 to 40 barges are designed in three styles: lead, middle, and trail; they fit together and function as a single barge unit. Such a tow can be 200 feet (61 meters) wide and over 1,000 feet (305 meters) long. Different tow configurations are shown in Figure 15-5. Exposure problems during an incident can vary considerably, depending on the method used.

Barges are defined as moderate, signifi-

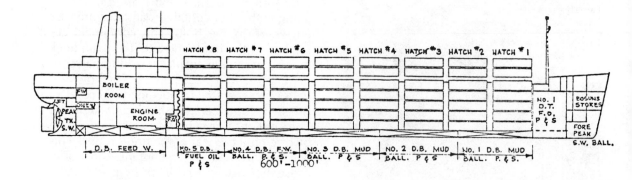

Figure 15-4. Diagram of the layout of a containerized ship. Note additional containers placed on the main deck over the hatch covers

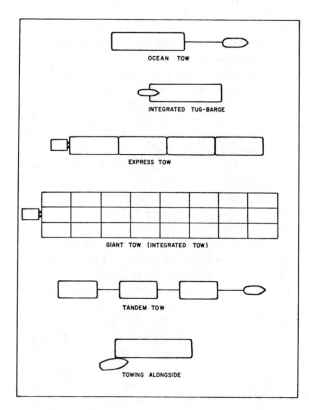

Figure 15-5. Different tow configurations in use

cant, or maximum, depending on the hazardous cargo they carry and the preventive measures required to keep the cargo from spilling or leaking. Regulations on protection against flooding, hull damage, and grounding are provided. The hull types are designed for containment and the requirements increase for collision and grounding damage as the cargo danger increases. After sustaining damage the barge must still meet temperature, pressure, and insulation requirements for safe carriage of the lading. A liquid cargo barge may carry chemicals, petrochemicals, or petroleum products such as fuel oil and gasoline. They come in varying sizes, depending on the waterways, and are divided by oiltight bulkheads. Examples of barge types, containment, cargoes, and requirements are given in Table 15-1.

The various skin arrangements and the protection each provides can be seen in Figure 15-6. The barge cargo tanks can be identified by shape (see Figure 15-7), pressure of contents, temperature, internal coatings, or whether they are built integral or independent of the barge.

Table 15-1. Barge Types and Requirements

Barge Type	Grounding Requirement	Collision Requirement	Flooding Requirement	Degree of Containment	Typical Cargoes
I	Must be able to run aground without over-stressing the hull structure.	Space required between cargo tanks and bow and sides of barge.	Must remain afloat if damaged at transverse bulkhead, with compartments flooded on both sides of damaged bulkhead.	Maximum	Chlorine (liquefied) Ethylene oxide (liquefied) Liquefied natural gas (LNG) Motor fuel antiknock compounds.
II	Same as Type I.	Space between cargo tank and bow and sides of barge required, but space may be smaller than in Type I barge.	Must remain afloat if damaged anywhere except at a transverse bulkhead.	Significant	Liquefied petroleum gases Anhydrous ammonia Camphor oil Ethyl ether Vinyl chloride
III	No requirement.	No requirement.	No requirement.	Moderate	Most acids Carbon tetrachloride Chloroform Most petroleum products, including fuel oils and gasoline.

Tankers

The tanker or tankship is a vessel specifically constructed to carry bulk liquid cargoes in its own tanks. Figure 15-8 illustrates a style of oil tanker construction commonly in use. The newer larger tankers vary somewhat, with all of the superstructure aft. Tankers can be easily distinguished from dry cargo ships by the lack of cargo handling gear and rigging. The tanker ship has pipes, manifolds, catwalks, vents, and hose handling equipment (Figure 15-9).

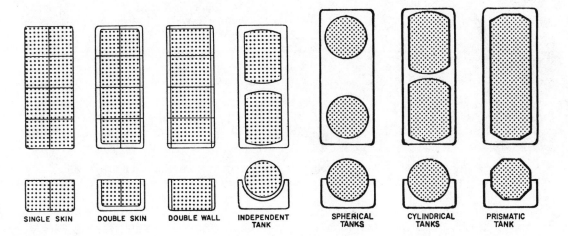

SINGLE SKIN   DOUBLE SKIN   DOUBLE WALL   INDEPENDENT TANK   SPHERICAL TANKS   CYLINDRICAL TANKS   PRISMATIC TANK

Figure 15-6. Four categories of tank barge construction

Figure 15-7. Three shapes of independent tanks on barges

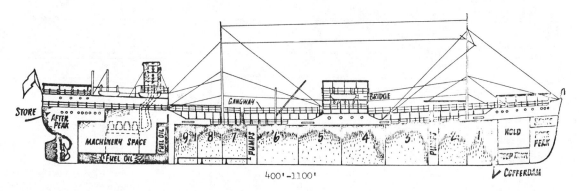

Figure 15-8. Diagram of style of tanker construction commonly in service

Figure 15-9. Deck arrangement, showing catwalk and piping obstructions that fire fighters will encounter on a tanker. The bridge is in the background

The cargo area is divided by two longitudinal bulkheads into three tanks: starboard, center, and port. A number of transverse bulkheads divide the area into sets of tanks numbered from fore to aft (front to back or bow to stern). The diagram in Figure 15-10 shows the tank arrangement and numbering system. For example, the tanks numbered 5P,

5C, and 5S would be the fifth set (port, center, starboard) from the bow, although some ships number from the stern. The location of tank bulkheads can be determined on the main deck by the rows of rivets. Older tankers will have the cargo tanks extend to the bottom of the ship, but newer ones are built with double bottoms. There are no decks within the cargo section. The bulkheads are oiltight and watertight. Pipes extend through the bottom of the tanks to the main deck where connections are made for loading and unloading. Pumps may be in an aft pump room or in watertight compartments which extend across the ship and to the bottom. The engine room is aft the cargo tanks below the deck areas. The living accommodations, navigation areas, and ship offices are in the deckhouse which may be at one location or separated.

Tankers carrying flammable or combustible liquids and liquefied gases are designed to keep ignition sources away from the cargo areas. This is done by physically separating tanks from areas where ignition sources might exist, restricting the accumulation of hazardous vapors, and building decks and bulkheads that are watertight and fire resistant to form a barrier to flame, heat, and smoke spread. The

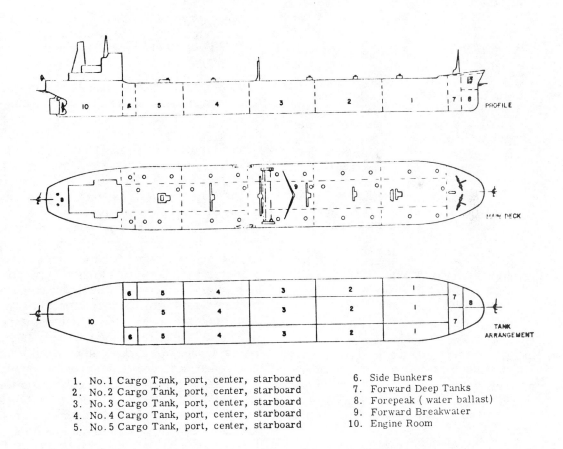

1. No. 1 Cargo Tank, port, center, starboard
2. No. 2 Cargo Tank, port, center, starboard
3. No. 3 Cargo Tank, port, center, starboard
4. No. 4 Cargo Tank, port, center, starboard
5. No. 5 Cargo Tank, port, center, starboard
6. Side Bunkers
7. Forward Deep Tanks
8. Forepeak (water ballast)
9. Forward Breakwater
10. Engine Room

Figure 15-10. Tank arrangement on tanker ships and numbering system employed to identify the tanks

regulations vary with the Coast Guard Classification of Oils, which are as follows:

Classification of Oils

For the purpose of regulating tank vessels, the regulations of the U.S. Coast Guard divide petroleum products into flammable liquids, combustible liquids, and flammable liquefied gases. These definitions are different from those used for truck and train throughout.

Flammable Liquids are liquids which will give off flammable vapors at or below 80 degrees F, or the liquid will flash in contact with flame when its temperature is at or below 80 degrees F. Flammable liquids are divided into three grades as follows:

Grade A—any flammable liquid having a Reid vapor pressure of 14 pounds or more. The Reid method of measuring vapor pressure consists of placing a small amount of liquid into a container with a pressure gauge and, after closing tight, heating the liquid to a temperature of 100 degrees F and measuring the vapor pressure, which is expressed in pounds per square inch absolute. This group includes: casinghead, natural gasoline, very light naphtha, butane blend gasoline, etc.

Grade B—any flammable liquid having a Reid vapor under 14 pounds and over 8½ pounds. This group includes most motor gasolines.

Grade C—any flammable liquid having a Reid vapor pressure of 8½ pounds or less and a flash point of 80 degrees F or below. This group includes most of the crude oils, "cut black" asphalt, creosote, benzol, toluol, alcohol, aviation gasoline, some jet fuel, etc.

Combustible Liquids are liquids which will give off flammable vapors only above 80 degrees F or liquids having a flash point above 80 degrees F. Combustible liquids are divided into two grades as follows:

Grade D—any combustible liquid having a flash point below 150 degrees F and above 80 degrees F. This group includes: kerosene, light fuel oils, very heavy crude oils, some jet fuels, etc.

Grade E—any combustible liquid having a flash point of 150 degrees F or above. This group includes: heavy fuel oils; "Bunker C;" diesel fuel; road oil; lubricating oil; asphalt; coal tar; and fish, animal, and vegetable oils.

Flammable Liquefied Gas means gas having a Reid vapor pressure exceeding 40 pounds which has been compressed and liquefied.

Separation of tanks can be by cofferdams constructed of two watertight bulkheads approximately 3 to 6 feet (0.9 to 1.8 meters) apart which can be flooded with water to form a fire break. The only openings into the cargo tanks are a manhole on the main deck and the tank vents which extend up the mast. This restricts the leakage of any vapors to an ignition source.

LNG tankers are somewhat different, mainly in the tank construction (Figure 15-11). The normal mild steel used in shipbuilding cannot be used because it loses its impact resistance and will crack easily under stress at low temperature of LNG. Spherical or prismatic tanks are constructed of stainless steel, high nickel steel, or aluminum alloy. Many have a capacity of 25,000 cubic meters. The tanks are of either freestanding or membrane construction. LNG tankers are equipped with special instrumentation to indicate the cargo level in the tanks, internal tank pressures, temperature in the cargo area, temperature in the secondary barrier area if one is provided, and gas detection equipment for several areas on the vessel. The LNG tankers are not allowed to vent off their product in U.S. ports.

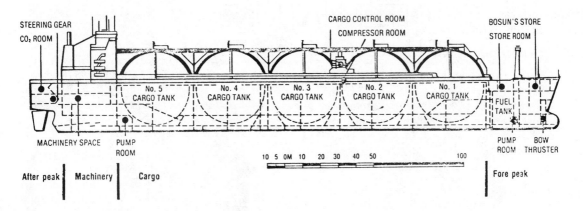

Figure 15-11. Inboard profile of LNG tanker

SHIPBOARD FIRE FIGHTING EQUIPMENT

All ships have fire fighting equipment on board. This ranges from first-aid appliances to total flooding fixed systems. Since ships differ in the equipment they have, a brief description will be given of the various resources. Each ship or its fire fighting plan will have to be checked for precisely what equipment it has. Barges are usually not provided fire fighting equipment.

Fire Main

The main fire fighting tool aboard ship is the fire main and its attached hydrants. The main may be single or looped and of 5-inch or 6-inch diameter. The major limitation is the size of the pumps supplying it. Most ships will have two pumps at separate locations for reliability. The pump size depends on the type of service the ship is in, its size, and the arrangement of the fire main system. Branch lines to hydrants are of 1½-inch or 2½-inch diameter. Hydrants must be located so that all parts of the weather decks, cargo decks, living quarters, store areas, machinery spaces or working spaces normally accessible during navigation can be reached by two streams of water. One stream must be from a single length of hose and the entire machinery area must be reached by single hoses from different outlets.

Fireboats or land-based companies can augment the supply to the ship on fire through a connection similar to a standpipe connection. This bolted flange, which provides a port facility hosethread to be attached to the fire main, is known as the International Shore Connection (Figure 15-12). Ships generally have at least one, and the connection is accessible from either side of the vessel. This enables fire departments to connect to the fire main system on a ship and supply water from land-based companies or fireboats. Fireboats and companies responding to ship fires should not only carry these adapters, but also know from preplanning where to attach them to ships frequenting the port.

Carbon Dioxide Systems

Carbon dioxide systems are widely used on ships for fire protection. Multiple-bottle storage areas provide adequate supplies for them (Figure 15-13). Carbon dioxide has several advantages; it causes no damage to cargo or machinery, is usable on electrical equipment, leaves no residue, can be used on a ship without power since it is a pressurized system, and will spread to all parts of the space it is re-

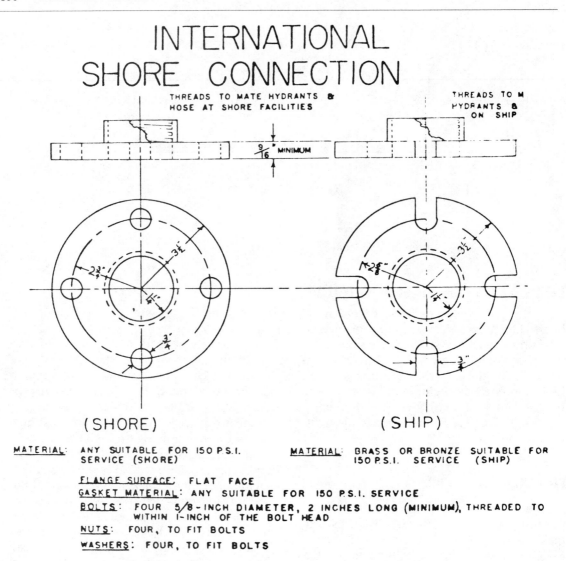

Figure 15-12. International shore connection to connect water supply from land-based companies or fireboats

leased in. The main disadvantages are the limited quantities of carbon dioxide, its ineffectiveness on oxidizing materials, and its inability to penetrate tightly packed storage such as containerized shipments.

Two systems are used on ships where hazardous materials are found. The first is a cargo or cargo tank system. Carbon dioxide is

released manually into the cargo hold, which has been tightly sealed. This may not totally extinguish the fire, but will hold it in check until a port is reached and the hold can be properly unloaded with adequate fire fighting equipment standing by. A cargo tank system is not required, but may be placed on a vessel to protect fuel storage or on tankers in

Figure 15-13. Fixed carbon dioxide systems used on ships for fire control

lieu of a steam smothering or deck foam system. These systems must release the required quantity of carbon dioxide within 5 minutes. The amount of carbon dioxide required to protect an area is considered to be 1 pound (0.45 kilogram) for every 30 cubic feet (0.85 cubic meter) of space. The operating instructions will detail the minimum number of carbon dioxide bottles to be used in relation to the amount of liquid in the tank.

The second system for ships is a total flooding system. It is used in electrical and machinery control areas where immediate extinguishment is vital to the ship's safety. In order to limit damage and maintain the effectiveness of the carbon dioxide, 85 percent of the required quantity must be discharged in 2 minutes. This system requires two separate and deliberate acts to avoid unintentional release. One is to release the required amount of agent and the other to operate the stop valve or direction valve. If additional carbon dioxide must be applied to maintain the concentration, a delayed discharge total flooding system is used.

For holds with flammable liquids, a special carbon dioxide system is used which applies the gas, at a rate of 1 pound (0.45 kilo-gram) for every 22 cubic feet (0.62 cubic meters) of space, in 2 minutes through additional piping and nozzles.

Foam Systems

Foam on ships is produced by developing a foam solution at a central location and pumping it through fixed piping systems to handlines or monitor nozzles where it is aerated.

The foam used is mainly mechanical foam, although older vessels may have chemical foam. Since some tankers carry polar solvents in the form of ketones, esters, or ethers, a foam concentrate suitable for use with these water-soluble products will be necessary. A special polar solvent foam has been developed for shipboard use which has a catalyst to increase stability so that it does not have to be applied gently, but can be discharged through handline nozzles and fixed-delivery nozzles. If this type of foam is not available, the tactic for control is to apply enough water to dilute the product, which could be considerable, and cool the exposed metal areas surrounding the fire. High-expansion and aqueous film-forming foams may soon find applications for shipboard usage.

The two systems found on ships are fixed systems piped into areas needing additional protection and deck foam systems. Fixed systems are used on many ships in special areas such as machinery, boiler or pump rooms, and bridge areas. These systems are usually manually controlled. They deliver 1.6 gallons (6 liters) per minute of foam solution for every 10 square feet (0.9 square meter) protected for a period of 3 minutes through marine floor nozzles or overhead spray deflectors.

Deck foam systems are used on tanker vessels, often in place of an inert gas system to protect the cargo tanks. These systems deliver the minimum required rate for 15 minutes. The foam system is more reliable than a gas system because of the design considerations and the possibility of an open tank fire. The piping system and foam stations are designed so that broken pipes can be circumvented and the system maintained in operation. There is strong dependence on fixed monitors due to their capacities, reach, number of personnel required to operate, and the time involved in initiating delivery when compared to handlines. Delivery is to be from aft of the area protected. This enables effective fire fighting operations to work forward from the aft house as long as foam pumps and proportioning system are aft and working.

Steam Smothering Systems

Steam smothering systems are no longer installed on ships, but an older oil tanker or cargo ship may have this type of fire protection. They can be used to flood the cargo tank or hold, which is enclosed. The fixed-piping system is controlled from a steam smothering master control valve, which is located on the main deck and marked with a sign and arrow. From the main steam line, branches extend to the various tanks, holds, and compartments. Each is controlled by an individual valve. Cargo ships may have a fitting that must be placed in the branch line before steam can be admitted to the hold. This is called the hold-steam-injection-fitting and is used to eliminate accidental discharge into a hold. The valves, connections, and the main control should be located during preplanning. These systems must never be used on water-reactive hazardous materials. It is also important that sufficient steam is available for a large continuous application.

Water Spray Systems

Water spray systems installed on a vessel are there for complete extinguishment of the fire. The exception to this is a special purpose system installed on a liquefied petroleum gas carrier to protect tanks, shield living quarters, and reduce the quantity of heat absorbed by the tank or exposed structure. Water spray is preferred for pump rooms since there is no asphyxiation hazard and no cleanup as with foam. The systems are individually designed with nozzles, capacities, and discharge patterns engineered as needed. They are not used to protect tanks or holds. The control valve is of the deluge type and is located outside the protected area in most cases. Operating instructions are posted near the marked control valve. Neither automatic nor manual sprinkler systems will be found on ships carrying hazardous materials.

Halogenated Systems

Halogenated extinguishing systems are not widely used on ships. At present only bromotrifluoromethane has been approved for fixed systems in limited application. The future may find these systems on tankers, cargo holds for container vessels, and total flooding systems for large engine rooms. Precautions will be necessary when entering areas using these agents because of the natural and heat-decomposed vapors. Ship fire fighting plans should be checked for the presence of a halogenated system.

FIRE FIGHTING

A fire on a ship may occur while steaming, at an anchorage, when mooring, or when unloading or loading at a pier, wharf, or dock. The location will determine if it is accessible to land based fire companies or only by and from ships. Those areas with fireboats find them most useful at fires on inaccessible ships or for operations on the water side of a tied-up ship or barge. All fireboat and land based company personnel responding should have a thorough knowledge of shipboard fire fighting. Several drills and inspections should be held annually to keep posted on the various ships and their hazards.

At a ship fire the incident commander can obtain assistance from the ship's master, deck and engineering officer, the Port Officer, or marine superintendent. Additional help can come from the Captain of the Port, the Chief Wharfinger, and insurance representatives who can secure tugs, barges, and specialized equipment if necessary. Those areas without fireboats should have agreements with the Coast Guard and tugboat operators to supply boats equipped for fire fighting.

In directing a ship fire the incident commander should not order a ship to be moved. It is better to recommend this to the captain of the ship or seek the authority and approval of the Captain of the Port before this is done. With hazardous cargo it may be necessary to move the ship from a pier area to a mooring still accessible from land, but with far less land exposures. It also may in some cases become necessary to move a ship to a remote location out of the channel and a safe distance from all exposures because of the possibility of rupture, polymerization, explosion, or toxic fume and smoke exposure. A final alternative is that the vessel may have to be flooded and sunk, depending on the nature of the problem. This should only be done under the direction of the Captain of the Port. In this case, movement to reduce the navigational hazard to a minimum is extremely important. Other considerations are the depth for ease of recovery, the condition of the bottom to avoid hull damage, and limitation of capsizing.

If the hull or watertight bulkheads must be opened, advice and preferably approval should be secured from the ship's captain. Extreme care must be taken not to damage vital machinery or electrical circuitry. When the watertight integrity of the ship is concerned, the Captain of the Port should be informed.

Stability

The careless use of excessive quantities of water during shipboard fire fighting operations may seriously impair the stability of the vessel. This can lead to loss of life, injuries, damage in excess of the fire loss, or even the total loss of the ship.

The incident commander must be concerned with the stability of a ship during fire fighting and secure advice from the senior officer of the ship on how to correct any list caused by water applied to extinguish the fire. The incident commander is not responsible for correcting the list or antisinking measures, but must take all precautions to avert further damage.

The stability depends upon the design of the ship, whether it is empty or loaded and primarily on how much additional weight is placed on board and where. The ability of a ship to remain in a stable position is controlled by the interaction of two opposing forces, gravity and buoyancy, diagrammed in Figure 15-14. The center of gravity G is the point where the weight of the ship and the cargo is concentrated. It remains in a constant position until the weight distribution on the ship is changed and then moves up and down along the centerline of the vessel, always acting with a downward force. The center of buoyancy B is the center of the force of the body of water displaced by the ship. It is the center of the immersed part of the ship and acts with an upward force perpendicular to the surface of the water. It moves laterally.

The addition of weight at a high point within the ship causes the center of gravity to rise and the vessel to list with decreased stability. The center of buoyancy shifts laterally to the lower side (Figure 15-15). If sufficient water is added to bring the center of gravity—the downward force—outside the center of buoyancy—the upward force—the ship will capsize.

There are various methods of preventing a ship from capsizing. These include the transfer of fuel, shifting of cargo, counterflooding, or pumping out flooded areas. The incident commander must remain cognizant

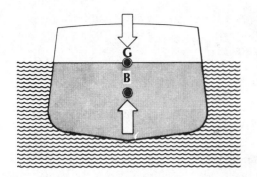

Figure 15-14. Interaction of two opposing forces, center of gravity G and center of buoyancy B, maintains stability of a ship

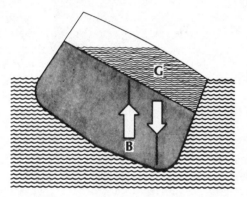

Figure 15-15. Ship lists as center of gravity moves to the outside of the center of buoyancy

of a number of points concerning ship stability. These include: the location of any flooded compartment or hold and its effect on buoyancy and gravity; whether the ship is empty or loaded, which can be seen by how high in the water the vessel sits; any openings that would admit water if the ship listed, such as a side cargo loading door; and the chance of injury should a hawser break due to the ship listing away from a dock. Dangerous lists can be caused on empty ships from handlines or master streams flooding areas of the main deck or above; this water should be removed immediately. Stability must always be a concern when flooding a hold. Although flooding should be done quickly if required, ample counter moves should be taken. Lower holds can be flooded effectively, but seldom is it practical to flood the tween decks. The areas above this should never be flooded.

Problems Encountered

A number of problems arise in marine fire fighting that are somewhat different from other fire situations. The physical size of the ship, especially a supertanker, can make circumstances difficult. Many ships are over 1,000 feet (305 meters) long. The fire may occur several decks below. This creates a subcellar fire that could be several stories below water level in an enclosed metal box. Fire spread is possible through a number of vertical and sometimes horizontal shafts, which must be discovered and protected. There will be a heavy fire load. Hazardous materials may be in bulk, tanks, or containerized and included among other combustible materials.

Several tactical problems must be considered. Approach may be from only one side, which could easily be downwind. Accessibility to the fire area and adjacent holds, cofferdams, or compartments which may be internal exposures may be difficult. Exposures on the sea and land sides may be large, and contain hazardous materials. Their protection may require a sizable commitment. The ventilation of the fire area can be extremely difficult, which could limit interior operations, allow heat buildup and fire spread, and pose severe hazards to personnel operating in the area.

It is important to consider and evaluate what has already been done to control the fire by the crew of the ship, how the supply of air to the fire can be reduced, and how to remove fire fighting water from the ship. If lower holds or bottom tanks are empty, the possibility of filling for increased stability should be discussed with the engineer of the ship. As in all hazardous material incidents, an avenue of escape must be maintained, especially when working below decks. Marine fire fighting units should be provided with good orders as to their task, because they will have difficulties with winds, tides, heat, and smoke as they maneuver to do their job.

The lack of oxygen in ship holds, the products of combustion, and buildups in closed hatch areas, along with carbon dioxide which may have been used by the ship's crew or fumigants in use, make the use of positive pressure self-contained breathing apparatus imperative. Hoselines or lifelines should accompany interior operations to guide lost fire fighters out of the vessel.

Cargo Hold Fires

The first step is the location of the hold involved. The next duty, as with all hazardous material fires, is to identify what is burning and the possibility of the fire extending. Due to the construction of ships there is considerable conduction and radiation of heat. The pipe ducts, air shafts, and other means of convection from the fire area will need to be found and cut off. The ventilation system must be shut down to avoid pulling heat and smoke to other areas of the ship. Once these avenues of travel are known, a fire area can be established and boundaries defined within which the fire will be contained. To maintain the fire in this area, hoselines must be stretched to exposed bulkheads, deck ventilators should be covered or secured to reduce the supply of air to the fire, hatches should be closed, and electrical circuits deenergized.

To assist in these operations, the incident commander should obtain the help of the ship's engineering or senior officer with the fire plan, blueprint, and the cargo map. If the ship has a foreign crew, an interpreter will be necessary for communications both on a shipboard visit for planning or during an incident.

After all loading or unloading operations

have ceased, all persons not directly involved in fire fighting should be put on shore.

Due either to a lack of hatch openings or the location of corridors in relation to the fire area, entry to the involved area may necessitate descending to the level of the fire at a remote location and then approaching the fire area. Heavy smoke may occur anywhere, depending on the construction of passageways, corridors, and ducts. Smoke ejectors may be helpful in moving smoke out of compartments and holds. The boundaries of the fire area must be frequently checked to control fire spread, even below the fire, for dropping materials. The first lines should be placed between the fire and the area of greatest exposure or hazard.

Water must be applied to the fire area from the most advantageous positions in order to gain maximum efficiency and reduce water use. If interior operations must be abandoned, master streams will have to be used from vessels or land based companies for control. If the fire cannot be controlled, an effort must be made to account for all personnel before withdrawing from the ship.

If access cannot be gained to a hold through a hatch inspection hole or the removal of the hatch cover, heavy streams and distributors can be used from the deck. Holes may have to be cut in the deck plate to work lines.

In order to complete overhaul, the cargo will eventually have to be removed. This is particularly true of lower hold fires. Assistance for cargo removal should be requested early. Whenever possible use two sources of water in extinguishment operations. To enhance hoseline operations, ventilate the fire area when it can be done directly to the outside without affecting the approaching fire fighting teams. If the first boundaries established cannot be held, set up at the next boundary area to confine the fire. Remember to remove accumulating water to assure the stability of the ship. Only as a last resort, consider flooding a hold with water. If necessary, this can be done by removing the hatch cover and using open butt hose lines.

Cargo Tank Fires

The greatest hazard on modern tankers is a fire spreading to the cargo area, which could be full of flammable or combustible petroleum products or chemicals or filled with the vapors from the former load. The capacity of the tanker can range from 250,000 to 1 million barrels of product. The tanks are much safer from fire when full, sealed, and vented than when empty. Once again, the importance of determining the commodity before proceeding is emphasized.

The attack of a fire in the tanker compartment takes two approaches. If the tank is intact, close off all air supplies through tank tops, ullage hose covers, Butterworth plates, and other tank openings. A small hole in a ruptured tank may be closed with a thoroughly wetted canvas. Then, the ship's fire protection system, carbon dioxide, steam or foam, should be used in the cargo area, being certain that valves in branch lines to uninvolved areas not adjacent to the fire are closed. Any exposed deck, bulkhead, or structural member nearby should be rapidly cooled with water fog.

If the tank has ruptured or is opened in some way, foam should be used for extinguishment. This is not appreciably different from fighting an oil storage tank fire on land. Concentrate as much foam delivery as possible on the fire for fast, effective extinguishment. Water may first need to be applied to cool the surface of the burning liquid to reduce foam breakdown. Take care to avoid overflowing the tank. If necessary, use an alcohol-type foam. Don't wash away a foam blanket or break it down with a fog line to attempt to cool surrounding metal. The incident commander should make sure early in the situation whether enough foam is available and, if not, have the supply officer secure adequate quantities.

Approach to a tanker may require the use of master streams and heavy lines to break up surrounding fires or sweep the sides of the ship and superstructure before boarding is possible. Board remote from the main fire area and attack with the wind. Protect all boarding ladders in case of the need for unexpected withdrawal.

The foam application should be concentrated on one tank at a time. A gentle application to permit an even flow with little agitation of the product is best. Direct the foam against an obstruction or tank wall and let it flow back over the surface. Provide fog protection to the personnel on the foam lines.

If foam supplies are not available, water fog can be quite effective on high flash point oils. The fog can cool the oils and form an emulsion over the surface to cut off the air. On low flash point volatile liquids, the water fog can completely cover the burning surface and dilute the vapor-air mixture so it will not support combustion. On large areas where the streams do not completely cover the area there will be a back flash to an extinguished area if there is any movement of the streams. With both high and low flash point fuels, water fog will slow combustion and absorb radiant heat.

Deck Fires

Since a number of tanker and barge fires occur during product transfer, a fire on the deck must be covered. The product overflowing, flowing from a burst hose, or leaking from a pipeline must be identified. Then the source of fuel should be shut down. On a tanker with a deck foam system this should then be used to extinguish the fire. Apply the foam correctly to get an even spread without scattering the fuel over the deck or dock. Water fog will have to be used on a barge or for a running spill fire where the source cannot be quickly controlled. Foam, if available, can be placed ahead of a flowing fuel fire to form a barrier and contain the fire. Water fog should be used again in this case to protect personnel and cool metal structures, but must be used with care not to destroy the integrity of the foam blanket. Any openings to cargo tanks or pumping areas should be closed immediately to keep the fire from obtaining further fuel sources.

Fires on the Water Surrounding a Vessel

These can best be handled by using foam lines from the ship applied to the side of the hull causing the foam to flow out from the ship and protect the ship. If foam is not available, water streams can be used to set up a current that will carry the fuel away from the vessel.

When applying either foam or water to a floating fire, the wind and current must be considered. If they are going in the same direction, application should be from the windward side and above the oil. If the wind and current are in opposite directions, the side of application will depend on which is stronger. The object is to apply the foam or water so that it flows down on the oil and carries it away from the ship.

Another vessel providing fire fighting in this case should position itself to drive the fire away from the exposed ship. Foam or water can be used, depending on availability and how quickly extinguishment is necessary. Care must be used to prevent the burning fuel from getting under docks or wharves or near other ships. Adjacent vessels should prepare to get under way or tugs should be summoned to move them.

Liquefied Flammable Gas Fires

These fires are essentially those involved with LPG and LNG. The procedures for handling LPG fires are generally known and have been previously covered in this text. A shipboard foam system would not necessarily be an efficient extinguishing device.

Fire fighting tactics for LNG are somewhat different from those for LPG. The first method of extinguishment is to stop the flow of fuel. If a fire is impinging on a tank, another vessel, or exposure, adequate protection should be provided. A fire will most probably be a spill-type without impingement. In this case extinguishment can be accomplished with dry chemical, carbon dioxide, or halogenated extinguishing agents. Foam and water have not been found effective on LNG. Water is, however, useful in providing exposure protection and in dispersing LNG vapors. Since large quantities of the above extinguishing agents are not readily available, a large spill will most likely be control burned while protecting exposures. The major hazard is the tremendous vaporization that could take place with a large spill. This could cause a tremendous flash fire over a large area, set fire to any exposures in its path, and cause injury or death to people in the area.

Barges

Fires on barges of flammable liquids, chemicals, and gases are handled like cargo tank and deck fires or fires in permanent storage facilities for similar products. Accessibility can be a problem and care must be taken not to sink the vessel or drive the fire into an exposure. With a large tow, the barges may have to be separated to confine the fire. Protection must also be provided so that the vessel does not get away from its mooring and become a floating fire, which could collide downstream and spread the fire.

Hazardous materials fires in marine transportation take on many aspects similar to other modes, but are influenced by the capacities and construction of the vessels hauling them. Fighting fires aboard a ship takes special training and knowledge and when coupled with the added problems of the hazardous commodities becomes a dangerous job requiring skilled technicians. If hazardous materials transverse or come into your city by marine transportation, plan ahead and apply the many lessons illustrated throughout the text.

QUESTIONS FOR REVIEW AND DISCUSSION

1. What documents will help identify hazardous materials on a ship and where they are found?

2. What are the five major topic areas to be covered in preplanning a ship? What do you consider the most important concern in each area and why?

3. Describe the differences between a tanker ship and a tank barge.

4. What are the three major fire fighting systems found on new ships? When would you use each if all three were on a vessel involved in a hazardous material incident?

5. What problems are encountered in shipboard fire fighting?

6. What precautions must be taken in fighting a hazardous material fire in the hold of a freighter?

CASE PROBLEM

During the night of Wednesday, December 18, the *Kirsten Sue*, a containerized vessel, docked about midnight. At four in the morning, a nightwatchman, making his round of the pier, smelled smoke. He investigated further and found the smoke coming from the *Kirsten Sue* (Figure 15-16). He boarded the ship, but could not locate the man on watch. The smoke was increasing; he located an alarm station and activated it to awaken the crew. The fire appeared to be in the Number 1 hold. The hatch had additional containerized units on top of it.

As the crew came forth, the watchman discovered that he could not communicate with the foreign crew except to point to the smoke pouring from the hold. The captain began to shout orders and the crew moved toward the hold with a 1½-inch hose line from a forward fire station.

They had some trouble due to the smoke, warming deck plate, and weather, which was 28°F (15.6°C) with the wind coming in from the northeast at 15 miles per hour (24 kilometers) with gusts to 20 miles per hour (32 kilometers). When they opened the access hole in the hatch cover, the heat drove them back.

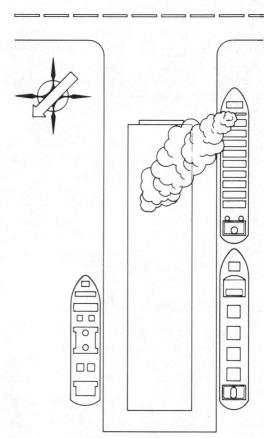

Figure 15-16. The *Kirsten Sue*

The watchman, realizing the fire was beyond the capability of the crew and developing a growing concern of the possibility of the fire spreading to the pier, left the vessel for his office telephone to call the fire department.

At 4:20 A.M., the fire department dispatched a full box assignment of four engines, two trucks, a squad, and a battalion chief to the pier for a ship fire. The fireboat was included as one of the four engines.

Waterfront Way, which passed in front of the pier, had a 16-inch (40.6 cm) main on a fully looped system with hydrants 400 feet (122 meters) east and 500 feet (152 meters) west. Drafting from the pier or Waterfront Way is severely limited due to the lift.

After some confusion with the Swedish Captain of the vessel, the battalion chief secured the dangerous cargo manifest and determined that the third container down in Hold Number 1 contained 55-gallon (208 liters) drums of vinylidene chloride. He had no way at this time of determining if the product was involved or exposed.

1. Several problems confronted the chief as he arrived. Describe each: accessibility, water supply, communications, identification of cargo, shipboard fire fighting systems.

2. What precautions should be taken with this product? As the battalion chief, what operations would you initiate to obtain access to the hold? Are technical assistance and outside resources necessary at this time? Explain your decision.

3. The ship has a fire main system and a carbon dioxide flooding system. Would you use these systems in the fire attack? If so, how could they be used? If not used, explain why and your alternative operations.

4. What exposure problems must be confronted? Is the stability of the vessel a concern? What can be done to maintain stability?

ADDITIONAL READINGS

Fire Fighting Manual For Tank Vessels, CG-329, U.S. Coast Guard, Department of Transportation, Washington, D.C., January 1, 1974.

Guide to Fixed Fire Fighting Equipment Aboard Merchant Vessels, NVC 6-72, U.S. Coast Guard, Department of Transportation, Washington, D.C., August 22, 1972.

Liquefied Natural Gas, Views and Practices, Policy and Safety, CG-478, U.S. Coast Guard, Department of Transportation, Washington, D.C., February 1, 1976.

Maguire, Hugh M., "You Have Got a Hell of a Ship Fire in Elliot Bay!" *Fire Command*, September 1973.

A Manual for the Safe Handling of Inflammable and Combustible Liquids and Other Hazardous Products, CG-174, U.S. Coast Guard, Department of Transportation, Washington, D.C., June 1, 1975.

Marine Fire Fighting Manual, Bureau of Training, Department of Fire, City of Long Beach, California, 1973.

Marine Fire Fighting Practices, Alaska Department of Education, Fire Service Training, Juneau, Alaska, October 1973.

Savage, Kent M. "Marine Gas Hazards Control," *Fire Command*, December 1974.

"Seattle's Shipboard Fire Fighting Training Is Comprehensive," *Fire Chief*, Chicago, March 1976.

16
Fixed exterior storage facilities

Storage facilities for hazardous chemicals come in a wide variety of sizes and shapes, each designed for specific reasons. The locations of these facilities can be either underground or above ground (Figure 16-1).

Associated with each storage facility are the piping, valves, fire protection devices, gauges, safety devices, as well as loading and unloading features.

Emergency response personnel have to be familiar with the types of storage tanks in their area along with the products stored and the auxiliary equipment. This can only be accomplished through an on-site visit. As stated so often throughout the text, the time to become familiar with a facility is before the incident, not while it is an on-going operation.

Figure 16-1. Above ground storage tanks

TYPES OF STORAGE TANKS

Bulk storage tanks of liquid or gas products can be classified in a number of ways, depending on the purpose of the classification:

1. Product the storage tank contains, such as gasoline, crude oil, or fuel oil

2. Method used for joining the steel plates that make up the tank, such as bolted, united, or welded

3. General purpose of the tank, such as production or storage

4. General configuration of the tank, such as horizontal, floating roof, or cone-shaped roof.

Emergency response personnel must be familiar with the different terms so that when preplanning the facility, they can take the technical information provided to them and rework it into one standard breakdown. For the purposes of this text the general configuration of the tank will be used as the classification method.

Storage Tank Safety Devices

In an ordinary storage tank there is a space above the liquid which contains a vapor/air mixture. As the product is added at the bottom, the liquid level rises and compresses the vapor/air mixture. Sufficient venting devices must be provided to prevent the building up of pressure during the tank filling operation. If the product is added too quickly or the relief vent fails to open, overpressurization and structural damage can occur.

Conversely, when the product is being pumped out, air must be allowed to enter the tank or a vacuum will be created in the vapor/air space. If the product is pumped out faster than air is admitted, the vacuum can cause an implosion of the tank walls.

In ordinary tanks as the product is added, the vapor/air mixture escapes into the atmosphere. The loss of vapor presents both an economic and an environmental hazard. Various techniques, as explained in the following paragraphs, are used to recover this mixture. In addition, the vented mixture of vapor and air is flammable, so care must be exercised to prevent it from traveling to an ignition source.

Flame arresters are installed in the vapor space to lower the temperature of the vapor below its self-ignition point. Flame arresters are usually metal tubes, but metal plates and screens can also be used.

Above ground storage tanks are exposed to the weather. This means that in the sunlight or warm weather the product heats up and more vapor is produced. This also creates an increase in pressure, which must be handled by venting. The opposite, of course, is also true. At night or when the temperature drops, some vapor condenses to liquid and leaves a vacuum in the vapor space. Air must be admitted to prevent internal collapse.

Emergency relief vents are also necessary to rapidly reduce pressure under sudden condition changes such as a fire. On some tanks, however, a weakened seam between the roof and shell is used. Overpressurization then causes a separation of the roof from the walls.

Horizontal tanks do not have a weakened seam which can be used for venting. If one of these is subject to overpressurization, the most likely failure will occur at the ends of the tank. (Refer to the discussion of a BLEVE in Chapter 8.)

Underground storage tanks experience the same problems enumerated above, except that the weather will not have as great an effect on the temperature of the product. The ground temperature stays relatively more constant than the air temperature. In addition, because it is buried, exposure to a fire is highly unlikely.

Above Ground Vertical Storage

The basic above ground storage tank is a cylinder, made out of steel plate, and welded together. The majority of these tanks are designed to withstand pressures from only 0 to 0.5 pounds psig.

The most common roof for these tanks is called the cone roof. The roof plates are cut to a tapered form so that the finished roof has a rise of 1 inch in 12 inches. Each joint between the roof plates and between the roof and staves is made with a single row of bolts. The usual method of supporting the roof is to use a center post resting on the tank bottom and a system of steel rafters extending from the center post to the tank shell.

Storage capacity for these types of tanks

can range from 20,000 gallons (760,000 liters) to over 10 million gallons (380,000,000 liters). Obviously, these can present serious problems to emergency response personnel.

Above Ground Floating Roof Tank

Control of the vapor/air space is of critical economic and environmental importance. One way to control this space is to use a floating roof tank.

In this type of installation, the roof of the tank actually floats on the top of the product (Figure 16-2). The top, therefore, rides up and down inside the tank shell as the liquid level changes.

However, because there is no support for the side wall at the roof level, special reinforcing must be installed around the circumference near the top. This will prevent buckling due to wind stresses when the product level is low.

In its basic form, the floating roof can be thought of as a pan, slightly smaller than the tank, floating on top of the product. Flexible "shoes" are used to close the space between the roof and the side wall to prevent the escape of product. The latest development in floating roof tanks is a "pontoon" (Figure 16-3), which provides stability as well as sealing capability.

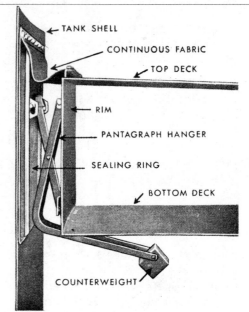

Figure 16-3. Floating roof tank pontoon seal. *Courtesy of the University of Texas, Petroleum Extension Service*

The floating roof tank shown in Figure 16-3 is also a double-decker. In this way the product touches the bottom deck which has no contact with the sun. This helps cut down on evaporation type losses.

The seal of the floating roof tank is provided by a "sealing ring" kept pressed against the side wall by a weight attached to a pantagraph. The actual space between the roof and the sealing ring is covered by a piece of fabric.

Except for the small volume under the seal between the side of the roof and the sealing ring, there is no vapor space. There is no vapor discharge when adding product nor is there any in rush of air when pumping product out of the tank. One final advantage is that corrosion is greatly reduced because the actual exposure to air necessary to produce it is so limited.

One great problem with this type of storage tank is that a heavy rainfall, or snow accumulation, can cause the roof to sink into the

Figure 16-2. Floating roof tank

product. This, in turn, can leave a large surface area of flammable liquid exposed.

Fire service officers must take special
note of this, because pouring large-diameter
streams on a floating roof tank in an attempt
to keep it cool can cause the roof to sink. Care
must be exercised to ensure that cooling water only hits the sides and does not land on the
roof. Some of these tanks are equipped with
drainpipes to conduct away the rainwater
gathered on the roof, as shown in Figure 16-4.
The drainpipe actually goes to the outside and
expands and contracts as the roof moves.

When a floating roof tank is empty, the
roof rests on pipe legs attached to the roof
(Figure 16-4).

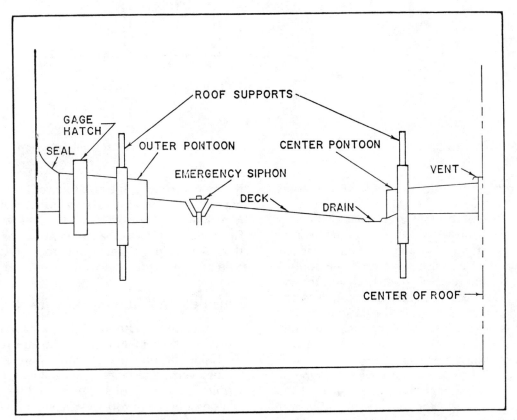

Figure 16-4. Section of a floating roof tank. *Courtesy of the University of Texas, Petroleum Extension
Service*

Outside Breather Roof Storage Tank

An outside breather tank utilizes a modi-
fication of the basic cone roof design while
the shell and bottom designs are the same.
However, the roof supports in the breather
tank installation are designed so that when
the roof is resting on them it is in the form of
an inverted cone (Figure 16-5).

When the liquid level is low or the vapor
pressure is reduced, the tank roof drops and
rests on the supports (Figure 16-5). As product
is added, the internal pressure increases and
the roof rises above the supports. Additionally, if the temperature on the inside should
increase, the pressure will also raise the roof.

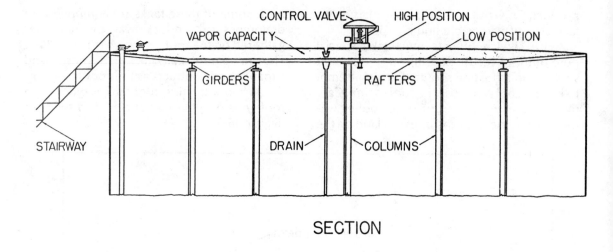

CONTROL VALVE HIGH POSITION
VAPOR CAPACITY LOW POSITION
GIRDERS RAFTERS
STAIRWAY DRAIN COLUMNS

SECTION

Figure 16-5. Breather roof tank design. *Courtesy of the University of Texas, Petroleum Extension Service*

The roof, therefore, is being kept in a raised position solely by the internal pressure.

Travel of the roof is limited to that which a normal cone roof would occupy. If the pressure should build up above the level that can be handled by raising the roof, the control valve (Figure 16-5) will open to relieve the excess pressure. The vapor capacity due to expansion is also shown in the illustration.

As long as the tank is kept nearly full and the vapor created does not exceed the volume created by the roof travel, no vapor will be lost to the outside. However, this type of tank is not designed to be used as a working tank where large volumes are moved in and out continuously because it must be kept close to full point to operate efficiently. As a result it is used primarily for full standing storage.

One maintenance problem is that the roof metal actually bends as pressure changes occur. This can lead to fatigue and structural failure.

Outside Balloon Roof Storage Tank

The balloon roof storage tank operates in the same manner as the breather roof except that the roof is larger. This larger area, created by making the roof overhang the tank walls (Figure 16-6), allows for a larger vapor space. All operating principles of this roof are the same as for the breather roof.

Outside Lifter Roof Storage Tank

On a lifter roof storage tank, the roof is an entirely separate member from the shell (Figure 16-7). The roof moves within a series of guides a vertical distance of several feet. This permits a larger vapor space than either the balloon or breather roof design, yet does not subject the metal structure to bending.

There are two ways of sealing the tank walls to the roof to prevent vapor interchange. One involves the use of a liquid seal made by having a metal skirt extend downward from the roof into a liquid-filled trough built on the upper part of the tank shell (Figure 16-8).

The other method uses a flexible curtain of treated fabric which will not pass the product vapors. The fabric forms a continuous ring around the top of the tank, with one edge fas-

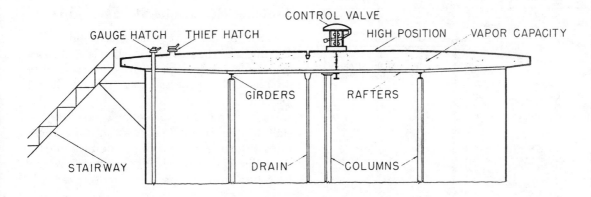

SECTION

Figure 16-6. Balloon roof tank design. *Courtesy of the University of Texas, Petroleum Extension Service*

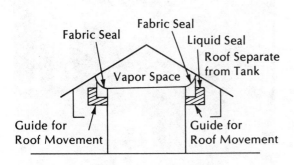

Figure 16-7. Lifter roof tank. *Courtesy of the University of Texas, Petroleum Extension Service*

tened to the movable roof and the other to the stationary tank shell (Figure 16-9). Weather protection is necessary for both types of seals.

Outside Spherical Storage Tanks

The spherical shaped storage tanks (Figure 16-10) are designed for efficient storage at higher pressures. Since a rounded surface can withstand higher pressure for the same thickness of material, that design is used where working pressures up to 25 psig are required.

Since this is a pressurized vessel, relief valves are provided at the roof level to prevent excess buildup of pressure.

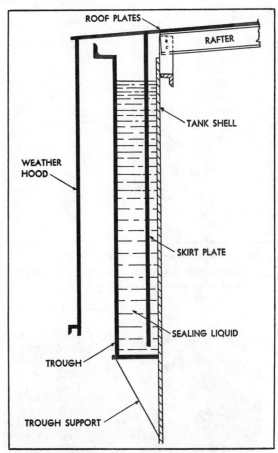

Figure 16-8. Lifter roof tank with liquid seal. *Courtesy of the University of Texas, Petroleum Extension Service*

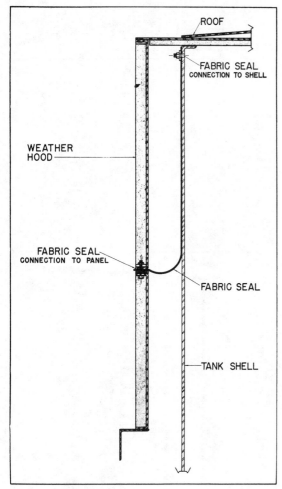

Figure 16-9.   Lifter roof tank with fabric seal. *Courtesy of the University of Texas, Petroleum Extension Service*

Figure 16-10.   Spherical shaped storage tanks

Outside Horizontal Storage Tanks

Horizontal storage tanks are utilized for storing both liquids (Figure 16-11) and gases (Figure 16-12) as well as pressurized (Figure 16-12) and nonpressurized (Figure 16-11) products.

The same type of relief valves are installed on horizontal tanks as on the vertical ones. Pressurized tanks can be determined by the elliptical or hemispherical heads which help distribute the pressures more evenly.

Figure 16-11.   Nonpressurized horizontal storage tanks—liquid

Figure 16-12.   Pressurized horizontal storage tank—gas

Underground Storage Tanks

Both liquids and gases can be stored underground. Liquefied petroleum gas can be stored in underground tanks both as a single tank or in multiple tanks for use by large municipal utilities. Figure 16-13 shows some of the tank vents for 1,000,000 cubic feet (30,000 cubic meters) of propane used by a gas company for inserting into the natural gas lines during peak periods of demand. These underground facilities also have the gauges and controls above ground (Figure 16-14).

Gasoline is another common commodity that is stored below ground, particularly at service stations. In Figure 16-15, the pump has been struck but the excess flow valve closed, limiting the quantity of liquid spilled. The resulting fire was extinguished with a single dry chemical extinguisher. Remote shutoff switches for shutting down the pump should the excess flow valve not close are also provided. Tank vents prevent the buildup of pressure when filling the tanks and will overflow the product should the tank be filled too much.

Figure 16-15. Gasoline pump spill fire

Figure 16-13. Underground storage of propane

Figure 16-14. Underground propane storage gauge

Problems with underground storage arise due to corrosion. Undetected leaks can then flow great distances and cause potential fire problems. This has been partially overcome with the use of fiberglass tanks, which resist leakage to a large extent. However, facilities using underground storage must keep track of the amount of product delivered versus the amount sold, plus that in inventory, to ensure that none is being lost through leakage.

Since piping from the storage tank to the pumps or delivery system is also underground, that is another potential area for leaks. Owners must ensure that installation of the tanks and pipes is done in accordance with the applicable codes and industry standards.

Small Storage Tanks

Storage tanks come in a large variety of sizes, containing a varied list of products, to meet the needs of the user. Propane can be found in small cylinders for use as a small heating torch, to propane cylinders for use on trailers, to compressed oxygen used for welding (Figures 16-16 and 16-17).

Industrial storage also has to be considered, particularly 55-gallon drums (Figure 16-18). Almost all types of hazardous materials can be found in these types of storage.

Small pressurized vessels have either relief valves and/or fusible plugs to prevent the buildup of pressure (Figure 16-19). However, small, nonpressurized vessels do not have any way to relieve a buildup of internal pressure.

Remember, any closed container has the potential to BLEVE. It is not necessary for the material to initially be under pressure. Once the container is exposed to heat, as it is under fire conditions, and remains closed, there is a potential for a violent rupture when the internal pressure exceeds the strength of the container. Also note that nothing has been said about the product within the container. Violent rupture occurs whether or not the product is a flammable liquid.

Once rupture has occurred, the product contained within the drum adds to the hazard. Be it flammable, corrosive, or a poison, it is now spread all over the initial area, compounding emergency response handling of the incident.

For further information on the explosion hazard of a closed container called a BLEVE, refer to Chapter 5.

Specialized Storage

Explosives are stored in specially designed magazines. These storage facilities are constructed so as to prevent break-ins, buildup of heat, and possible sparks which could ignite the material.

Many products have specialized storage containers to handle specific hazardous properties. For example, a substance marked flammable solid—dangerous when wet must be

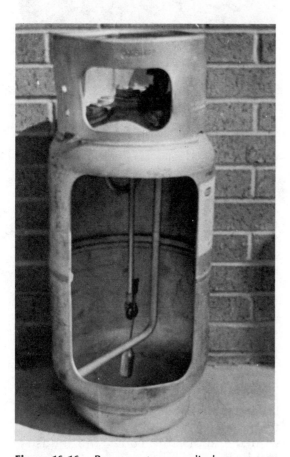

Figure 16-16. Propane storage cylinder

Figure 16-17. Oxygen storage cylinder

Figure 16-19. Propane cylinder relief valve

Figure 16-18. 5-gallon drum of corrosive material

protected from allowing moisture to enter. The phosphorus pentasulfide in Figure 16-20 is a flammable solid in a special outside stor-

age container. Emergency response personnel must become familiar with the various types in their response area before an incident occurs.

Figure 16-20. Flammable solid storage facility

PHYSICAL LAYOUT

The physical layout of a facility can range from a single small cylinder to many hundreds of large, vertical tanks. The larger the facility, the more complex becomes a hazardous material incident.

Base and Supports

Vertical tanks must be built on a sound, well-compacted base. Any shifting of the underlying ground can create stresses in the metal that can lead to leaks and even structural failure.

Draining must be provided to prevent the accumulation of water at the base. Water standing at the base can lead to rust, corrosion, and in the colder areas, freezing. Again, this could weaken the structural elements and cause leaks or failures.

Structural supports for horizontal tanks also present a problem. When built of steel and left unprotected, the structural supports can be weakened during a spill fire, and may cause the tank to drop, rupture, and add more fuel to the fire. Steel supports for any flammable liquid tanks should be protected by fire resistive material with a two-hour rating.

Saddles made of concrete are often used to support horizontal tanks as shown in Figure 16-21. These maintain their integrity for a longer time than the unprotected steel. Of course there is still the BLEVE potential, even with the protected supports, so emergency response personnel must use care when approaching the incident.

With the continuing energy problems, many types of tanks are pressed into service for storage. Figure 16-21 also shows a horizontal storage tank using wood for support. Under fire conditions, collapse of the tanks can be predicted. Emergency response personnel must be alert to these makeshift operations.

Sometimes storage tanks are placed directly on the ground (Figure 16-22). While the danger of collapse is eliminated, many other problems are created. Corrosion due to the metal in contact with the moist ground is always a problem. In addition, should a fire occur, the product would completely surround the storage tanks and flame could easily directly impinge on the vapor space. Further, it would be difficult for fire service hose streams to sweep the product out of the way.

Figure 16-21.   Concrete saddle for supporting horizontal tanks (left) and makeshift wooden support (right)

Figure 16-22.   Storage tanks directly on the ground

Containment Facilities

Spillage from a hazardous materials storage tank can occur from several causes:

1. Overfilling
2. Leak from structural stress
3. Leak from valves or piping
4. Massive failure of tank due to catastrophe (fire, earthquake, flood, or tornado)

Spillage for any of these reasons must be controlled. One way of controlling the flow of product is to channel the flow to impoundment basins. Here it could burn without threatening any exposures nor doing harm to the surrounding environment. However, because large tanks are usually located in very industrialized areas, there generally isn't room for large impoundment basins.

The other solution is to build a dike around the individual tank or a group of tanks (Figure 16-23). This wall would contain a spill and keep it from spreading. Exposures would then be protected and damage to the environment from the spill could be lessened. However, not all facilities have a diked area, and emergency response personnel must be prepared to cope with such an occurrence.

The materials used for the dikes can be earth, steel, or concrete. They must be strong enough to withstand the head pressure created by the liquid filling the diked area.

If a large dike is used to enclose several tanks, the enclosed area can be broken up by low walls into smaller units. In this way small or minor spills will not needlessly expose the other tanks.

Figure 16-23. Diked area surrounding vertical storage tank

The size of the dikes is dependent upon the type of product contained in the tanks. Products that are not subjected to boilover (refined oil for example) should have sufficient diked area to contain the volume of the largest tank. This requirement provides several safety factors for the following reasons:

1. The largest tank has the spill.
2. The tank is completely full.
3. The tank will not drain below the height of the dike wall because the head pressures would equalize.

Dikes for products such as crude petroleum, which are subject to boilover, are usually designed to contain the volume of all the tanks enclosed. The top of the dike should also have a deflector to prevent the liquid wave from washing over the top in case of a sudden release.

Obviously, one problem that fire service personnel are confronted with is that in a spill and fire, water and foam are applied to the burning product. This adds to the liquid volume and care must be exercised to ensure that the dike is not overflowed.

Another problem potential is the accumulation of rainwater in the diked area. If this water were allowed to stand, the usable volume of the dike would be reduced. Then, in the event of a spill, the product could overflow the dike.

To overcome the problems of rainwater and water from fire fighting, dikes should be equipped with drains with control valves. The valves would normally be kept closed so that in the event of a spill, the product would not leave the diked area. Rainwater could be removed periodically by opening the valves. In the event of a fire, if the flammable liquid floats on the water, the dike can be drained very carefully of the water on the bottom.

Finally, there are times when fire fighting must take place from outside the diked area. If the dike wall is too high, access for fire fighting will be difficult or impossible. In addition, if workers are inside the diked area when a spill occurs, the wall should be low enough to allow quick egress.

One of the obvious things required of emergency service personnel is to preplan the storage facility. Officers must know the location of the drain valves and piping control valves for the various tanks. Alternate control points must be determined in advance in case the spill or fire overruns the valve locations. For example, Figure 16-24 shows the control valves located directly in the diked area. If the tank filled this dike, control of the flow would have to occur from another point in the facility.

Figure 16-24. Storage tank control valves in diked area

Determining Amount of Product

The amount of product in the storage tanks must be determined by emergency response personnel. This will be used in determining the potential problems during an incident. Supervisory personnel must know how to determine product levels for the storage facilities within their jurisdiction.

Among the methods that can be used are:

1. Remote digital readout at the control room.
2. Gauging rods to determine liquid height and be converted to volume.
3. Float-operated gauge.
4. Pressure-operated gauges.
5. Use of infrared scanning device which shows differences in temperature of the metal. (Product height in tank can be converted into volume.)[1]

HANDLING AN INCIDENT

When an incident occurs in a storage facility, emergency response personnel must:

1. Consult preplan.
2. Size-up incident.
3. Attack the fire.
4. Proceed with rescue.

The need for and items to be included in a preplan are discussed in detail in an earlier chapter. If that information has been followed and a preplan exists, the incident commander can begin to size-up the problem.

Size-Up

Step 1. Identify the product which created the incident. Information must be obtained from plant personnel because it is possible that the same tank is used for a variety of products.

Step 2. Determine whether more than one product is stored in the tank. (Is it compartmented?)

Step 3. Refer to the various reference sources to determine the hazards, physical properties, and extinguishing methods and agents for the identified product:

1. Effect on humans (special protective equipment, evacuation)
2. Effect on the environment (streams, ground water, air)
3. Specific gravity
4. Water solubility
5. Water reactivity
6. Flash point
7. Reactivity problems
8. Explosive limits
9. Polymerization
10. Extinguishing agents or effective cover which will reduce hazardous vapors
11. Radioactive hazards

Step 4. Check location of the tanks in reference to the exposures. Exposures include buildings, other tanks, and overhead electric lines.

Step 5. Determine the types of storage tanks, their safety features, shutoff valves, and dike drain valves.

Step 6. Check availability of resources (personnel, equipment, water, extinguishing agent).

Step 7. Consider weather conditions that affect the attack (wind direction and speed, rain, temperature)

Once size-up is completed, the incident commander must decide whether to:

1. Attack the fire
2. Contain the fire and let it burn up the fuel
3. Withdraw emergency response personnel

If the decision is made to attack the fire, the procedures are outlined in the following paragraphs.

Attack

If the information obtained during the initial size-up indicates an attack is warranted, the incident commander must begin it immediately. However, keep in mind that as the attack continues and new information is received, the commander can reevaluate and revise the tactics and strategy being used. On the basis of these data a containment or a withdrawal can be initiated.

1. Evacuate the downwind area of the vapor cloud.

2. Have all personnel approach from upwind. Make sure all are equipped with the appropriate protective equipment.

3. Keep all unnecessary personnel including spectators at least one mile away.

4. If there is a gas leak without a fire, use hose streams to disperse the vapor. Then, under cover of the streams go in and shut off the control valves. Make sure a backup hose line from a separate water source is available. If the flow cannot be shut down, then the vapor cloud must be dispersed with hose streams. Use caution however, in case the combination of product and water forms a hazardous substance. In that case runoff must be contained by diking. Remember to keep personnel, civilians, and apparatus away from the vapor cloud.

5. If there is a leak with fire do not extinguish the fire until the leakage is stopped. Using the cover of hose streams, with a back-up line, the control valve should be shut down. Keep the exposures cool at the point of flame impingement. At large fires radiant heat is also a problem and water must be applied directly to the exposures

to keep it cool. Remember to approach horizontal tanks from the sides. (Refer to Chapter 8 for a further discussion.)

6. Listen for the operation of the relief valve. As pressure increases, the pitch of the noise also increases. This should be an indication that withdrawal is necessary.

7. Large tank storage fires will require protection of the exposures, particularly other tanks, with large quantities of water. Unmanned monitor streams can be used to great advantage under these conditions.

8. Extinguishment can be tried using special agents and techniques. Incident commanders will probably need to set up a supplemental source of supply of the extinguishing agent. Dikes will have to be kept from overflowing.

9. Based upon the type of construction of the tanks, the attack will vary. Care must be exercised not to compound the problem by failing to take this into account.

Rescue Inside Storage Tanks

A rescue inside a storage tank can result from:

1. Explosion or fire
2. Presence of toxic liquids, vapors, or dusts
3. Deficiency of oxygen
4. Physical hazards such as slipping, falling, or falling objects

Explosion or Fire. Even after a tank has been freed of vapor, flammable mixtures again may be formed by the admission of vapor or a liquid petroleum from some other source. Vapors or liquid petroleum may enter the tank through unblanked lines or leaks in the bottom of the tank. Vapors may evolve within the tank from sludge (BS&W), sidewall scale, or residue; or from oil trapped in hollow roof supports, foam chambers, pontoons, or heating coils; or from oil-soaked wooden structures or other absorbent materials.

Toxic Liquids, Vapors, or Dusts. The sludge in the tanks (which may continue to contain lead) should always be considered hazardous. When such tanks are entered, the precautions necessary to protect workers against a toxic hazard from tetraethyllead must be followed. Hydrogen sulfide (H_2S), a very toxic gas, may be found in significant concentrations in and around tanks used for crude oil, gas oil, and distillates. All tanks which contain these stocks should be considered dangerous.

Oxygen Deficiency. If petroleum vapor in strong concentration is breathed, a reaction results that produces a stage of excitement leading to unconsciousness similar to that produced by alcohol or chloroform. Although with rest and fresh air there may be recovery in a few hours, all physical reactions that result from such vapor inhalation should promptly be reported to a physician.

Atmosphere (air) within a tank that registers on the vapor indicator more than 14 to 20 percent of the lower explosive limit is considered unsafe for breathing, even for a short period, because of the toxic nature of petroleum vapors.

Air within a tank that has been closed for an extended period, even though the tank has been cleaned and is empty, may become deficient in oxygen because of rusting (oxidation) of the metal of the tank. No one should enter such a tank before it has been ventilated, unless wearing protective self-contained breathing apparatus that provides an independent air supply.

Physical Hazards

1. A discharge of steam or petroleum product into a tank.
2. Objects that have been dropped or that fall from the upper part of a tank.
3. Falling from scaffolding, stairways, or ladders.

4. Falling from or through the roof of a tank.
5. Tripping over hose lines or over other objects.
6. Slipping on tank floors.
7. Colliding with structural tank supports or piping.
8. Structural failure of a tank or its appurtenances.
9. Insufficient light.

Rescue Procedures. The following rescue procedures should be used. They are excerpted from "Crude Oil Tanks."[2]

1. Tanks must be tested to determine the amount of hydrocarbon vapors present, and for the presence of hydrogen sulfide in tanks which contained crude oil.
2. Tests for hydrocarbon vapor must be made with a vapor indicator. Anyone who makes such tests should be thoroughly instructed in the handling and reading of vapor indicators, and should be satisfied that the instruments used are in proper working condition. Tests should be made at various points within the tank. If ventilating equipment is in operation, samples should be taken which are representative of the tank atmosphere.
3. First tests of the atmosphere should be made at openings where vapors are leaving the tank. When the vapor indicator registers not more than 14 to 20 percent of the lower explosive limit, the tester should then enter the tank and make further tests. Tests for vapor should be repeated at frequent intervals and, if the indicator shows the presence of vapors above this limit, everyone in the tank should leave promptly. *No one should re-enter until the atmosphere within the tank is again at or below this limit, unless wearing self-contained breathing apparatus.*
4. The presence of hydrogen sulfide can be

detected by the way this gas blackens moistened lead–acetate paper. Also, hydrogen–sulfide detectors are available to test an atmosphere for this gas. Anyone who uses these instruments should be thoroughly instructed in the handling and reading of them. The instrument should be checked to assure that it is functioning properly before it is used.

5. When working inside a tank, floor sumps should be covered or otherwise guarded to prevent falls. Personnel should step over pipe or hose lines, never on them. Tools or other equipment should never be dropped or thrown from the upper part of the tank.

6. Precautions should be taken to assure that the skin does not come in contact with oil. However, in the event such contact cannot be avoided, the oil should be washed from the skin with soap and water as quickly as possible.

7. In low concentrations, hydrogen sulfide is detectable by its characteristic rotten-egg odor. However, it is important to recognize that the smell cannot be depended upon to forewarn of dangerous concentrations, as exposure to the gas dulls the sense of smell.

   Excess exposure to hydrogen sulfide causes death by paralysis of the respiratory system. In milder doses it also attacks the eyes. Repeated short accidental exposures may lead to some chronic irritation of the eyes, nose, and throat; but this usually disappears soon after removal from exposure. The effects of exposure to hydrogen sulfide are not cumulative in action.

   Hydrogen sulfide is a flammable gas, and burns in air. The flammable limits are 4.3 to 46 percent by volume in air. Therefore, the usual precautions against sources of ignition must be observed when this hazard exists. Remember, because hydrogen sulfide may be present in the *vicinity*

of tanks which contained fresh crude oil, personnel in this area should wear *adequate self-contained breathing apparatus.*

8. All sources of ignition should be eliminated from the area where flammable vapors may be present. Signs should be posted warning that sources of ignition should be kept out of the area.

   Any equipment which may provide a source of ignition should not be permitted within the vicinity of the tank until the area has been tested and found vapor-free. Then, if the equipment is used, it should be placed well away from the tank, preferably to the windward, in order to minimize the ignition hazard. No work should be done if the direction of the wind might carry vapors into areas where they might produce a hazardous condition, nor when an electrical storm is threatening or in progress.

   No artificial lights, other than flashlights or low-voltage battery-operated lanterns, should be used inside the tank until the tank has been vapor-freed. Portable lights used outside the tank in the path of possible vapor travel should be of the explosionproof type, and should be connected to approved extension cords equipped with approved connectors and switches. Such equipment, when used, should be thoroughly inspected to be sure that it will not be a source of ignition.

9. Ventilation while rescue is in progress is critical. Toxic concentrations and vapors within their flammable limits must be removed. Consult the workers on the scene to use their mechanical ventilation devices. If the emergency units' equipment is used, then make sure all equipment is an explosionproof design.

Ordinary (natural) ventilation may be also used to remove tank vapors. All roof and shell manhole covers should be removed—in that order—so that air will circulate freely through the tank. Caution should be exercised when bottom manhole covers are

removed because of escaping petroleum vapors.

FIRE FIGHTING WITH FOAM

When extinguishing a flammable liquid fire special agents are necessary. These water additives are called foam and can be divided into five classifications:

1. Protein
2. Aqueous film forming foam (AFFF)
3. Fluoroprotein
4. Synthetic
5. Chemical

Protein

Protein-based foam is formed from the decomposition of vegetable or animal proteins within metallic salts added for strength. There are two basic types, 6 percent and 3 percent, which when proportioned with water produce 6 percent and 3 percent solutions by volume. This means that when using a 3 percent concentrate, 97 percent water is added to make the foam. In the same manner when using a 6 percent concentrate, the water portion is 94 percent. Air is then added to the mixture to expand the solution into the finished foam product. The expansion ratio varies between the manufacturers from 1:7 and 1:12. Foam with this type of air ratio is known as low-expansion foam.

A five-gallon can of 3 percent concentrate foam will require 161.7 gallons of water. This number can be determined as follows:

$$\frac{\text{Quantity of foam}}{\text{Percentage of concentrate}} = \text{total solution}$$

Total solution − quantity of foam

= quantity of water

In the preceding example the quantity of foam is 5 gallons and the percentage of concentrate is 3 percent. This yields:

$$\frac{5}{0.03} = 166.7 \text{ gallons total}$$

166.7 − 5 gallons of concentrate
= 161.7 gallons of water

Now, having a total solution of 166.7 gallons and using a 1:10 expansion (average) ratio, the finished foam product would be

$166.7 \times 10 = 1667$ gallons of finished foam

The 3 percent foam has several advantages over the 6 percent variety. These are:

1. Longer application times
2. Smaller storage space

Aqueous Film Forming Foams (AFFF)

Aqueous film forming foam uses several chemicals called fluorocarbon surfactants to function. These groups of chemicals operate by altering the physical properties of water (reducing the surface tension) so that it is able to spread and float across the surface of a hydrocarbon fuel even though it is more dense and should sink. Because of this property one manufacturer has called the foam light water. AFFF is also available in 3 percent and 6 percent concentrations.

It is important to note that when AFFF is used on metal tanks or vessels where the metal has heated up, it tends to break down. This lowers its ability to stop burnback and may cause problems when used on storage tank fires.

Fluoroprotein Foam

A further refinement of AFFF was the combining of the fluorocarbon surface active agents with protein foam. This combination is called fluoroprotein foam. Variable physical properties can be achieved by varying the proportions of the two products.

The fluoroprotein foam was developed specifically to achieve an acceptable degree of compatability between protein foam and Purple-K powder. This was necessary because by itself protein foam was not compatible with the Purple-K powder.

The low expansion foams, protein, AFFF, and fluoroprotein extinguish fires involving flammable liquids by:

1. Excluding air from the flammable vapors
2. Eliminating vapor release from the fuel surface
3. Separating the flames from the fuel surface.

Synthetic Foam

Another class of fire fighting foams is the synthetic detergent type. These are sometimes referred to as high expansion foams because of their expansion ratios in the order of 1:15 and 1:20. They use ordinary, air-aspirating equipment.

These foams flow well, are capable of rapid flame breakdown, and can gain control of hydrocarbon fires at low to moderate solution application densities.

Ultra-high-expansion foam is produced in specialized equipment by driving a high-volume air stream, by means of a fan, through a metal or cloth grid which is continuously sprayed with a synthetic foam solution at a predetermined rate and concentration. Expansion ratios of 1 to 800 have been achieved, depending upon the type and chemical composition of the foam as well as its stabilizers.

While not normally considered applicable for hydrocarbon fires, tests were conducted at the National Aviation Facilities Experimental Center in Atlantic City, New Jersey. These tests used ultra-high-expansion foams at discharge rates from 100 to 500 gpm on 60-foot diameter, JP-4 aviation fuel fires. The results of these experiments

> ... indicate that ultra-high expansion foam is capable of obtaining progressive and effective control of JP-4 fuel fires at solution application rates from 0.04 to 0.10 gpm/ft². However, the vulnerability of this type of foam to disruption by wind in exposed outdoor conditions and the limited vapor-securing characteristics, restrict its use as a primary fire fighting agent.[3]

One 5-gallon can of an ultra-high-expansion foam can produce about one-third of a million gallons of foam solution. This would be enough to cover a football field 12 inches deep.

Chemical Foam

Chemical foam is formed by the reaction of an alkaline with an acid in the presence of a stabilizer. The two products to mix come in either powder form or already in solution as a liquid. The acid solution is aluminum sulfate and the alkaline is sodium bicarbonate.

This foam is not used very much any more because of the inconvenience of mixing plus the time delay required for application. In addition, getting the foam on the fire requires special application tools, so that another delay is inserted.

Getting Foam into the Water Stream

Mechanical foam (all of the previously discussed foam except chemical) must be induced into the hose stream in the correct proportion. This can be accomplished in several ways.

The in-line eductor either at the pumper or near the nozzle uses the water flow to create a vacuum and bring foam up into the stream (Figure 16-25). A metering valve on the eductor is set to the proportion of concentrate being used (3 percent or 6 percent) to bring the correct amount into the stream. Air is mixed at the nozzle and the foam is expanded as it leaves the nozzle. Some in-line eductors also have a bypass that allows the flow of foam to be shut off and the line used as a regular attack line.

Some apparatus have proportioners built into the pumper. Special foams tanks are added with the proportioner being set for the type of foam carried. Then, only one valve has to be opened at the pump panel to discharge foam from a line.

An in-line compressed air system em-

ploys an air compressor to inject air into a proportioned solution of water and foam liquid concentrate. This is mixed within the piping system to produce foam.

The balanced pressure system uses a separate foam concentrate pump to move the foam from the tank to the apparatus discharge. Unused concentrate is returned to the tank. The proportioner is installed at the pumper discharge outlets.

Alcohol-Type Foams

Regular protein type foams will not work on a certain class of flammable liquids called polar solvents. This inability to extinguish polar solvent fires results from the fact that they are soluble in water. Therefore, their ability to blanket is greatly reduced. A special "alcohol-type" protein foam, resistant to dilution, is therefore necessary.

In addition to alcohol, some other polar solvents are esters, organic solvents, acetone, formic acid, and methanol.

Foam Application

Foam must be applied as gently as possible. If a straight stream is being used, the foam should be banked off of the wall or other obstruction as shown in Figure 16-25. It can also be bounced onto the surface of the spill by hitting the ground in front of the spill (Figure 16-26).

If a foam nozzle has a spray attachment, it is possible to gently apply the foam directly to the surface. The object is to prevent the mixing of foam and fuel.

Remember, do not disturb the foam blanket by directing a stream of water into it. Flashback can occur and a reignition is possible.

Successful use of foam is also dependent upon the rate at which it is applied. Application rates are determined by the gallons of foam solution reaching the fuel surface every minute. Remember, if the application rate is

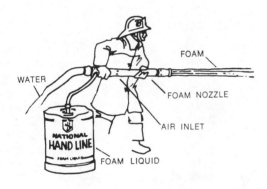

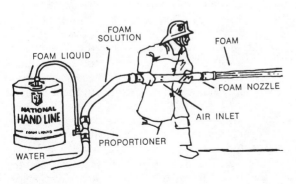

Figure 16-25. In-line foam eductors

Banking foam off wall.

Striking ground in front of fire.

Figure 16-26.   Foam application techniques

less than the minimum recommended, time required to extinguish will be prolonged or, if it is too low, the fire may not be controlled. Table 16-1 provides some suggested application rates.

Table 16-1.   Foam Application Rates

Type of Foam	Nozzle Pressure, psi	Flow, gpm	Square Foot Coverage	gpm per sq ft
Protein	100–150	60	375	0.16
Protein	100–150	95	600	0.16
AFFF	100–150	60	600	0.10
AFFF	100–150	95	950	0.10

Once you have determined your application rates, it is necessary to calculate how much foam will be needed for a one-hour supply. The following example will illustrate how to perform the calculations.

Example: A 2,000 square foot spill fire is going to be extinguished with 3 percent pro-

tein foam, applied at the rate of 95 gpm. How many gallons of solution are needed on hand for a one-hour supply?

Needed supply = 2,000 square feet
× 0.16 gpm/sq ft × 60 minutes × 003
= 576 gallons

Using Foam to Extinguish Storage Tank Fires

Mr. N. R. Lockwood of the Mobil Oil Corporation presented a paper entitled "The Protection of Refinery Properties" at a meeting at the Applied Physics Laboratory.[4] The paper so clearly presents the capability of using foam on a storage tank fire that it is presented here:

> The net result is that today we recommend subsurface foam injection. With this system we're able to go into the product line with the foam. That has a number of advantages. The possibility of losing that line through some fire accident is fairly remote. It's a large line. It is easy to reach, and so we don't have to enter the dike except possibly to open the tank product line valve ...
>
> We have reached the conclusion that where tanks 200 feet and over are concerned, the possibility of getting the fire out by surface application through foam chambers or foam towers is fairly remote. In a tank of that size the updraft and air turbulence makes it very difficult to apply foam to the burning surface with monitors. Another drawback to surface-applied foam is that the flame exposed foam loses its property of sealing or forming a coherent blanket after traveling about 100 feet on the burning surface from its point of application. Therefore, we believe that subsurface injection is the proper answer because we can inject foam, and it will reach the burning surface at various points within the tank as determined by the design of the system. Subsurface injection, of course, can be supplemented by surface

application methods.

On our floating roof tanks, we either use portable or fixed systems. On the large open top tanks of over 150 feet diameter, lots of oil companies, including Mobil, put on fixed systems for rim protection. Basically, a rim fire in an open top tank is a relatively easy fire to extinguish, and it can be extinguished by using 1½" foam hose streams from the windgirder. The fire usually progresses around the seal at a relatively slow rate, and its intensity, of course, is not anywhere near what one gets from a full surface fire in a tank. Rim fires can be easily approached by fire fighters. However, when the tank gets to be around the 200 foot diameter size, you can see from a logistics viewpoint it requires a lot of hose to get around the rim of the tank. Therefore, on those tanks we used fixed systems, and the design requirements of the fixed system are covered in the appendix of NFPA No. 11 . . .

With a covered-type floater with a pan-type, aluminum pontoon, or a fiberglass-type floating roof within the tank, we feel that foam should be provided for the full tank surface. Basically, if fire fighters are available, we would not do that, but if you do not have a complete truck and equipment available to foam the entire surface, it's a good idea to put on a fixed system.

During the period of initial fill of a covered floater, the area below the floating roof vents into the area above and into the space below the fixed roof. During that period, the area between the two roofs can approach the flammable range. During that period and for about 18 to 24 hours afterwards, depending on the volatility of the stock and the ambient temperature and the wind, the area between the two roofs may be in the flammable range before ventilation can take place and the space becomes too lean to burn.

Most tank fires are caused by external fires due to spills, etc., and are due to the heat of an external fire that causes the roof of a cone roof tank to rupture at the weakest seam. Then you have a full surface fire. In the case of a floating roof tank, especially with the open top tank, all you would expect would be a rim fire. So our recommended procedure is to extinguish ground fires first . . .

We prefer the use of monitors for fighting rim fires or dike fires. You have seen pictures of how large capacity monitors can be used. We use them similar to the way one handles a crash fire, in that you start working from one point with monitors, and work your way around until you have foamed the entire diked area . . .

There are certain limitations to foam. Low expansion foams of the conventional type that we use, by this I mean both fluoroprotein and AFFF are employed in 3% concentration at 4 to 8 expansion. We do not use foams on gases or liquefied gases with boiling points below ambient temperature. High expansion foams can be used on liquid natural gas. Their function is not to extinguish the fire but to control the fire. Foam shouldn't be used on electrical equipment. It should not be used on materials that react violently with water such as metallic sodium. We're not particularly concerned about that in refineries. If foam is to be used in conjunction with dry chemical powder, a compatible type of dry chemical must be employed or else a compatible type of foam concentrate must be chosen. Foam should not be used on hot oils or burning asphalts or burning liquids where the liquid temperature is above the boiling point of water unless it is applied at a very slow rate. It will cause violent frothing and slopover at fast rates of addition.

Foam can cause problems which we call "boilover." When you have a "boilover," which occurs with crude oil, a rather extensive fire ensues. What happens is that with crude, when you have a material with a wide boiling range, the light ends in the mixture become heated up from the fire. They tend to volatize and burn off, and the heavy ends in the mixture tend to sink, and as they sink they cause more of the light ends to vaporize and come to the surface and burn. This circulation produces a downward progression of heated oil called a "heat wave" which continues down at the rate of 2 to 3 feet per hour. As this proceeds through the depth of liquid in the tank, its entire contents reaches 300° to 400°F. After a certain period of burning it is probably questionable whether one could successfully extinguish a tank of crude where you have the full surface involved. The only way that foam can exist on the surface of such a fire is to cause the crude oil to "slopover" the sides of the tank by applying foam a little bit at a time. In effect this is sort of a "controlled slopover," which cools the surface of crude oil in the tank. Such actions may eventually permit foam to extinguish crude oil fires. Fires of this type happened in the old days when we had wooden roof tanks. Today, with our modern steel tanks, the possibility of this happening is extremely remote. Usually, where crudes are stored in cone roof tanks in the temperate zones where the daily temperatures are relatively high, the vapor space is too rich to burn so ignition doesn't readily occur. We recommend that floating roofs be used for storage of crude and in that case all that might occur is a rim fire. Experiences show that even when the floating roof sinks, the possibility of a boilover in a modern tank is extremely remote.[4]

STORAGE INCIDENTS

Bayonne, New Jersey, Oil Terminal Fire,[5] January 6, 1973

New packing flanges were being installed on a pump used for number 6 fuel oil. In order to complete the installation an arc-welding outfit was being used. This ignited the oil-soaked packing at the flange.

In addition to the fire at the flange, there was still some number 6 fuel oil in the dike from a previous spill. The valves to the other tank were closed. Unfortunately, the valve for number 2 tank could not be reached.

The ground fire weakened the unprotected steel pipe feeding the number 2 tank. The line sagged and broke, thus providing more fuel to the fire. As this ground fire intensified, the weak roof seam tanks experienced a build-up of internal pressure. However, the seams opened, relieved the pressure and no product was lost.

The ground fire was extinguished using foam lines attacking from three sides.

Would your department have sufficient mutual aid to handle the water supply problems? Where would sufficient supply of foam come from? Could you have prevented the rupture of the weak seam roof welds?

Baton Rouge, Louisiana, Chlorine Leak,[6] December 10, 1976

A chlorine cloud moving at the rate of 10 miles per hour caused the evacuation of 10,000 people. The leak from a stationary chlorine tank involved approximately 100 tons of the gas. It took 8 hours for the tank to empty and another 8 hours for the cloud to dissipate.

Where would you keep 10,000 displaced citizens for 16 hours? How would you feed and clothe them and arrange for sanitary facilities during that time? How would you communicate with them to keep them informed? How would you protect the vacated area? What provisions for a medical emergency have you made? Who would you get to assist in the evacuation?

QUESTIONS FOR REVIEW AND DISCUSSION

1. Describe the types of tanks used to store gases and liquids. Include the types of supports used to secure the tanks.

2. Prepare a survey form of your local area, detailing the types of tanks used for storage.

3. Describe the various types of safety devices used on storage tanks.

4. Explain why tanks with movable roofs present different problems to fire fighting personnel.

5. Select a storage facility in your area and prepare a physical layout of the complex.

6. What type of containment methods are used in storage facilities? What problems can each method present to emergency response personnel?

7. Prepare a procedure that emergency response personnel can follow when handling an incident at a storage facility.

8. Explain the advantages and disadvantages of using each of five major types of foam.

9. If twenty 5-gallon (380 liters total) cans of 6% foam are used on a fire, how much water will be required to deliver it?

10. Why is it necessary to use a special foam on alcohol fires?

11. Describe the technique for applying foam to a spill or fire.

12. A 2,000-square-foot surface area of a flammable liquid is burning. A foam of AFFF is to be applied. At what rate would you apply the foam and how much will be needed for a 1-hour application?

CASE PROBLEM—STORAGE

Right in the center of the business district of Stove Pipe, Oklahoma, there is a small-tank farm and loading rack. There are three horizontal tanks, each containing 35,000 gallons of various types of gasoline. The supports for the tanks are unprotected steel. Uphill from the horizontal tank is a 250,000 gallon tank for heating oil.

On a Saturday morning in July at 10:30 A.M. a truck at the filling rack overflowed. The gasoline ran underneath the horizontal tanks and ignited. The valve on the horizontal tank has not been shut off so the flow of gasoline continues. The temperature is 88°F (31°C).

The downtown area is composed of retail establishments which are crowded with Saturday shoppers. The ground is level all around the storage area, but gasoline has begun to flow towards a large department store and toward the storm sewers.

The volunteer fire department has 35 fire fighters available to bring three 1,000 gpm pumpers, a utility truck, and and ambulance. There is an excellent hydrant system in the town, with a flow of 6,000 gpm in the vicinity of the fire.

Mutual aid companies within a 15-mile radius can supply 60 fire fighters, 7 pumpers (at least 1,000 gpm), 1 ladder truck, and 4 ambulances. The airport, 12 miles away, has a crash vehicle available to respond if requested.

The foam supply with the Stove Pipe Fire Department consists of ten 5-gallon cans of AFFF foam and a single eductor.

1. Describe in detail how you would handle this incident.

NOTES

1. I. M. Levitt, "Users Report," *The Probeye View*, Hughes Aircraft Company, July/August 1977, p. 1.

2. "Crude Oil Tanks," The University of Texas, 2nd ed., December 1974 (Austin, Texas), pp. 100–101.

3. G. B. Geyer, "History, Research, and Foam Developments," in *Proceedings of a Seminar on Fire-Fighting Foams—Their Characteristics and Uses in Fuel Conservation*, The Johns Hopkins University, Applied Physics Laboratory, February 6, 1974 (Silver Spring, Md.), pp. 18–19.

4. N. R. Lockwood, "The Protection of Refinery Properties," in *Proceedings of a Seminar on Fire-Fighting Foams—Their Characteristics and Uses in Fuel Conservation*, The Johns Hopkins University, Applied Physics Laboratory, February 6, 1974 (Silver Spring, Md.), pp. 41–45.

5. F. X. Donovan, "Streams Limit Tank Collapse in Jersey Oil Terminal Fire," *Fire Engineering*, September 1973, pp. 128–129.

6. "Baton Rouge Plant Leaks Deadly Gas Across Mississippi," *The Washington Post*, December 11, 1976, p. A3.

Glossary

Absolute Pressure The true pressure. It equals atmospheric plus gauge pressure and is abbreviated psia.

Absolute Zero The lowest point on the Kelvin scale, at which there is a total absence of heat.

Absorbent Material Used to soak up liquid hazardous materials. Example: Commercial bagged clay, kitty litter, Zorbal.

Absorption The taking in of toxic materials by contact with the skin.

Acaricide A pesticide used to control spiders, ticks, and mites.

Acid or Ammonia Suits Special protective clothing that prevents toxic or corrosive substances or vapors from coming in contact with the body.

Active Ingredient The chemical that has pesticidal action. Active ingredients are listed in order on a pesticide label as percentage by weight or as pounds per gallon of concentrate. *See* Inert Ingredient.

Acute Poisoning Poisoning by a single exposure to a toxic chemical.

AFFF Aqueous Film Forming Foam. An extinguishing agent designed to flow on a burning liquid.

Air Bill A shipping paper prepared from a bill of lading which accompanies each piece of an air shipment.

Air Lift Axle A single air-operated axle which, when lowered, will convert a vehicle into a multiaxle unit, providing the vehicle with a greater load-carrying capacity.

Air-Reactive Materials that will ignite at normal temperatures when exposed to air.

Air Spring A flexible air-inflated chamber in which the air pressure is controlled and varied to support the load and absorb road shocks; formerly called air bag.

Alcohol Foam Blankets fires in the same manner as conventional foam, but it is intended for use with liquids which are soluble in water, such as alcohol and acetone. It must be applied more carefully than regular foam because the mechanical strength of the bubbles is less.

Alpha Rays Made up of very large particles which are the same as the nucleus of the helium atom.

Asphyxiating Materials Substances that can cause death through displacing the oxygen in the air.

Atmospheric Pressure The pressure caused by the weight of the air elevated above the earth's surface. At sea level it equals 14.7 pounds per square inch (psi).

Atom The smallest particle of an element that can exist.

atto Metric prefix for 10⁻¹⁸, 0.000 000 000 000 000 001, abbreviated a.

Autoignition Temperature Same as ignition temperature.

Axle Weight Amount of weight transmitted to the ground by one axle or the combined weight of the two axles in a tandem assembly.

Bactericide A pesticide used to control bacteria.

Baffle An intermediate partial bulkhead that reduces the surge effect in a partially loaded tank.

Beta Rays Smaller than alpha rays, beta radiation is made up of electrons.

Bipyridyls A group of synthetic organic pesticides that include the herbicide paraquat.

BLEVE Boiling Liquid Expanding Vapor Explosion. A container failure with a release of energy, often rapidly and violently, also accompanied by a release of gas to the atmosphere and propulsion of the container or container pieces.

Blow Down Valve A manually operated valve whose function is to quickly reduce tank pressure to atmosphere.

BMCS Bureau of Motor Carrier Safety, under the Federal Highway Administration of DOT. Responsible for establishing regulations for in-service motor vehicles and their operators (drivers).

Boiling Point The temperature at which a liquid boils, given in either °C or °F at a pressure of 760 mm of mercury (Hg), one atmosphere, or 14.7 psia (absolute).

Boyle's Law When the temperature and mass of a gas are kept constant, the product of the pressure and volume is equal to a constant.

Brake Hoses A flexible conductor for the transmission of fluid pressure in the brake system.

British Thermal Unit (Btu) The amount of heat necessary to raise one pound of water, one degree fahrenheit in temperature from 63°F to 64°F.

Bulk Container A cargo container, such as that attached to a tank truck or tank car, used for transporting materials in bulk quantities.

Bulkhead A structure used to protect against damage caused by shifting cargo and/or to separate loads.

Bung A cap or screw used to cover the small opening in the top of a metal drum or barrel.

Carbamate A synthetic organic pesticide containing carbon, hydrogen, nitrogen, and sulfur.

Carbon Dioxide A gas stored in cylinders and applied through a fixed or semifixed system, or from a portable extinguisher. It is useful for inerting a closed area or for putting out small local fires.

Cargo Manifest A shipping paper that lists all of the contents being carried by the transporting vehicle or vessel.

centi Metric prefix for 10⁻², 0.01, abbreviated c.

Certification Label A label permanently affixed to the forward left side of the trailer stating that the vehicle conforms with all applicable Federal Motor Vehicle Safety Standards in effect on the date of original manufacture.

Charles' Law If the volume of a gas is kept constant and the temperature is increased, the pressure increases in direct proportion to the increase in absolute temperature.

CHEMTREC CHEMical TRansportation Emergency Center. Provides assistance during a hazardous materials emergency:
1. Relays action information in regard to the specific chemical.
2. Will contact manufacturer or other expert for additional information or onsite assistance.

Chronic Poisoning Poisoning that is a result of repeated exposure to sublethal doses over a period of time.

Class A Combustibles Ordinary combustibles, which leave a residue after burning.

Class B Combustibles Flammable liquids and gases.

Class C Combustibles Class A or B fires that occur in or near electrical equipment.

Class D Combustibles Combustible metals that are easily oxidized.

Class A Explosive A material or device that presents a maximum hazard through detonation.

Class B Explosive A material or device that presents a flammable hazard and functions by deflagration.

Class C Explosive A material or device that contains restricted quantities of either Class A or Class B explosives or both, but presents a minimum hazard.

Class A Poison A poisonous gas or liquid of such nature that a very small amount of the gas, or vapor of the liquid, mixed with air, is dangerous to life.

Class B Poison Any substance known to be so toxic to humans that it poses a severe health hazard during transportation.

Cleanout Fitting Fitting installed in the top of a tank to facilitate washing of the tank interior.

COFC Container-on-Flatcar.

Combination The change that occurs when two chemicals are combined and the result is a different chemical.

Combustible Liquid Any liquid that has a flash point at or above 100°F (37.78°C) and below 200°F (93.33°C).

Combustible Metal Metal that will burn.

Combustion A rapid oxidation or chemical combination, usually accompanied by heat or light.

Command Post Central control point for the incident. Information flows to and from the command post. Special resource people are based at the command post.

Common Name of Pesticide Well-known made-up name accepted by the Environmental Protection Agency to identify the active ingredients in a pesticide. It is listed under the active ingredient statement on the label.

Compound A pure substance composed of two or more elements.

Compressed Gas Any material or mixture having the container pressure exceeding 40 psi at 70°F (21.11°C) or, having an absolute pressure exceeding 104 psia at 130°F (54.44°C).

Compressed Gas in Solution A nonliquefied gas that is dissolved in a solvent but at high pressures.

Condensation Going from the gaseous to the liquid state.

Conduction Heat transfer through the movement of atoms within a substance.

Connection Box Contains fittings for trailer emergency and service brake connections and electrical connector to which the lines from the towing vehicle may be connected. Formerly called junction box, light box, bird box.

Consignee The person who is to receive a shipment.

Consist A rail shipping paper similar to a cargo manifest. It will contain a list of the cars in the train in order and may list those cars carrying hazardous materials and their location on the train.

Container An article of transport equipment that is:

1. Of a permanent character and strong enough for repeated use.
2. Specifically designed to facilitate the carriage of goods by one or more modes of transport without intermediate reloading.
3. Fitted with devices permitting its ready handling, particularly its transfer from one mode to another.

The term *container* does not include vehicles. Also referred to as: freight container, cargo container, intermodal container.

Container Chassis A trailer chassis having simply a frame with locking devices for securing and transporting a container as a wheeled vehicle.

Container Ship A ship specially equipped to transport large freight containers in horizontal, or, more commonly, in vertical container cells. The containers are loaded and unloaded usually automatically by special cranes.

Container Specification Number A number found on a shipping container preceded by the initials DOT, which indicate that the container has been built according to federal specification.

Contaminant A pesticide or other toxic material found as a residue in or on a substance where it is not wanted.

Control Agent Any material used to contain or extinguish a hazardous material or its vapor.

Convection Heat transfer from one place to another by actual motion of the hot material.

Corrosive Material Any liquid or solid that causes destruction of human skin tissue or a liquid that has a severe corrosion rate on steel or aluminum.

Crossover Line Installed in a tank piping system to allow unloading from either side of tank.

Cryogenics Substances having temperatures below –150°F (–101.1°C).

Cubic Capacity Useful internal load-carrying space usually expressed in cubic feet, cubic yards, or cubic meters. Also referred to as available cube, or simply cube.

Curie The amount of radioactive material that will give 37 billion disintegrations per second.

Dangerous Cargo Manifest A cargo manifest used on ships, which lists all the hazardous materials on board, including their location.

deca Metric prefix for 10^1, 10.0, abbreviated da.

deci Metric prefix for 10^{-1}, 0.1, abbreviated d.

Decomposition Breaking down of a substance to a less complex form. This can be accomplished by the introduction of heat, through the addition of neutralized chemicals, or through biodegradation.

Deflagration The intense burning rate of some explosives; black powder being one example.

Defoliant A herbicide used to remove unwanted plant growth without killing the whole plant.

Detonation A wave that passes along the body of an explosive, instantaneously converting the explosive into gas.

Diluent Any liquid or solid material used to carry or dilute an active ingredient.

Dip Tube Installed for pressure unloading of product out the top of the tank.

Disinfectant A pesticide that controls germs.

Distillation Going from the liquid to gaseous to liquid state.

Dome The circular fixture on the top of a tank car, which contains valves and relief devices.

Dosimeter Designed for measuring accumulated exposure doses of gamma radiation to emergency response personnel.

DOT Department of Transportation; the administrative body of the executive branch of the federal government responsible for transportation policy, regulation, and enforcement. *See also* FHWA and NHTSA, and BMCS.

Doubles Trailer combination consisting of a truck tractor, semitrailer, and a full trailer coupled together. Formerly called double-trailer or double-bottom.

Dry Chemical A special fire-extinguishing chemical, sodium or potassium bicarbonate or monosodium phosphate powder, usually available from a semi-fixed or portable extinguisher.

Dummy Coupler A fitting used to seal the opening in an air brake hose connection (gladhands) when the connection is not in use; a dust cap.

Element The simplest form of a substance and the basic building block of chemistry.

Emergency Shutoff Lever An automatic or manually operated safety valve control that stops the flow of a liquid.

Emergency Valve A self-closing tank outlet valve.

Emergency Valve Operator A device used to open and close emergency valves.

Emergency Valve Remote Operator A secondary closing means, remote from tank discharge openings, for operation in event of fire or other accident.

Emulsifiable Concentrate Material mixed with a solvent (usually petroleum) which forms an emulsion when mixed with water for application.

Emulsion A mixture in which one liquid is suspended as tiny drops in another liquid, such as oil in water.

Endothermic Reaction Heat is absorbed during the reaction.

EPA Registration Number The number that appears on the pesticide label to identify the individual pesticide product. May appear as "EPA Reg. No."

Etiologic Agent A microorganism, or its toxin, that causes or may cause human disease.

Excess Flow Valve A safety valve designed to shut off the flow of a liquid when the flow rate exceeds a set rate.

Exothermic Reaction Reaction that produces heat.

Explosive A material capable of burning or detonating suddenly and violently.

Explosive Limits Same as Flammable Limits.

Extremely Flammable A liquid pesticide that has a flash point of 20°F (-6.67°C) or lower, determined by closed cup or Seta flash test.

femto Metric prefix for 10^{-15}, 0.000 000 000 000 001, abbreviated f.

FHWA Federal Highway Administration; the DOT division concerned with highway construction and usage. Other similar divisions of DOT relate to air, rail, and water transportation.

Fifth Wheel A device used to connect a truck tractor or converter dolly to a semitrailer in order to permit articulation between the units. It generally is composed of a lower part consisting of a trunnion, plate, and latching mechanism mounted on the truck tractor (or dolly) and a kingpin assembly mounted on the semitrailer.

Fifth Wheel Pickup Ramp A steel plate designed to lift the front end of a semitrailer to facilitate engagement of the kingpin into the fifth wheel.

Fill Opening An opening in top of a tank used for filling the tank. Usually incorporated in a manhole cover.

Flame Impingement The points where flames contact the surface of a container.

Flammable Any material that is capable of being easily ignited and of burning with extreme rapidity.

Flammable Compressed Gas Any flammable material or mixture in a container having a pressure exceeding 40 psi at 100°F (37.78°C).

Flammable Gas Any compressed gas that will burn.

Flammable Limits The range of gas or vapor concentrations (percent by volume in air) that will burn or explode if an ignition source is present. Limiting concentrations are commonly called the *lower explosive limits* (LEL) and the *upper explosive limit* (UEL). Below the LEL the mixture is too lean to burn, and above the UEL it is too rich to burn.

Flammable Liquid Any liquid having a flash point below 100°F (37.78°C) as determined by tests prescribed in the Federal Regulations. Also, a liquid pesticide that has a flash point between 20°F to 80°F, determined by closed cup or Seta flash test. *Note*—This is the EPA definition for pesticide labeling. It varies from the NFPA and DOT definitions.

Flammable Solid Any solid material, other than an explosive, which is liable to cause fires through friction, absorption of moisture, spontaneous chemical changes, retained heat from manufacturing or processing, or which can be ignited readily and when ignited burns vigorously and persistently.

Flashing Liquid-tight rail on top of a tank that contains water and spillage and directs it to suitable drains. May be combined with DOT overturn protection.

Flashing Drain Metal or plastic tube that drains water and spillage from flashing to the ground.

Flash Point The lowest temperature at which a liquid will give off sufficient flammable vapor for ignition to occur. Values can be determined by the open cup (OC) or the closed cup (CC) method. In general, the open cup value is about 10° to 15°F (−12.22° to −9.44°C) higher than the closed cup valve.

Freezing Point The temperature in °C or °F at which a liquid solidifies.

Full Protective Clothing Clothing that will prevent gases, vapors, liquids, and solids from coming in contact with the skin. Full protective clothing includes the helmet, self-contained breathing apparatus, coat and pants customarily worn by fire fighters (turnout or bunker coat and pants), rubber boots, gloves, bands around legs, wrist and waist. As well as covering for neck, ears, and other parts of the head not protected by the helmet, breathing apparatus, or face mask.

Fungicide A pesticide that controls or inhibits fungus growth.

Fusible Plug A safety relief device in the form of a plug of low-melting metal. The plugs close the safety relief device channel under normal conditions and are intended to yield or melt at a set temperature to permit the escape of gas.

Gamma Rays A form of electromagnetic radiation similar to X rays.

Gauge Pressure The pressure read on a gauge, which does not take atmospheric pressure into account. The abbreviation for this pressure reading is psig.

giga Metric prefix for 10^9, 1 000 000 000.0, abbreviated G.

Gladhands Fittings for connection of air brake lines between vehicles. Are also called hose couplings, hand shakes, or polarized couplers.

Gram The weight of one cubic centimeter of water at 4°C.

Gram-Calorie One gram-calorie is the amount of heat necessary to raise one gram of water one degree centigrade from 14.5°C to 15.5°C.

Granules Dry, coarse particles of some porous material (clay, corncobs, walnut shells) into which a pesticide is absorbed.

Gross Weight The weight of a trailer together with the weight of its entire contents.

Half-Life The length of time it takes for one-half of a given amount of a radioactive substance to change into the next element.

Hatch Plan A schematic drawing of the location of all cargo on a ship.

Hazard Class A group of materials as designated by the Department of Transportation that share a common major hazardous property; for example, radioactivity, flammability.

Hazardous Material A substance or material in a quantity or form that may pose an unreasonable risk to health and safety, or property when stored, transported, or used in commerce. *Note:* This is the Department of Transportation definition.

Head The front and rear closure of a tank shell.

Heating Tube A tube installed inside a tank used to heat the contents. Also may be called 'fire tube.'

Heat of Fusion The quantity of heat that must be supplied to a material at its melting point to convert it completely to a liquid at the same temperature.

Heat of Vaporization The quantity of heat that must be supplied to a liquid at its boiling point to convert it completely to a gas at the same temperature.

hecto Metric prefix for 10^2, 100.0, abbreviated h.

Hemispherical Head A head that is half a sphere in shape. Used on MC-331 high-pressure tanks.

Herbicide Pesticide that controls plant life.

High-Expansion Foam Detergent-base foam that expands in ratio of 1,000 to 1. It is low-water-content foam.

Hitch A connecting device at the rear of a vehicle used to pull a trailer with provision for easy coupling.

Hopper Sloping panels at bottom of tank that direct dry bulk solids to the outlet piping.

Hose Tube A housing used on tank and bulk commodity trailers for the storage of cargo-handling hoses. Also called hose troughs.

Hypergolic Materials Materials that ignite on contact with one another.

ICC Interstate Commerce Commission. An independent federal government agency in the Executive Branch (not affiliated with DOT) charged with administering acts of Congress affecting rates and routes for transport of interstate commerce.

Ignition Temperature The minimum temperature in °F or °C required to ignite gas or vapor without a spark or flame being present. Values provided in reference texts are only approximate because they change substantially with changes in geometry, gas, or vapor concentrations and in the presence of catalysts.

Impermeable Cannot be penetrated by liquid or vapor.

Incipient Fires Fires in the beginning stages.

Individual Container A cargo container, such as a box or drum, used to transport materials in small quantities.

Inert Ingredient Material in a pesticide formulation that generally has no pesticidal activity, but could be flammable or combustible.

Ingestion The taking in of toxic materials through the mouth.

Inhalation The taking in of toxic materials by breathing through the nose or mouth.

Inorganic Chemistry The branch of chemistry dealing with compounds that do not contain carbon.

Irritating Material A liquid or solid substance that upon contact with fire or when exposed to air gives off dangerous or intensely irritating fumes, but not including any poisonous material, Class A.

Isolation Perimeters An area around a pesticide incident within which only necessary personnel with full protective gear are allowed.

Jacket A metal cover that protects the tank insulation.

Jackknife Condition of truck tractor-semi-trailer combination when their relative positions to each other form an angle of 90° or less about the trailer kingpin.

Kilo Metric prefix for 10^3, 1,000, abbreviated k.

Kingpin Attaching pin on a semitrailer that mates with and pivots within the lower coupler of a truck tractor or converter dolly while coupling the two units together.

Kingpin Assembly *See* Upper Coupler Assembly.

Labels Four-inch-square (26-centimeter-square) diamond markers required on individual shipping containers that are smaller than 640 cubic feet (19.2 cubic meters).

LD 50 The dose of a pesticide active ingredient taken by mouth or absorbed by the skin which is expected to cause death in 50 percent of the test animals. The lethal dose is measured in milligrams per kilogram of body weight.

Liquefied Gas Partially liquid at a temperature of 70°F (21.11°C).

Liter The volume of one kilogram of water at 4°C.

Lower Explosive Limits See Flammable Limits for definition.

Manhole Openings usually equipped with removable, lockable covers and large enough to admit a person into a tank trailer or dry bulk trailer.

Manifold Used to join a number of discharge pipelines to a common outlet.

mega Metric prefix for 10^6, 1 000 000.0, abbreviated M.

Melting Point The temperature in °F or °C at which a solid becomes a liquid.

Meter A measure of length based on the spectrographic color line of the element krypton.

micro Metric prefix for 10^{-6}, 0.000 001, abbreviated μ.

milli Metric prefix for 10^{-3}, 0.001, abbreviated m.

Miscible Two or more liquids that can be mixed and will remain mixed under normal conditions.

Mixture A combination of chemicals that contains two or more substances that do not lose their individual identities.

nano Metric prefix for 10^{-9}, 0.000 000 001, abbreviated n.

Nematicide A pesticide used to control nematodes. Nematodes are microscopic round-worms.

Newton The force required to impart an acceleration of one meter per second to a mass of one kilogram.

NFPA 704 National Fire Protection Association Pamphlet 704, which describes a system for marking hazardous materials as to their health hazard, flammability, and reactivity.

NHTSA National Highway Traffic Safety Administration of DOT; responsible for establishing motor vehicle safety standards and regulations for *new* vehicles. Formerly called National Highway Safety Bureau (NHSB).

Nitrophenols Synthetic organic pesticides containing carbon, hydrogen, nitrogen, and oxygen that are used as wood preservatives, fungicides, or disinfectants. Affect liver and central nervous system in the human body.

Nonflammable Gas A compressed gas that is not classified as flammable.

Nonliquefied Gas One that is entirely gaseous at a temperature of 70°F (21.11°C).

NOS or NOIBN These notations designate *Not Otherwise Specified* or *Not Otherwise Identified By Name* and will appear on shipping papers when the materials conform to a hazardous material definition, but are not listed by generic name in the regulations. Example: Flammable NOS.

Oil Hazardous Materials-Technical Assistance Data System (OHM-TADS) Organization within the Environmental Protection Agency (EPA) that provides information on some hazardous substances to emergency teams responding to spills.

Organic Chemistry The branch of chemistry that deals with compounds containing carbon.

Organic Peroxide An organic derivative of the inorganic compound hydrogen peroxide.

Organochlorine Compounds (Chlorinated Hydrocarbons) Synthetic organic pesticides that contain chlorine, carbon, and hydrogen. These pesticides affect the central nervous system. Examples: DDT and Endrin.

Organophosphates Synthetic organic pesticides that contain carbon, hydrogen, and phosphorous. Toxic to humans because they prevent proper transmission of nerve impulses. Examples: Parathion and Malation.

ORMS Other Regulated Materials that do not meet the definitions of hazardous materials but possess enough hazard characteristics in transport that they require some regulation.

Outage The space left in a vessel filled with a flammable liquid.

Outlet Valve The valve farthest downstream in a tank piping system, to which the discharge hose is attached.

Outrigger Structural load-carrying members attached to and extending outward from the main longitudinal frame members of a trailer.

Overturn Protection For fittings on top of a tank in case of rollover. May be combined with flashing rail or flashing box.

Oxidizer Releases oxygen under decomposition and will support combustion.

Oxidizing Material A substance that yields oxygen readily to stimulate the combustion of organic matter.

Package Markings The descriptive name, instructions, cautions, weight, or specifica-tion marks required to be placed on outside of containers of hazardous materials.

Pascal The pressure obtained when a force of one newton acts on an area of one square meter.

Pesticide Chemical or mixture of chemicals used to destroy, prevent, or control any living organism considered to be a pest.

Physical Properties Properties of a material that relate to the physical states common to all substances: that is, a solid, a liquid, or a gas.

Pickup Plate A sloped plate and structure of a trailer, located forward of the kingpin and designed to facilitate engagement of fifth wheel to kingpin.

pico Metric prefix for 10^{-12}, 0.000 000 000 001, abbreviated p.

Piggyback Transport Type of shipping in which bulk containers from one mode, such as highway transportation, are placed on flat cars or container ships for transportation by another mode, such as rail or marine.

Pintle See Hitch.

Pipeline A way to transport a product.

Placards Ten-and-3/4-inch (273.0 mm)-square diamond markers required on the transporting vehicle such as a truck or tank car or a freight container 640 cubic feet (19.2 cubic meters) or larger.

Polarized Couplers Fittings for connecting of air brake lines between vehicles. The service and emergency couplings are unilateral and will not mate with each other.

Precipitate Result when a solution of one chemical is combined with the solution of another chemical and an insoluble material is produced.

Pressure A force applied over a given area.

psia See Absolute Pressure for definition.

PSTN Pesticide Safety Team Network. Regional teams of the National Agricultural Chemical Association designed to assist with pesticide incidents.

Pumpoff Line A pipeline that usually runs from the tank discharge openings to the front of the trailer. Most pumps are mounted on the tractor.

Pyrophoric Capable of igniting spontaneously when exposed to dry or moist air at or below 130°F (54.44°C).

Radiation Heat transfer by electromagnetic waves.

Radioactive Capable of spontaneously emitting ionizing radiation.

Radioactive Material Any material, or combination of materials, that spontaneously emit ionizing radiation, and have a specific activity greater than 0.002 microcuries per gram.

Radioisotopes Artificially radioactive elements.

Reactive Capable of or tending to react chemically with other substances.

Reactivity The degree of ability of one substance to undergo a chemical combination with another substance.

Reducing Agent A fuel that becomes chemically changed by the oxidizing process.

Refrigeration Unit Cargo space cooling equipment.

Reid Vapor Pressure Equilibrium pressure exerted by vapor over the liquid at 100°F (37.78°C) as expressed in pounds per square inch absolute, as defined in 46 CFR 30.10-59.

Relay Emergency Valve A combination valve in an air brake system, which controls brake application and which also provides for automatic emergency brake application should the trailer become disconnected from the towing vehicle.

Rotary Gauge A gauge for determining the liquid level in a pressurized tank.

Running Lights Marker, clearance, and identification lights required by regulations.

Rupture Disk A safety device in the form of a metal disk that closes the relief channel under normal conditions. The disk bursts at a set pressure to permit the escape of gas.

Safety Relief Valves A device found on pressure cargo tanks containing an operating part that is held in place by spring force. Valves open and close at set pressures.

Sandshoe A flat steel plate that serves as ground contact on the supports of a trailer and used instead of wheels, particularly where the ground surface is expected to be soft.

Self-Accelerating Decomposition Temperature The temperature above which the decomposition of an unstable material proceeds by itself, independently of the external temperature.

Semitrailer A truck trailer equipped with one or more axles and constructed so that the front end and a substantial part of its own weight and that of its load rests upon a truck tractor.

Sheer Sections A safety feature, such as a valve or joint, built onto a cargo tank, which is designed to fail or break completely to prevent a failure or break of the tank itself.

Shipping Papers A shipping order, bill of lading, manifest, or other shipping document issued by the carrier.

Sliding Fifth Wheel A fifth wheel assembly capable of being moved forward or backward on the truck tractor to vary load distribution on the tractor and to adjust the overall length of combination.

Slope Sheet Panels located at each end of payload compartment that direct product by gravity to hoppers.

Soil Contamination Contamination of the ground area where a pesticide spill or fire occurs, or where contaminated runoff water flows.

Solution Mixture of one or more substances in another in which all ingredients are completely dissolved.

Solvent A liquid that will dissolve a substance to form a solution. Some examples of solvents are water, petroleum distillate, xylene, or methanol.

Specific Gravity The ratio of the weight of a volume of the product to the weight of an equal volume of water. In the case of liquids of limited solubility, the specific gravity will predict whether the product will sink or float on water; for example, if the specific gravity is greater than 1, the product will sink, and if the specific gravity is less than 1, the product will float.

Spontaneously Combustible Solid materials that can decompose in the absence of air, generally causing igniting or deflagrating.

Standard Transportation Commodity Code (or STCC Code Number) A listing of code numbers in general use by carriers for categories of articles being shipped.

Stress A state of tension put on or in a shipping container by internal chemical action, external mechanical damage, or external flames or heat.

Sublimation Going from the solid to gaseous to solid state.

Sump The low point of a tank at which the emergency valve or outlet valve is attached.

Switch List A list of the cars in a train used by railroad crews in a yard when they are making up a train.

Tandem Two-axle suspension.

Technical Assistance Personnel or printed materials which provide technical information on the handling of hazardous materials.

Technical Pesticide Highly concentrated pesticide, which is to be combined with other materials to formulate pesticide products.

Temperature The condition of an object that determines whether heat will flow to or from another object.

tera Metric prefix for 10^{12}, 1 000 000 000 000.0, abbreviated T.

Threshold Limit Value The concentration of a toxic substance that can be tolerated with no ill effects.

TIR Transport International Routier, translated: International Transport by Road. This is the name of a customs convention (or agreement) that exists among many countries, principally European. It permits vehicles or containers, properly approved and certified, to be sealed under Customs direction in one country and be transported across borders of member countries without reinspection until arrival at final destination, where the seal is then removed under Customs supervision.

TOFC Trailer-on-Flatcar; also referred to as a piggyback.

Tow Bar A beam structure used to maintain rigidly the distance between a towed vehicle and the towing vehicle.

Toxic Materials that can be poisonous if inhaled, swallowed, or absorbed into the body through cuts or breaks in the skin.

Triaxle Three-axle construction in which at least two of the axles are equally spaced at approximately 48 to 50 inches (120 to 125 centimeters) apart and the third axle may be spread or equally spaced.

Truck A self-propelled vehicle carrying its load on its own wheels and primarily designed for transportation of property rather than passengers.

Truck Tractor A powered motor vehicle designed primarily for drawing semitrailers, and constructed so that it carries part of the trailer weight and load.

Truck Trailer A vehicle without motor power, primarily designed for transportation of property rather than passengers, and to be drawn by a truck or truck tractor.

Unstable Materials that are capable of rapidly undergoing chemical changes or decomposition.

Upper Coupler Assembly Consists of upper coupler plate, reinforcement framing, and fifth wheel kingpin mounted on a semitrailer. Formerly called upper fifth wheel assembly.

Upper Explosive Limits *See* Flammable Limits.

Vapor Density This is actually a specific gravity rather than a true density, because it equals the ratio of the weight of a vapor or gas (with no air present) compared to the weight of an equal volume of air at the same temperature and pressure. Values less than 1 indicate that the vapor or gas tends to rise, and values greater than 1 indicate that it tends to settle. However, temperature effects must considered.

Vaporization Going from a liquid to a gaseous state or going from a solid to a gaseous state.

Vapor Pressure The equilibrium pressure of the saturated vapor above the liquid, measured in millimeters of mercury (760 mm Hg= 14.7 psia) at 20°C (68°F) unless another temperature is specified. Conversion is done as follows:

$$psi = \frac{mm\ Hg \times 14.7}{760}$$

Water Foam Either mechanical or chemical foam is produced by a special foam nozzle or by a fixed system. It is used to form a blanket over the surface of burning liquids. It is effective only with liquids which are not appreciably soluble in water.

Water Fog A finely divided mist produced by either a high- or low-velocity fog nozzle. It is used for knocking down flames and cooling hot surfaces.

Water Pollution Contamination of water by a pesticide or other unwanted material.

Water-Reactive Materials, generally flammable solids, that will react in varying degrees when mixed with water or in contact with humid air.

Water Solubility The ability of a liquid or solid to mix with or dissolve in water.

Waybill The shipping paper prepared by the railroad from a bill of lading. Waybills generally accompany a shipment and are carried by the conductor in the caboose of the train.

Wettable Powder A finely ground pesticide dust that will mix with water to form a suspension for application. This formulation may not burn, but may release toxic fumes under fire conditions.

Wood Preservatives Pesticides used to treat any wood to prevent insect damage, dry rot, or other damage.

INDEX

INDEX OF INCIDENTS